Mathematics in Industry

The European Consortium for Mathematics in Industry 43

The *ECMI* subseries of the *Mathematics in Industry* series is a project of *The European Consortium for Mathematics in Industry*. *Mathematics in Industry* focuses on the research and educational aspects of mathematics used in industry and other business enterprises. Books for *Mathematics in Industry* are in the following categories: research monographs, problem-oriented multi-author collections, textbooks with a problem-oriented approach, conference proceedings. Relevance to the actual practical use of mathematics in industry is the distinguishing feature of the books in the *Mathematics in Industry* series.

Martijn van Beurden · Neil V. Budko ·
Gabriela Ciuprina · Wil Schilders ·
Harshit Bansal · Ruxandra Barbulescu
Editors

Scientific Computing in Electrical Engineering

SCEE 2022, Amsterdam, The Netherlands, July 2022

Springer

Editors
Martijn van Beurden
Eindhoven University of Technology
Eindhoven, The Netherlands

Gabriela Ciuprina
Politehnica University of Bucharest
Bucharest, Romania

Harshit Bansal
Eindhoven University of Technology
Eindhoven, The Netherlands

Neil V. Budko
DIAM, Numerical Analysis
Delft University of Technology
Delft, Zuid-Holland, The Netherlands

Wil Schilders
Department Mathematics and Computer Science
Eindhoven University of Technology
Eindhoven, The Netherlands

Ruxandra Barbulescu
INESC-ID
Lisbon, Portugal

ISSN 1612-3956 ISSN 2198-3283 (electronic)
Mathematics in Industry
The European Consortium for Mathematics in Industry
ISBN 978-3-031-54516-0 ISBN 978-3-031-54517-7 (eBook)
https://doi.org/10.1007/978-3-031-54517-7

This Springer imprint is published by the registered company Springer Nature Switzerland AG
The registered company address is: Gewerbestrasse 11, 6330 Cham, Switzerland

Paper in this product is recyclable.

Preface

The 14th International Conference on Scientific Computing in Electrical Engineering was held from 11–14 July 2022, in Amsterdam, the Netherlands. The conference took place at the Centre for Mathematics and Computer Science (CWI), Amsterdam Science Park, Amsterdam, the Netherlands. It was a festive event, marking the 25th anniversary of SCEE, as the first conference was held in Darmstadt in 1997.

The conference topics were:

- Computational Electromagnetics: Modelling and parameter extraction, discretization and solution methods, Applications: antennas, microwave, interconnects and on-chip passive structures.
- Circuit Simulation and Design: Reduced order modelling, numerical integration techniques, TCAD/EDA tools and techniques, Applications: radio frequency, power electronics, optical networks.
- Coupled Problems: Field-circuit coupled problems, Multi-physics: substrate coupling, coupling with electrical, thermal and mechanical problems, Applications: co-simulation, electromagnetic compatibility, bio-engineering.
- Mathematical and Computational Methods: Inverse problems, optimization, multi-scale schemes, solutions methods for large linear systems, differential-algebraic equations, grid computing and parallel computing.

In the latter category, also the relatively new and popular topic of scientific machine learning was addressed, as quite a few researchers are now focussing on this theme, for example, with physics-informed neural networks (PINNs).

This conference edition had several invited/keynote speakers both from academia and industry and contributed presentations in lecture and poster format. SCEE 2022 was honoured by the presence of the following invited speakers:

- Ursula van Rienen (University of Rostock, Germany), Some Highlights from Computational Electromagnetics @ SCEE
- Ricardo Riaza (Universidad Politécnica de Madrid, Spain), A Projective-Based Formalism for Symmetric Modelling of Electrical Circuits
- Michael Günther (University of Wuppertal, Germany), Port-Hamiltonian Systems: A Useful Approach in Electrical Engineering?
- Idoia Cortes Garcia (Eindhoven University of Technology, the Netherlands/TU Darmstadt, Germany), Multiphysical Modelling and Co-Simulation of Superconducting Magnets in Accelerator Circuits
- Carolina Urzúa Torres (TU Delft, the Netherlands), Boundary Element Methods for Electromagnetic Scattering at Complex Geometries
- Fernando Henriquez (EPFL-Switzerland), RELU Neural Network Galerkin Boundary Element Method

Participants of SCEE 2022 in front of the CWI building in Amsterdam

Another feature of this conference was the Industry Morning, where three renowned speakers from industry gave very nice presentations on urgent topics within the electronics industry:

- Liesbeth Vanherpe (ASML, Eindhoven, the Netherlands), Scientific Computing at ASML
- Andras Poppe (SIEMENS Industry Software STS Strategic Innovation group, Hungary, Budapest University of Technology and Economics (BME), Department of Electron Devices, Hungary), Creating New Multi-Domain Digital Twins of LEDS with an Attempt to Describe Their Ageing for Predictive Maintenance Schemes
- Jörg Ostrowski (ABB), Research within ABB.

In addition to these talks, we had a total of 33 oral presentations and 26 poster presentations, completed with two special sessions: a meeting of the European project (Marie-Skłodowska-Curie EID) ROMSOC and a meeting of the ECMI Special Interest Group MSOEE.

On Wednesday evening, the SCEE standing committee, the program committee and the local organizing committee also had a meeting, followed by a lovely dinner with the invited speakers in restaurant "De Kas", a restaurant in a greenhouse that uses only their own grown products, and recently received a Michelin green star. A special highlight of the SCEE 2022 was the visit to the Van Gogh Museum.

After this excursion, the conference dinner took place in the Vondelpark3 restaurant, which is located in the heart of Amsterdam's most famous park, in the former Vondelpark pavilion. This venue is also used by Dutch broadcasting organization WNL for their Sunday morning talk show. During the dinner, in the midst of a warm atmosphere, many ideas and new research directions were discussed in parallel to the enjoyment of good food and wine.

For us, organizing SCEE 2022 took quite some effort. As many of you would know, the 14th edition of the conference was first scheduled to take place in Darmstadt, Germany. Due to strict COVID-19 regulations, the standing committee of SCEE, in close consultation with the Darmstadt organizers, decided to choose a different location. It was decided that the conference would be hosted again in March 2022 in the Netherlands, like in 2020, but now in Amsterdam. Thanks to the efforts of Wil Schilders, who managed to gather a team of organizing committee and avoid postponing the conference by a period of two years. However, due to COVID-19-related measures in the Netherlands, and similar problems in other European countries, in the first months of this year, we had to postpone the conference till 11–14 July 2022 in anticipation that the situation would be better then. We were finally able to have a great and enjoyable in-person conference in the summer of 2022. Over the past year, a lot of hard work has been put into getting the proceedings published in 2023. We thank the reviewers and the SCEE program committee members for their assistance during the reviews of the abstracts and the papers for the proceedings.

September 2023

Martijn van Beurden
Neil Budko
Gabriela Ciuprina
Wil Schilders
Harshit Bansal
Ruxandra Barbulescu

Organization

Local Organizing Committee

Wil Schilders (Chair)	TU Eindhoven, the Netherlands
Martijn van Beurden	TU Eindhoven, the Netherlands
Neil Budko	TU Delft, the Netherlands
Gabriela Ciuprina	Politehnica University of Bucharest, Romania
Ruxandra Barbulescu	Politehnica University of Bucharest/INESC-ID Lisbon, Romania/Portugal
Harshit Bansal	TU Eindhoven, the Netherlands

Program Committee

Wil Schilders (Chair)	TU Eindhoven, the Netherlands
Martijn van Beurden	TU Eindhoven, the Netherlands
Neil Budko	TU Delft, the Netherlands
Gabriela Ciuprina	Politehnica University of Bucharest, Romania
Georg Denk	Infineon, Germany
Herbert de Gersem	TU Darmstadt, Germany
Michael Günther	University of Wuppertal, Germany
Stefan Kurz	Bosch, Germany
Ulrich Langer	Johannes Kepler University Linz, Austria
Jan ter Maten	University of Wuppertal, Germany
Jörg Ostrowski	ABB, Switzerland
Ursula van Rienen	University of Rostock, Germany
Vittorio Romano	University of Catania, Italy
Ruth Vazquez Sabariego	KU Leuven, Belgium
Sebastian Schöps	TU Darmstadt, Germany
Caren Tischendorf	Humboldt University of Berlin, Germany

Standing Committee

Ursula van Rienen (Chair)	University of Rostock, Germany
Gabriela Ciuprina (Secretary)	Politehnica University of Bucharest, Romania

Wil Schilders	Treasurer, TU Eindhoven, the Netherlands
Michael Günther	University of Wuppertal, Germany
Jörg Ostrowski	ABB, Switzerland

Sponsors

TU Eindhoven
ROMSOC
CWI
Platform Wiskunde Nederland
4TU.AMI
ABB
NDNS+
ECMI's MSOEE

Acknowledgement

We would like the thank Eindhoven University of Technology, viz. the Centre for Analysis, Scientific Computing and Applications (CASA) within the Department of Mathematics and Computer Science and the Electromagnetics (EM) group within the Department of Electrical Engineering, and Delft University of Technology, Department of Applied Mathematics (DIAM), for their help and support in the organization of the SCEE 2022 Conference.

We are also grateful for the financial support from the Centre for Mathematics and Computer Science (CWI), Platform Wiskunde Nederland, the Applied Mathematics Institute of the four Universities of Technology in the Netherlands (4TU.AMI), the mathematics cluster NDNS+ (Nonlinear Dynamics of Natural Systems), the European Marie-Curie-Skłodowska Industrial Doctorate project Reduced Order Modelling, Simulation and Optimization of Coupled System (ROMSOC), ECMI's special interest group MSOEE, and finally ABB.

Last but not least, we would like to thank all the members of the standing committee and the program committee, who helped us very much in preparing and running the conference. The careful reviewing process was only possible with the help of the members of the scientific committee who were handling the reviewing process. The anonymous referees did a wonderful job that helped the authors to improve the quality of their contributions.

Finally, we express our gratitude to our colleagues from Springer Heidelberg for continued support and patience during the preparation of this volume.

ACKNOWLEDGMENT

Contents

Circuit Simulation and Design

Harmonic Balance with Small Signal Perturbation

Kai Bittner, Martin K. Steiger, and Hans Georg Brachtendorf(✉)

University of Applied Sciences of Upper Austria, 4232 Hagenberg, Austria
kai.bittner@zeiss.com,
{Martin.Steiger,Hans-Georg.Brachtendorf}@fh-hagenberg.at

Abstract. Investigating perturbations of a periodic steady state of an electric circuit is of interest e.g. for small signal responses, noise analysis or the generation of X-parameter models. We present a method based on Harmonic Balance, to compute the Fourier coefficients of the circuit response for a small signal perturbation of the input. The relation to two-tone Harmonic Balance is investigated and it is shown that under suitable conditions the perturbation method can approximate the full two-tone solution at extremely lower costs. The method is tested on a Gilbert mixer circuit.

1 Introduction

Many problems require the computation of the distortion of the periodic steady state (PSS) of a circuit if a small signal perturbation is applied. For instance, if the small signal response itself is of interest [1]. Furthermore, noise analysis [2] is based on the injection of small random signals. Another application is the generation of X-parameters [3,4], which provide a behavioral model for a neighborhood of the PSS.

Here, we present a frequency domain method, which is based on Harmonic Balance (HB) [5]. If HB is suitable for the simulated circuit then the perturbed solution is obtained easily, with only little extra cost, compared to the HB for the unperturbed PSS. In Sect. 2 we reformulate the problem as an infinite system of equations for the Fourier coefficients. Simplifications for real-valued signals are given in Sect. 3. In Sect. 4 we introduce a discretization scheme, which results in a linear system of equations, for a truncated sequence of Fourier coefficients. The relation to multi-tone HB is described in Sect. 5. In Sect. 6 we describe how our method can be utilized for the computation of X-parameters. The method is tested on a Gilbert cell mixer circuit in Sect. 7.

2 Small Signal Distortion of a Periodic Steady State

Consider the circuit equations from modified nodal analysis

$$\tfrac{d}{dt}q\big(x(t)\big) + f\big(x(t)\big) + s(t) = 0, \tag{1}$$

where $x(t) \in \mathbb{R}^n$ is the vector of unknowns, $f(x) \in \mathbb{R}^n$ the vector sums of static currents entering each node, $q(x) \in \mathbb{R}^n$ the vector sums of charges and magnetic fluxes, and $s(t) \in \mathbb{R}^n$ is the vector of time-dependent sources.

M. van Beurden et al. (Eds.): SCEE 2022, MI 43, pp. 3–10, 2024.
https://doi.org/10.1007/978-3-031-54517-7_1

Let $x(t)$ be a PSS, i.e., $x(t)$ solves (1) and for all t

$$s(t) = s(t+T),\; x(t) = x(t+T).$$

A perturbation $\tilde{s}(t) = s(t) + \Delta s(t)$ will lead to the perturbed solution $\tilde{x}(t) = x(t) + \Delta x(t)$. For small $\Delta s(t)$ and $\Delta x(t)$ one obtains by linearization of q and f an approximated version of the circuit equations, namely

$$\tfrac{d}{dt} q\big(x(t)\big) + \tfrac{d}{dt}\big(C(t)\,\Delta x(t)\big) + f\big(x(t)\big) + G(t)\,\Delta x(t) + s(t) + \Delta s(t) = 0, \qquad (2)$$

where the Jacobians $C(t) := \frac{dq}{dx}\big(x(t)\big)$ and $G(t) := \frac{df}{dx}\big(x(t)\big)$ are T-periodic matrix-valued functions. Taking the difference of (2) and (1) one obtains the linear, time-variant differential algebraic equation

$$\tfrac{d}{dt}\big(C(t)\,\Delta x(t)\big) + G(t)\,\Delta x(t) + \Delta s(t) = 0 \qquad (3)$$

for the perturbation Δx of the PSS $x(t)$.

Now we consider a harmonic perturbation $\Delta s(t) = \hat{s}\, e^{i(m\omega + \Delta\omega)t}$, where $\omega = \frac{2\pi}{T}$ is the angular frequency, the amplitude $\hat{s} \in \mathbb{C}^N$ is small, and $\Delta\omega \in \mathbb{R}$ is a frequency offset which can be chosen such that $|\Delta\omega| \le \frac{\omega}{2}$ for suitable $m \in \mathbb{Z}$. With the Fourier expansions

$$C(t) = \sum_{k\in\mathbb{Z}} C_k\, e^{ik\omega t}, \quad G(t) = \sum_{k\in\mathbb{Z}} G_k\, e^{ik\omega t}, \quad \Delta x(t) = \sum_{k\in\mathbb{Z}} X_k e^{i(k\omega + \Delta\omega)t}$$

one obtains from (3)

$$\frac{d}{dt}\Big(\sum_{\ell\in\mathbb{Z}} e^{i(\ell\omega+\Delta\omega)t} \sum_{k\in\mathbb{Z}} C_{\ell-k} X_k\Big) + \sum_{\ell\in\mathbb{Z}} e^{i(\ell\omega+\Delta\omega)t} \sum_{k\in\mathbb{Z}} G_{\ell-k} X_k + \hat{s}\, e^{i(m\omega+\Delta\omega)t} = 0, \qquad (4)$$

For the expansion of Δx we have to assume that the homogeneous equations (i.e. for $\Delta s(t) = 0$) have only the trivial solution. This means the circuit does not contain an oscillator. Equating coefficients in (4) now yields

$$i(\omega\ell + \Delta\omega) \sum_{k\in\mathbb{Z}} C_{\ell-k} X_k + \sum_{k\in\mathbb{Z}} G_{\ell-k} X_k + \hat{s}\,\delta_{\ell,m} = 0, \qquad \ell \in \mathbb{Z}, \qquad (5)$$

where $\delta_{\ell,m}$ is the Kronecker delta.

3 Real-Valued Signals

In practice, we can assume that $q(x)$ and $f(x)$ are real-valued functions. This implies that $C(t)$ and $G(t)$ are real-valued, too, with Fourier coefficients satisfying $C_{-k} = \overline{C_k}$ and $G_{-k} = \overline{G_k}$. Since the stimulus $s(t)$ is in practice also a real-valued (typically sinusoidal) signal, it is of interest how the solution of (3) can be used for the solution of real-valued problems. It turns out that this requires only little extra effort.

In a preliminary step, solving (3) for the conjugate complex $\overline{\Delta s(t)}$ leads in (5) to the equations

$$i(\omega\ell - \Delta\omega) \sum_{k\in\mathbb{Z}} C_{\ell-k} X_k^* + \sum_{k\in\mathbb{Z}} G_{\ell-k} X_k^* + \bar{\hat{s}}\,\delta_{\ell,-m} = 0,$$

with the solution coefficient X_k^*. Substituting $\ell \to -\ell$ and $k \to -k$ we obtain due to $C_{-k} = \overline{C_k}$ and $G_{-k} = \overline{G_k}$ that

$$\overline{i(\omega\ell + \Delta\omega) \sum_{k\in\mathbb{Z}} C_{\ell-k} \overline{X_{-k}^*} + \sum_{k\in\mathbb{Z}} G_{\ell-k} \overline{X_{-k}^*} + \hat{s}\,\delta_{\ell,m}} = 0,$$

i.e. $X_{-k}^* = \overline{X_k}$ and thus the corresponding solution of (3) for $\overline{\Delta s(t)}$ is

$$\Delta x^*(t) = \sum_{k\in\mathbb{Z}} \overline{X_k}\, e^{-i(k\omega+\Delta\omega)t} = \overline{\Delta x(t)}.$$

Therefore, $\mathrm{Re}(\Delta x(t))$ and $\mathrm{Im}(\Delta x(t))$ are the solutions for the real-valued perturbations $\mathrm{Re}(\Delta s(t))$ and $\mathrm{Im}(\Delta s(t))$, respectively.

4 Discretization

Since (5) is an infinite system one needs to approximate it by a finite system of equations. Because one can expect the Fourier coefficients to decay for $|k| \to \infty$, we set $X_k = 0$, $|m_1 - k| > K$ for some $m_1 \in \mathbb{Z}$. To determine the remaining coefficients X_k, $k = m_1 - K, \ldots, m_1 + K$ one chooses $2K+1$ equations from (5), namely

$$i(\omega\ell + \Delta\omega) \sum_{k=m_1-K}^{m_1+K} C_{\ell-k} X_k + \sum_{k=m_1-K}^{m_1+K} G_{\ell-k} X_k + \hat{s}\,\delta_{\ell,m} = 0, \quad \ell = m_2 - K, \ldots, m_2 + K, \tag{6}$$

for some $m_2 \in \mathbb{Z}$. A possible choice would be $m_1 = m_2 = m$, to adapt to the center frequency. However, if the computation has to be repeated for several values of m, fixed values of m_1 and m_2 may be used to speed up computations, since only one *LU*-factorization has to be performed.

The regularity of the system matrix is equivalent to the fact that there is only the trivial solution to the homogeneous system. Since we have assumed this property for the original system (5), it should typically hold for a sufficiently good approximation (4).

The matrices C_k and G_k can be computed numerical integration, namely the trapezoidal rule, i.e.,

$$G_k = \frac{1}{T} \int_0^T G(t)\, e^{-ik\omega t}\, dt \approx \frac{1}{TN} \sum_{\ell=0}^{N-1} G\left(\tfrac{\ell T}{N}\right) e^{-2\pi i k\ell/N},$$

where $N > 2K+1$ to avoid aliasing. Therefore, an efficient computation of G_k, C_k, and $x_\ell := x(\frac{\ell T}{N})$ can be done by employing the Fast Fourier Transform, if a suitable N is chosen, e.g. a power of two.

In contrast to the method in [1], where the linear, time-variant system (3) is solved in the time domain by classical time stepping methods, the presented method works in the frequency domain. It is well suited if the matrix sequences (C_k) and (G_k) are decaying fast, which implies a fast decay of (X_k), too. This is a case if the PSS $x(t)$ is nearly sinusoidal, i.e., if the PSS can be computed by an HB efficiently, than the described perturbation method will perform very well, too.

5 Relation to Two-Tone Harmonic Balance (HB)

Two-tone signals are of the form $x(t) = \hat{x}(t,t)$, where

$$\hat{x}(t_1,t_2) = \hat{x}(t_1+T_1,t_2) = \hat{x}(t_1,t_2+T_2), \quad t_1,t_2 \in \mathbb{R}.$$

They have a bi-variate Fourier expansion of the form

$$x(t) = \sum_{k,\ell\in\mathbb{Z}} \hat{X}_{k,\ell}\, e^{2\pi i(k/T_1+\ell/T_2)t}.$$

With the substitution $k \to k - n\ell$ and $X_{k,\ell} = \hat{X}_{k-n\ell,\ell}$ this becomes

$$x(t) = \sum_{k,\ell\in\mathbb{Z}} X_{k,\ell}\, e^{i(k\omega+\ell\Delta\omega)t}, \tag{7}$$

where $\omega = \frac{2\pi}{T_1}$ and $\Delta\omega = \frac{2\pi}{T_2} - n\omega$. Typically, $n \in \mathbb{Z}$ is chosen to obtain a small frequency offset $\Delta\omega$, i.e., $|\Delta\omega| \le \frac{\omega}{2}$, which corresponds to the setting in Sect. 2. By a multi-rate HB [6,7] one can compute the coefficients $X_{k,\ell}$ for the PSS of a circuits driven by a two-tone signal.

The two-tone solution of (1) with the real-valued source term

$$\tilde{s}(t) = s(t) + \hat{s}\, e^{i(n\omega+\Delta\omega)t} + \overline{\hat{s}}\, e^{-i(n\omega+\Delta\omega)t}$$

with a T_1-periodic single-tone signal $s(t)$, can be approximated using the solution of (5) for sufficiently small $\hat{s}$. Due to linearity one solves for $s(t)$ by a single-tone HB, then solves (5) for $\Delta s(t) = \hat{s}\, e^{i(n\omega+\Delta\omega)t}$. The solution for $\overline{\Delta s(t)} = \overline{\hat{s}}\, e^{-i(n\omega+\Delta\omega)t}$ follows immediately from Sect. 3. By linear combination one obtains the solution

$$x(t) = \sum_{k\in\mathbb{Z}} \sum_{\ell=-1}^{1} \tilde{X}_{k,\ell}\, e^{i(k\omega+\ell\Delta\omega)t}, \tag{8}$$

where $\tilde{X}_{k,0}$ is the result of a single-tone HB with source $s(t)$, while $\tilde{X}_{k,1} = X_k$, $\tilde{X}_{k,-1} = \overline{X_{-k}}$ (cf. Sect. 3) are obtained from the subsequent perturbation approach (5). Obviously, for small $\hat{s}$ the expansion (8) will be a good approximation of the two-tone signal (7).

A general multi-tone analysis can be performed by computing the perturbed solution for several harmonic perturbations $\hat{s}_\ell\, e^{i(k_\ell\omega+\ell\Delta\omega_\ell)t}$ separately, and using the linearity for superpositions of the harmonics.

Note, that the computational cost of the perturbation approach is much smaller than for the full two-tone HB since it requires only the computation of a single-tone PSS with an essentially smaller system of equations to be solved. The cost for the final step of solving the linear system (5) equals essentially the cost of one Newton step in the preciding single-tone HB.

6 Extraction of X-Parameter Models

X-parameters [3,4] are a generalization of S-Parameters to describe the relation of power waves in electronic circuits or devices. While S-Parameter are used as behavioral model for linear systems, X-parameters describe the behavior of a non-linear network in the neighborhood of a PSS. For an N-port system with a large signal incident wave of fixed amplitude at port 1 the X-parameter model reads as

$$B_{p,k} = X^{(FB)}_{p,k}(|A_{1,1}|, f_0)\, P^k$$
$$+ \sum_{\substack{q=1,\ell=1 \\ (q,\ell)\neq(1,1)}}^{q=N,\ell=K} X^{(S)}_{p,k,q,\ell}(|A_{1,1}|, f_0)\, A_{q,\ell}\, P^{k-\ell} + X^{(T)}_{p,k,q,\ell}(|A_{1,1}|, f_0)\, \overline{A}_{q,\ell}\, P^{k+\ell},$$

where $A_{p,k}$ and $B_{p,k}$ denote the k-th Fourier coefficient of the incident and scattered wave at the p-th port, respectively. The first term describes the contribution of the large signal input $DC + A_{1,1}\, e^{2\pi i f_0} + \overline{A_{1,1}}\, e^{-2\pi i f_0}$, where $P = e^{i\,\arg(A_{1,1})}$ and the amplitude $|A_{1,1}|$ is fixed. The remaining terms contain variable small signal contributions.

Since the incident and scattered waves are related to voltages and currents at the ports by

$$a_p = \frac{U_p + I_p Z_p}{2\sqrt{\mathrm{Re}(Z_p)}} \qquad b_p = \frac{U_p - I_p \overline{Z_p}}{2\sqrt{\mathrm{Re}(Z_p)}}, \tag{9}$$

with the port impedance Z_p (often $50\,\Omega$), we can obtain the X-parameters of a given circuit with N ports as follows. The ports are connected to voltage sources with internal impedance Z_p. To obtain the large signal contribution $X^{(FB)}_{p,k}(|A_{1,1}|, f_0)$, a HB with input signal

$$V(t) = U_0 \cos(2\pi f_0 t)$$

at port 1, while all other sources are set to zero. The amplitude U_0 is chosen to get the proper value for $|A_{1,1}|$ as described e.g. in [4, Eq. (14)]. Using (9) we obtain

$$X^{(FB)}_{p,k}(|A_{1,1}|, f_0) = \frac{U_{p,k} - I_{p,k}\overline{Z_p}}{2\sqrt{\mathrm{Re}(Z_p)}},$$

where $U_{p,k}$ and $I_{p,k}$ are the k-th Fourier coefficient of voltage and current at the p-th port, respectively, which can be extracted immediately from the HB solution.

To compute the remaining X-parameters one can now use the perturbation method described in Sect. 2. We reformulate the problem as an infinite system of equations for the Fourier coefficients. Simplifications for real-valued signals are given in Sect. 3 and 4. First the matrices G_k and C_k are computed. Now, to determine $X^{(S)}_{p,k,q,\ell}$ and $X^{(T)}_{p,k,q,\ell}$ we solve (6) with $\Delta\omega = 0$ and ℓ chosen as the corresponding X-parameter index. The source term $\hat{s}$ is chosen as if all voltage sources are set to zero except for q-th port (as it is done for the computation of S-Parameters). Due to linearity, the voltage at q-th port can be any non-zero value, e.g. 1V is usually a good choice. Using again the relation (9) we obtain the X-parameters as

$$X^{(S)}_{p,k,q,\ell} = \frac{U_{p,k} - I_{p,k}\overline{Z_p}}{U_{q,\ell} + I_{q,\ell} Z_q} \qquad X^{(T)}_{p,k,q,\ell} = \frac{\overline{U_{p,-k}} - \overline{I_{p,-k}}\, Z_p}{U_{q,\ell} + I_{q,\ell} Z_q}.$$

One should take into account, that the system (6) has to be solved for the same ℓ for different ports q, i.e., the same matrix for several right hand sides. To save computation time, one should therefore first set up the system matrix for (6) (for each $\ell = 1, \ldots, K$), do an LU-factorization, and solve for each right hand side (port) by forward and backward substitution.

7 Numerical Test

We have tested the method on a Gilbert cell mixer [8, Fig. 2.8] with a local oscillator input of 100 MHz and an radio frequency (RF) input of 99.9 MHz yielding a frequency offset $\Delta\omega = -0.1$ MHz. The perturbation method as well as the 2-tone HB were performed with a cutoff of the Fourier series after $K = 31$ and $N = 64$ sampling points for the trapezoidal rule. The RF signal is treated as perturbed input and the amplitude is swept from 0.1mV to 0.4V. In Fig. 1 one can see the absolute value of Fourier coefficients of the output signal for local oscillator amplitude 1 V plotted against the RF amplitude. As one can see the coefficients $\tilde{X}_{0,1}$ and $X_{0,1}$ agree very well for RF input almost up to 0.1 V. The coefficient $X_{0,2}$ (only obtained by two-tone HB) is included as a measure of non-linearity.

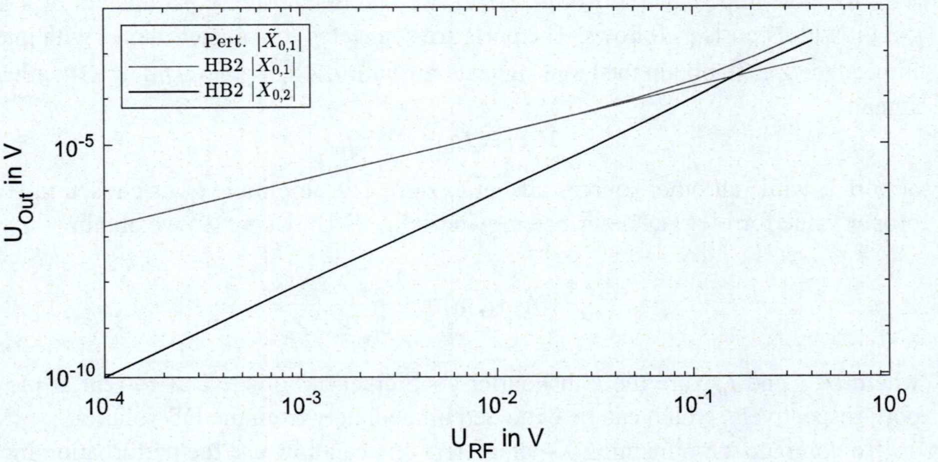

Fig. 1. Fourier coefficient $X_{0,1}$ for perturbation technique and two-tone HB, as well as $X_{0,2}$ for two-tone HB as reference

The above result is confirmed by the relative ℓ^2-error (i.e. in the Euclidean norm) over all coefficients (Fig. 2), obtained by comparison to a two-tone HB of high precision.

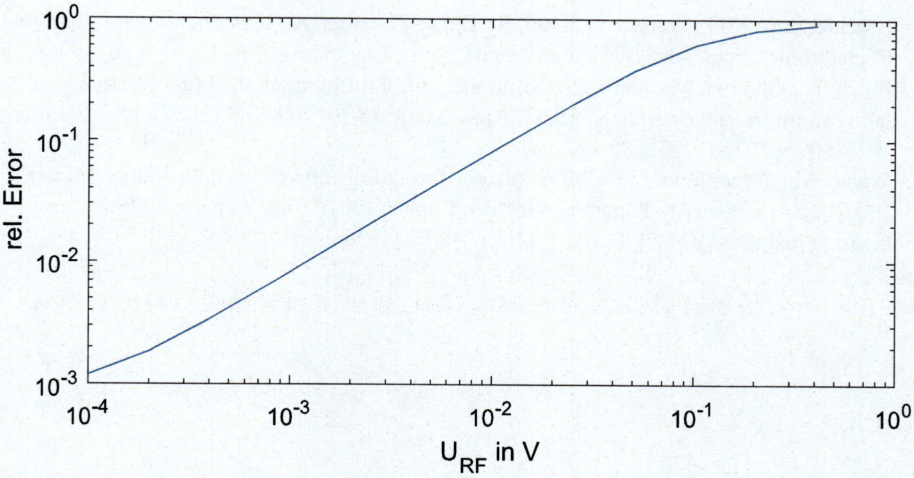

Fig. 2. ℓ_2-error of perturbation technique.

8 Conclusion

The presented HB perturbation method permits the efficient analysis of the distortion of a PSS by a small signal. The results allow a first assessment of the qualitative characteristics of RF circuits with moderate nonlinear behavior.

Acknowledgements. This project AMOR ATCZ203 has been co-financed by the European Union using financial means of the European Regional Development Fund (INTERREG) for sustainable cross boarder cooperation. Further information on INTERREG Austria-Czech Republic is available at https://www.at-cz.eu/at.

References

1. Okumura, M., Sugawara, T., Tanimoto, H.: An efficient small signal frequency analysis method of nonlinear circuits with two frequency excitations. IEEE Trans. Comput.-Aided Des. Integr. Circuits Syst. **9**(3), 225–235 (1990). https://doi.org/10.1109/43.46798
2. ter Maten, E.J.W., Fijnvandraat, J.G., Lin, C., Peters, J.M.F.: Periodic AC and periodic noise in RF simulation for electronic circuit design. In: Antreich, K., Bulirsch, R., Gilg, A., Rentrop, P. (eds.) Modeling, Simulation, and Optimization of Integrated Circuits, Birkhäuser, Basel, pp. 121–134 (2003)
3. Verspecht, J., Root, D.: Polyharmonic distortion modeling. IEEE Microwave Mag. **7**(3), 44–57 (2006). https://doi.org/10.1109/MMW.2006.1638289
4. Comberiate, T.M., Schutt-Ainé, J.E.: LIM2X: generating X-parameters in the time domain using the latency insertion method. IEEE Trans. Compon. Packag. Manuf. Technol. **4**(7), 1136–1143 (2014). https://doi.org/10.1109/TCPMT.2014.2318034
5. Kundert, K., Sangiovanni-Vincentelli, A.: Simulation of nonlinear circuits in the frequency domain. IEEE Trans. Comput.-Aided Design Integr. Circuits Syst. **5**(4), 521–535 (1986). https://doi.org/10.1109/TCAD.1986.1270223

6. Brachtendorf, H.G., Welsch, G., Laur, R., Bunse-Gerstner, A.: Numerical steady state analysis of electronic circuits driven by multi-tone signals. Electr. Eng. **79**(2), 103–112 (1996)
7. Pulch, R., Günther, M., Knorr, S.: Multirate partial differential algebraic equations for simulating radio frequency signals. Eur. J. Appl. Math. **18**(6), 709–743 (2007). https://doi.org/10.1017/S0956792507007188
8. Wang, M.: Reconfigurable CMOS mixers for radio-frequency applications. Master's thesis, Queen's University Kingston, Ontario, Canada (2010). https://qspace.library.queensu.ca/bitstream/handle/1974/5712/Wang_Min_201006_MASc.pdf

A Projective-Based Formalism for Symmetric Modeling of Electrical Circuits

Ricardo Riaza(✉)

Departamento de Matemática Aplicada a las Tecnologías de la Información y las Comunicaciones and Information Processing and Telecommunications Center
Escuela Técnica Superior de Ingenieros de Telecomunicación,
Universidad Politécnica de Madrid, Ciudad Universitaria s/n, 28040 Madrid, Spain
ricardo.riaza@upm.es

Abstract. We survey in this contribution some recent ideas involving the use of a homogeneous formalism to set up electrical circuit models. Broadly, the goal is to avoid any lack of generality in the modeling process by avoiding unnecessarily restrictive assumptions in the form of the characteristics of the circuit devices. We discuss how to use this approach in the framework of nodal analysis, aiming at the development of computationally efficient models. Except for some minor technicalities arising in the index analysis, the discussion is deliberately kept at a simple level.

1 Introduction

Choosing either a current-controlled or a voltage-controlled description for any given circuit device always entails both a lack of generality and of symmetry in symbolic circuit analysis. This is already the case in the most elementary stages of circuit theory, when one chooses between the forms $v = Ri$ and $i = Gv$ to write Ohm's law [16]. The same happens in the nonlinear context: choosing between any of the two nonlinear counterparts of such relations, namely $v = f(i)$ or $i = g(v)$ for whatever functions f or g, necessarily excludes some devices from the analysis. Focusing on specific forms of f or g which are known to admit a global inverse entails an obvious loss of generality. Similar remarks apply to capacitors, inductors and memristors, now involving the charge/flux variables in their characteristics.

A way to circumvent these limitations comes from choosing a homogeneous description for the devices (as a historical anecdote, worth noting is that homogeneous coordinates were created, in a completely different context, by A. F. Möbius in 1827 [14], exactly the year of publication of Ohm's book [16], what motivates an anachronic speculation about what the evolution of circuit theory might have been should Ohm have used this formalism). In the linear resistive setting, such a homogeneous standpoint would amount to writing Ohm's law as

$$Pv - Qi = 0 \tag{1}$$

for two real parameters P, Q which do not vanish simultaneously. The assumptions $P \neq 0$ or $Q \neq 0$ imply that the resistance or the conductance, respectively, are well-defined

M. van Beurden et al. (Eds.): SCEE 2022, MI 43, pp. 11–22, 2024.
https://doi.org/10.1007/978-3-031-54517-7_2

as $R=Q/P$ or $G=P/Q$. The variables $(P:Q)$ are defined only up to a nonzero constant and therefore define a pair of homogeneous coordinates of a point lying on a *projective line* $\mathbb{RP}$, which seems to provide the natural mathematical context to accommodate all possible resistance values (including zero and infinity) in a comprehensive manner. The same ideas apply to linear circuits in sinusoidal steady state, working in this case in the complex domain.

Still in the linear context and from the same homogeneous perspective, a way to reduce model dimensionality without losing generality comes from writing Ohm's law in parametric form, that is,

$$i = Pu \tag{2a}$$

$$v = Qu, \tag{2b}$$

where u is an adimensional variable from which both the current and the voltage can be explicitly computed. In turn, this provides a route to extend the approach to the nonlinear context, namely by writing the characteristics of nonlinear (resistive) devices as

$$i = \psi(u) \tag{3a}$$

$$v = \zeta(u), \tag{3b}$$

for certain functions ψ, ζ. Such a global parametric description is well-defined for a broad class of devices, as discussed in [21,22]. Current-controlled and voltage-controlled descriptions are of course included in (3), just by letting ψ or ζ be the identity map. However, (3) also accommodates, for example, hysteresis loops not admitting a global description in terms of either the current or the voltage. The form (3) is also of interest when, for the sake of generality, one wants to leave the nature of the device unspecified in the modeling process, at least up to a certain stage.

A systematic approach to linear and nonlinear circuit analysis stemming from these ideas is discussed in [20,21]. One of the challenges of this approach is to accommodate it within a computationally efficient framework, for simulation purposes. Needless to say, nodal analysis is the key tool to set up automatically circuit models with a reduced dimensionality, something which is essential in large-scale circuit analysis and simulation. In this direction, the main purpose of this contribution is to discuss some features of nodal models of nonlinear circuits in which *some* branches are given a homogeneous description of the form (3), for any of the reasons discussed in the previous paragraph. This will be addressed in Sects. 2 and 3: in particular, the latter section provides an index characterization of the resulting differential-algebraic circuit models. For simplicity, we will restrict the use of homogeneous descriptions to (some) resistive devices, by assuming that capacitors are voltage-controlled and inductors current-controlled. For space restrictions, other analytical aspects involving homogeneous descriptions of circuits will be discussed in less detail in Sect. 4. Finally, Sect. 5 compiles some concluding remarks.

2 Nodal Models

Digraphs and the Incidence Matrix. We refer the reader to [1,3–5] for background on graph and digraph theory and, in particular, for details on the claims which are here presented without proof. We assume throughout the document that the digraph underlying the circuit has n nodes (vertices) and m branches (edges), and also that it is connected and has no branches with just one incident node. Let us choose a reference node and map the remaining nodes and the branches onto the sets $\{1, \ldots, n-1\}$ and $\{1, \ldots, m\}$, respectively. In this setting, the entries of the *reduced incidence matrix* $A = (a_{jk}) \in \mathbb{R}^{(n-1)\times m}$ are defined as $a_{jk} = 1$ (resp. -1) if the k-th branch leaves (resp. enters) node j, and 0 otherwise. Any reduced incidence matrix of a digraph is totally unimodular (that is, the determinant of all square submatrices is either 1, -1 or 0), and the determinant of a square submatrix of order $n-1$ is ± 1 if and only if the branches corresponding to its columns define a spanning tree.

Nodal Analysis. After choosing a reference node, Kirchhoff laws can be easily described in terms of the corresponding reduced incidence matrix as $Ai = 0$, where i is the m-dimensional vector of branch currents, and $v = A^{\mathsf{T}}e$, where v and e stand for the vectors of branch voltages and node potentials, with dimensions m and $n-1$, respectively.

We will split the incidence matrix according to the nature of the circuit device lying on each branch: specifically, we will let A_c, A_l, A_g, A_v, A_i stand for the submatrices of A defined by the columns corresponding to capacitors, inductors, voltage-controlled resistors, voltage sources and current sources, respectively. Resistors with a homogeneous description will be later labeled with the subscript h. The same notational convention will apply to the components of the voltage and current vectors; that is, v_c, v_l, etc. will be defined from the components of the voltage vector which correspond to capacitors, inductors and so on.

As indicated in Sect. 1, we will assume for simplicity that capacitors are voltage-controlled by a smooth relation of the form $q_c = \eta(v_c)$, with $C(v_c)$ standing for the incremental capacitance matrix $\eta'(v_c)$. Analogously, inductors will be assumed to be defined by a smooth current-controlled characteristic reading as $\varphi_l = \phi(i_l)$; we will let $L(i_l) = \phi'(i_l)$ denote the incremental inductance matrix. Furthermore, $C(v_c)$ and $L(i_l)$ will be assumed to be nonsingular (invertible) for all values of v_c and i_l.

In the context of nodal analysis, it is very common to assume that resistors are voltage-controlled by a smooth relation of the form $i_g = \gamma(v_g)$. With this setup, the expressions provided above for Kirchhoff laws make it possible to write the nodal equations in the form

$$C(v_c)v_c' = i_c \tag{4a}$$
$$L(i_l)i_l' = A_l^{\mathsf{T}} e \tag{4b}$$
$$0 = A_c i_c + A_v i_v + A_l i_l + A_g \gamma(A_g^{\mathsf{T}} e) + A_i i_s(t) \tag{4c}$$
$$0 = v_c - A_c^{\mathsf{T}} e \tag{4d}$$
$$0 = v_s(t) - A_v^{\mathsf{T}} e, \tag{4e}$$

where $i_s(t)$ and $v_s(t)$ denote the excitations in the (assumed independent) current and voltage sources. By eliminating capacitor voltages and currents one easily gets the Modified Nodal Analysis (MNA) model, namely

$$A_cC(A_c^{\mathsf{T}}e)A_c^{\mathsf{T}}e' = -A_vi_v - A_li_l - A_g\gamma(A_g^{\mathsf{T}}e) - A_ii_s(t) \tag{5a}$$
$$L(i_l)i_l' = A_l^{\mathsf{T}}e \tag{5b}$$
$$0 = v_s(t) - A_v^{\mathsf{T}}e, \tag{5c}$$

widely used in nonlinear circuit simulation [7,8,12,18,19,24,25]. With some extra work, later results can be naturally extended to MNA models. However, we mostly restrict the analysis to models of the form (4) for the sake of simplicity.

Homogeneous Description of (Some) Resistors. The aforementioned voltage-control assumption on resistors may be unnecessarily restrictive in different practical situations. Even if this assumption may be reasonable for most resistors, it excludes resistive devices for which no global controlling variable exist and, more important, rules out the chance to leave the nature of the device unspecified up to a certain stage of the modeling process: for instance, the latter is important when one wants the model to account for *both* a voltage-controlled or a current-controlled device in a given circuit branch (think e.g. of an ideal switch). Along the lines presented in Sect. 1, a way to do so is to consider a second class of resistors (besides the voltage-controlled ones already introduced above) and give them a parametric description of the form

$$i_h = \psi(u_h) \tag{6a}$$
$$v_h = \zeta(u_h). \tag{6b}$$

Notice the use of the subindex h for these resistors and be aware of the fact that ψ and ζ are now vector-valued (these maps were used for a single device in Sect. 1). Also worth noting is that the family described by (6) includes current-controlled resistors as a particular case, obtained by setting $\psi(u_h) = u_h$. Under a smoothness assumption on such resistors and provided that there are no coupling effects, we will let P_j and Q_j (with j indexing the set of homogeneous resistors) stand for the corresponding entries of ψ' and ζ', respectively. We further assume that, for each homogeneous resistor, at least one of the parameters P_j or Q_j is not zero.

In this framework, a nodal model with the structure depicted in (4) takes the form

$$C(v_c)v_c' = i_c \tag{7a}$$
$$L(i_l)i_l' = A_l^{\mathsf{T}}e \tag{7b}$$
$$0 = A_ci_c + A_vi_v + A_li_l + A_g\gamma(A_g^{\mathsf{T}}e) + A_h\psi(u_h) + A_ii_s(t) \tag{7c}$$
$$0 = v_c - A_c^{\mathsf{T}}e \tag{7d}$$
$$0 = v_s(t) - A_v^{\mathsf{T}}e \tag{7e}$$
$$0 = \zeta(u_h) - A_h^{\mathsf{T}}e, \tag{7f}$$

whereas the MNA reads as

$$A_cC(A_c^{\mathsf{T}}e)A_c^{\mathsf{T}}e' = -A_vi_v - A_li_l - A_g\gamma(A_g^{\mathsf{T}}e) - A_h\psi(u_h) - A_ii_s(t) \tag{8a}$$
$$L(i_l)i_l' = A_l^{\mathsf{T}}e \tag{8b}$$
$$0 = v_s(t) - A_v^{\mathsf{T}}e \tag{8c}$$
$$0 = \zeta(u_h) - A_h^{\mathsf{T}}e. \tag{8d}$$

3 Index Analysis

The notion of the *index* is essential for the analysis and the numerical treatment of differential-algebraic systems such as (4), (5), (7) or (8) [11, 12, 19]. In particular, index-one systems are important because they admit (at least in a local sense) a state-space reduction in terms of the differential variables arising in the model, and also because they are better suited for numerical treatment than higher-index systems.

Under the assumption that $C(v_c)$ and $L(i_l)$ are nonsingular, the nodal models (4) and (7) display a *semiexplicit* form: this makes the index analysis a bit simpler than for the MNA counterparts (5) and (8). Indeed, for semiexplicit equations the index-one condition amounts to the nonsingularity of the matrix of partial derivatives of the functions defining the algebraic part of the system (namely, the equations which do not involve time derivatives) with respect to the algebraic variables (again, those for which no time derivative appears in the model).

Specifically, for the model (7), the algebraic variables are e, i_c, i_v and u_h, and the index-one condition is equivalent to the nonsingularity of the matrix

$$\begin{pmatrix} A_g G A_g^{\top} & A_c & A_v & A_h P \\ -A_c^{\top} & 0 & 0 & 0 \\ -A_v^{\top} & 0 & 0 & 0 \\ -A_h^{\top} & 0 & 0 & Q \end{pmatrix}, \tag{9}$$

where $G = \gamma'(v_g)$ is diagonal (that is, no coupling effects among resistors are allowed), and P and Q are diagonal matrices whose diagonal entries are defined by the corresponding parameters of the different homogeneous resistors. Note that in (9) we omit the dependence of G on $A_g^{\top} e$ and of P and Q on the homogeneous variables u_h.

Our main result (Theorem 1 below) presents an index-one characterization of (7) in terms of the structure of spanning trees in the circuit. This approach can be traced back to Kirchhoff's seminal paper [10], and is of particular interest for nonpassive problems, namely, those in which some of the conductances G_i and/or the ratios P_j/Q_j or Q_j/P_j for homogeneous resistors (remember that, for each j, at least one of the parameters P_j and Q_j does not vanish) become zero or negative. Note that the subindices of the individual devices correspond here to the global numbering of the digraph edges and not to their position in the G, P and Q matrices. In the same direction, we assume that each spanning tree T is defined by the index set of its constituting branches or *twigs*; that is, every such T amounts to a subset of $\{1, \ldots, m\}$ with $n-1$ elements which specify a spanning tree. The complement of T in $\{1, \ldots, m\}$ is written as $\overline{T}$ and stands for the index set of the cotree branches or *chords*. Note also that the absence of proper trees implicitly defines a null sum in (10) below, ruling out such configurations from the index-one setting.

Theorem 1. *The determinant of the matrix (9) equals the polynomial*

$$\sum_{T \in \mathscr{T}_p} \prod_{\substack{i \in T_g \\ j \in T_h \\ k \in \overline{T}_h}} G_i P_j Q_k, \tag{10}$$

where $\mathscr{T}_p$ stands for the set of proper trees (namely, those including all voltage sources and capacitors, and neither current sources nor inductors), and T_g, T_h denote the indices of T corresponding to twigs with voltage-controlled resistors and with homogeneous resistors, respectively, whereas $\overline{T}_h$ denotes the set of chords with homogeneous resistors.

Therefore, provided that $C(v_c)$ and $L(i_l)$ are nonsingular, the model (7) is index one exactly for the values of the variables $v_g = A_g^{\mathsf{T}}e$, u_h which do not annihilate (10).

Proof. For notational simplicity, let us join together capacitors and voltage sources under the subscript *cv*: we are therefore led to characterize the determinant of

$$\begin{pmatrix} A_g G A_g^{\mathsf{T}} & A_{cv} & A_h P \\ -A_{cv}^{\mathsf{T}} & 0 & 0 \\ -A_h^{\mathsf{T}} & 0 & Q \end{pmatrix}.$$

We will do so by looking at this matrix as the Schur complement [9] of the identity block in

$$\begin{pmatrix} 0 & A_{cv} & A_h P & A_g G \\ -A_{cv}^{\mathsf{T}} & 0 & 0 & 0 \\ -A_h^{\mathsf{T}} & 0 & Q & 0 \\ -A_g^{\mathsf{T}} & 0 & 0 & I \end{pmatrix}$$

and, in turn, the latter matrix as the one obtained after setting $P_0 = I$ and $Q_0 = 0$ in

$$\begin{pmatrix} 0 & A_{cv} P_0 & A_h P & A_g G \\ -A_{cv}^{\mathsf{T}} & Q_0 & 0 & 0 \\ -A_h^{\mathsf{T}} & 0 & Q & 0 \\ -A_g^{\mathsf{T}} & 0 & 0 & I \end{pmatrix}.$$

By denoting $\tilde{A} = (A_{cv}\ A_h\ A_g)$, $\tilde{P} = \mathrm{diag}(P_0, P, G)$, $\tilde{Q} = \mathrm{diag}(Q_0, Q, I)$, this matrix has the form

$$\begin{pmatrix} 0 & \tilde{A}\tilde{P} \\ -\tilde{A}^{\mathsf{T}} & \tilde{Q} \end{pmatrix}.$$

By multiplying the first $n-1$ columns by -1 and performing an obvious permutation of columns, one can easily check that

$$\det \begin{pmatrix} 0 & \tilde{A}\tilde{P} \\ -\tilde{A}^{\mathsf{T}} & \tilde{Q} \end{pmatrix} = (-1)^{(n-1)(\tilde{m}+1)} \det \begin{pmatrix} \tilde{A}\tilde{P} & 0 \\ \tilde{Q} & \tilde{A}^{\mathsf{T}} \end{pmatrix},$$

where $\tilde{m}$ is the total number of voltage sources, capacitors and resistors. Now, the latter determinant can be computed by means of a generalized Laplace expansion (see e.g. [9]) along the first $n-1$ rows. In light of the properties of determinants of the square submatrices of the incidence matrix and the diagonal form of $\tilde{P}$, it is not difficult to see that

$$\det \begin{pmatrix} \tilde{A}\tilde{P} & 0 \\ \tilde{Q} & \tilde{A}^{\mathsf{T}} \end{pmatrix} = \sum_{T \in \mathscr{T}_{cvr}} \left((-1)^{\sigma(T)} \det A_T \left(\prod_{j \in T} \tilde{P}_j \right) \det \left(\tilde{Q}_{\overline{T}}\ \tilde{A}^{\mathsf{T}} \right) \right)$$

where $\tilde{Q}_{\overline{T}}$ is the submatrix of $\tilde{Q}$ defined by the columns indexed by $\overline{T}$ and $\sigma(T)$ is the exponent which corresponds to the spanning tree T in the Laplace expansion, namely

$$1+\ldots+n-1+\sum_{j\in T} j.$$

Additionally, $\mathscr{T}_{cvr}$ denotes the family of spanning trees just including capacitors, voltage sources and resistors.

In turn, after a transposition and a permutation of rows one gets

$$\det\left(\tilde{Q}_{\overline{T}}\ \tilde{A}^{\mathsf{T}}\right) = (-1)^{(n-1)(\tilde{m}-n+1)} \det\begin{pmatrix} \tilde{A} \\ \tilde{Q}^{\mathsf{T}}_{\overline{T}} \end{pmatrix} = (-1)^{(n-1)(\tilde{m}-n+1)}(-1)^{\sigma(T)} \det A_T \prod_{k\in\overline{T}} \tilde{Q}_k,$$

where the last identity owes to the structure of the matrix $\tilde{Q}_{\overline{T}}$. Altogether, the different expressions obtained for the successive determinants yield

$$\det\begin{pmatrix} 0 & \tilde{A}\tilde{P} \\ -\tilde{A}^{\mathsf{T}} & \tilde{Q} \end{pmatrix} = \sum_{T\in\mathscr{T}_{cvr}} \left((-1)^{2[(n-1)(\tilde{m}+1)+\sigma(T)]-(n-1)n} (\det A_T)^2 \prod_{j\in T} \tilde{P}_j \prod_{k\in\overline{T}} \tilde{Q}_k \right).$$

Note that the exponent of -1 has the same parity as $n(n-1)$, which is necessarily an even number. The fact that $\det A_T = \pm 1$ then implies that the determinant above amounts to

$$\sum_{T\in\mathscr{T}_{cvr}} \left(\prod_{j\in T} \tilde{P}_j \prod_{k\in\overline{T}} \tilde{Q}_k \right).$$

Finally, we make use of the definition of $\tilde{P}$ and $\tilde{Q}$ to show that the latter expression yields (10). First, the conditions $P_0 = I$ and $Q_0 = 0$ for the entries which correspond to capacitors and voltage sources reduce the range of the sum to the family of proper trees (namely, those including *all* voltage sources and capacitors) since, otherwise, a voltage source or a capacitor in the cotree annihilates the corresponding term because of the condition $\tilde{Q}_k = 0$. For voltage-controlled resistors and by construction of $\tilde{P}$ and $\tilde{Q}$, each $\tilde{P}_i$-entry is simply the incremental conductance G_i, whereas the corresponding entry in $\tilde{Q}$ amounts to 1. This is responsible for the factors $\prod_{i\in T_g} G_i$ in (10). Finally, homogeneous resistors contribute the remaining factors in (10) simply because $\tilde{P}_j = P_j$ and $\tilde{Q}_k = Q_k$, again by the definition of $\tilde{P}$ and $\tilde{Q}$. This completes the proof. □

Some particular cases merit additional remarks: in the absence of homogeneous resistors, (10) essentially amounts to Maxwell's characterization of the admittance nodal matrix [13], since only the conductances within the twigs of proper trees are involved. By contrast, when all resistors are given a homogeneous description, then (10) amounts to

$$\sum_{T\in\mathscr{T}_p} \prod_{\substack{j\in T_h \\ k\in\overline{T}_h}} P_j Q_k, \tag{11}$$

just formulated in terms of the homogeneous parameters P, Q. In particular, from (11) one can also accommodate the case in which all the parameters P_j are nonzero, an assumption which allows for a local current-controlled description of all resistors: after dividing the (11) by the product of all P_j-parameters, this expression then amounts to the sum of chord resistance products extended over the set of proper trees. The example that follows should be of help in clarifying the expressions arising in all these contexts. Let us also emphasize that, in strictly locally passive problems (in which all incremental conductances are positive and, for each j, P_j and Q_j do not vanish and have the same sign), all terms in (10) have the same sign and therefore the sum is not zero, provided that there exists at least one proper tree. The latter condition is equivalent to the absence of VC-cycles and IL-cutsets, a topological condition which is well-known to characterize index-one configurations in strictly locally passive circuits [7,25].

Example. For illustration, consider the simple circuit depicted in Fig. 1. To keep the terms of the discussion as simple as possible, assume that both the capacitor and the inductor are linear, with nonzero capacitance and inductance C and L, respectively.

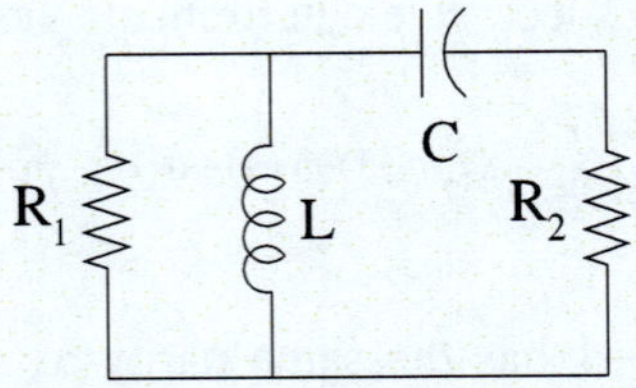

Fig. 1. An elementary circuit example.

As indicated above, in nodal analysis it is very often the case that resistors are assumed to be voltage-controlled. In our setting, this would mean that they are governed by the (assumed differentiable) characteristics $i_1 = \gamma_1(v_1)$ and $i_2 = \gamma_2(v_2)$. By directing the branches top-down or (the one accommodating the capacitor) towards the right, the nodal equations (4) take the form

$$Cv_c' = i_c \tag{12a}$$
$$Li_l' = e_1 \tag{12b}$$
$$0 = i_c + i_l + \gamma_1(e_1) \tag{12c}$$
$$0 = -i_c + \gamma_2(e_2) \tag{12d}$$
$$0 = v_c - e_1 + e_2, \tag{12e}$$

with the subindices 1 and 2 in the node potentials corresponding to the NW and NE nodes. The determinant of the matrix of partial derivatives of (12c)–(12e) with respect to the algebraic variables e_1, e_2, i_c is easily seen to be $G_1 + G_2$, with $G_j = \gamma'_j(e_j)$ for $j = 1, 2$. Notice that $G_1 + G_2$ corresponds to the sum of conductances in the circuit proper trees, shown in Fig. 2.

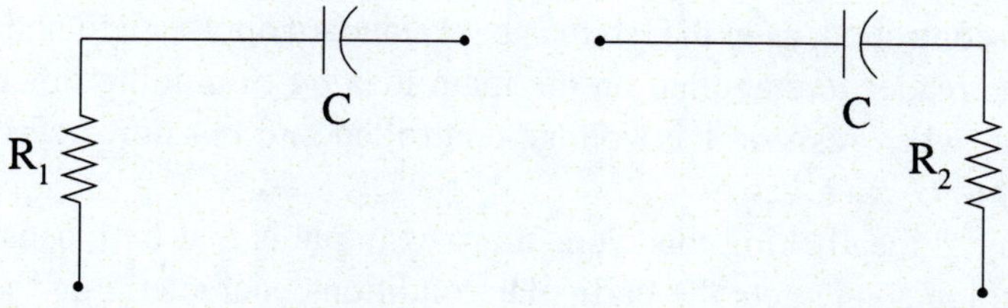

Fig. 2. Proper trees.

In this voltage-controlled setting, we may therefore guarantee that the circuit is index one if and only if $G_1 + G_2 \neq 0$. However, this assumption excludes some cases which may be of interest in practice. To avoid lacking generality, we may resort to a homogeneous description of the resistors and write $i_j = \psi_j(u_j)$, $v_j = \zeta_j(u_j)$ for $j = 1, 2$. Now the model (7) is defined by (12a)–(12b) together with the algebraic equations

$$0 = i_c + i_l + \psi_1(u_1) \tag{13a}$$
$$0 = -i_c + \psi_2(u_2) \tag{13b}$$
$$0 = v_c - e_1 + e_2 \tag{13c}$$
$$0 = -e_1 + \zeta_1(u_1) \tag{13d}$$
$$0 = -e_2 + \zeta_2(u_2) \tag{13e}$$

and some elementary computations show that the corresponding determinant is now

$$P_1 Q_2 + Q_1 P_2, \tag{14}$$

where $P_j = \psi_j{}'(u_j)$, $Q_j = \zeta_j{}'(u_j)$ for $j = 1, 2$.

The polynomial shown in (14) is homogeneous of degree one in each pair of variables (P_j, Q_j): in greater generality, so is (11). What we want to illustrate is that the expression just obtained accounts for all possible forms in the characteristics of resistors. For instance, the voltage-control assumption supporting (12) above essentially amounts to assuming $Q_1 \neq 0 \neq Q_2$, what makes it possible to divide (14) by $Q_1 Q_2$ (this is known as a *dehomogenization* of (14)) to get the aforementioned expression $G_1 + G_2$, with $G_j = P_j / Q_j$. But other cases are also of interest: for instance, the assumption that only the first resistor is voltage-controlled yields, by the same token, the expression

$$G_1 Q_2 + P_2,$$

which exemplifies the form (10) obtained in Theorem 1. The tree on the left of Fig. 2 accounts for the term $G_1 Q_2$ (since resistor 1 is a twig and resistor 2 is a chord) and, likewise, the tree on the right just yields the second term, namely P_2. Another case of interest results from the assumption that both resistors are current-controlled: here, after dividing (14) by $P_1 P_2$ one gets the expression $R_1 + R_2$, with $R_j = Q_j / P_j$; note that the R_j's arise as the *chord* resistances from each spanning tree, that is, R_1 comes from the tree on the right of Fig. 2 and R_2 from the one on the left. Worth remarking is that when both resistors are current-controlled, the condition $R_1 + R_2 = 0$ prevents the model from being index one, something that remains somehow hidden if devices are assumed to be voltage-controlled, as in the (so to speak) standard approach to nodal analysis. We encourage the reader to elaborate on the main idea by examining other cases, e.g. by checking that, when resistor 1 is voltage-controlled and resistor 2 current-controlled, (14) amounts to $G_1R_2 + 1$.

The essential idea behind this elementary example is that by means of dehomogenization techniques one gets the particular conditions characterizing the index in specific contexts from the general form shown in (14): we refer the reader to [20] for further uses of such techniques.

4 Other Applications of Homogeneous Models

In this section we discuss, in less detail, other applications of the homogeneous formalism. The first one involves DC circuits, in which operating points are computed by open-circuiting capacitors and short-circuiting inductors. Using the model (7), the equations describing operating points are easily seen to be

$$0 = A_l i_l + A_v i_v + A_g \gamma(A_g^{\mathsf{T}} e) + A_h \psi(u_h) + A_i I_s \tag{15a}$$

$$0 = -A_l^{\mathsf{T}} e \tag{15b}$$

$$0 = v_c - A_c^{\mathsf{T}} e \tag{15c}$$

$$0 = V_s - A_v^{\mathsf{T}} e \tag{15d}$$

$$0 = \zeta(u_h) - A_h^{\mathsf{T}} e, \tag{15e}$$

where I_s and V_s stand for the DC excitation terms in the sources. Provided that a DC operating point does exist, the nonsingularity of the matrix of partial derivatives of the right-hand side of (15) makes it unique, as a straightforward consequence of the inverse function theorem. This matrix of partial derivatives reads as

$$\begin{pmatrix} A_g G A_g^{\mathsf{T}} & A_l & A_v & A_h P & 0 \\ -A_l^{\mathsf{T}} & 0 & 0 & 0 & 0 \\ -A_c^{\mathsf{T}} & 0 & 0 & 0 & I_c \\ -A_v^{\mathsf{T}} & 0 & 0 & 0 & 0 \\ -A_h^{\mathsf{T}} & 0 & 0 & Q & 0 \end{pmatrix},$$

which is easily seen to be nonsingular if and only if so is

$$\begin{pmatrix} A_g G A_g^{\mathsf{T}} & A_l & A_v & A_h P \\ -A_l^{\mathsf{T}} & 0 & 0 & 0 \\ -A_v^{\mathsf{T}} & 0 & 0 & 0 \\ -A_h^{\mathsf{T}} & 0 & 0 & Q \end{pmatrix}.$$

Notice that this matrix has the same form as (9), and therefore the characterization of index-one configurations stated in Theorem 1 should have a dual result characterizing nondegenerate operating points in the homogeneous framework.

Other Uses of the Homogeneous Formalism. From a different perspective, the intrinsic symmetry provided by the homogeneous description of devices can be extended, at least for linear circuits, to handle both voltage and current sources in a unifying framework. This idea can be pursued further as to accommodate Thévenin and Norton equivalent circuits within a comprehensive setting [23].

Projective-based techniques in circuit modeling also find applications in power system modeling, fault diagnosis and the qualitative analysis of nonlinear circuits: see [17, 20, 22] and references therein. The very idea behind (2) suggests that homogeneous descriptions should be useful in the modeling of switching circuits, a topic which is in the scope of future research. Possibly, other application fields may benefit from this formalism.

5 Concluding Remarks

To sum up, homogeneous description of devices are of interest when trying to avoid any lack of generality in the circuit modeling process. Additionally, one may retain the computational advantages of the nodal approach by using homogeneous descriptions only for specific devices which for whatever reason demand so. Needless to say, the idea can be naturally extended to reactive devices and also to memristors. Controlled sources and coupling effects may also be accommodated in this framework along the lines suggested in [21]. Future work may extend the index analysis here presented to MNA models, as well as to index-two configurations and also to other families of circuit models, including distributed models involving partial differential-algebraic equations [2, 6, 15, 24, 25].

References

1. Bapat, R.B.: Graphs and Matrices. Springer, Heidelberg (2010). https://doi.org/10.1007/978-1-84882-981-7
2. Bartel, A., Baumanns, S., Schöps, S.: Structural analysis of electrical circuits including magnetoquasistatic devices. App. Numer. Math. **61**, 1257–1270 (2011)
3. Bollobás, B.: Modern Graph Theory. Springer, New York (1998). https://doi.org/10.1007/978-1-4612-0619-4
4. Bondy, J.A., Murty, U.S.R.: Graph Theory. Springer, Heidelberg (2008)
5. Chen, W.-K.: Graph Theory and its Engineering Applications. World Scientific, Singapore (1997)
6. Diab, M., Tischendorf, C.: Splitting methods for linear circuit DAEs of index 1 in port-Hamiltonian form. In: van Beurden, M., Budko, N., Schilders, W. (eds.) SCEE 2020, pp. 211–219. Springer, Cham (2021). https://doi.org/10.1007/978-3-030-84238-3_21
7. Estévez-Schwarz, D., Tischendorf, C.: Structural analysis of electric circuits and consequences for MNA. Int. J. Circuit Theory Appl. **28**, 131–162 (2000)
8. Günther, M., Feldmann, U.: CAD-based electric-circuit modeling in industry. Surv. Math. Ind. **8**, 97–129, 131–157 (1999)
9. Horn, R.A., Johnson, C.R.: Matrix Analysis. Cambridge University Press, Cambridge (1985)
10. Kirchhoff, G.: Ueber die Auflösung der Gleichungen, auf welche man bei der Untersuchung der linearen Vertheilung galvanischer Ströme geführt wird. Annalen der Physik und Chemie **72**, 497–508 (1847). English translation: IRE Tr. Circuit Theory **5**, 4–7 (1958)
11. Kunkel, P., Mehrmann, V.: Differential-Algebraic Equations. Analysis and Numerical Solution. EMS (2006)

12. Lamour, R., März, R., Tischendorf, C.: Differential-Algebraic Equations: A Projector Based Analysis. Springer, Heidelberg (2013). https://doi.org/10.1007/978-3-642-27555-5
13. Maxwell, J.C.: A Treatise on Electricity and Magnetism. Clarendon Press, Oxford (1892)
14. Möbius, A.F.: Der Barycentrische Calcul. J. A. Barth (1827)
15. Nedialkov, N., Pryce, J.D., Scholz, L.: An energy-based, always index ≤ 1 and structurally amenable electrical circuit model. SIAM J. Sci. Comput. **44**, B1122–B1147 (2022)
16. Ohm, G.S.: Die Galvanische Kette, Mathematisch Bearbeitet. T. H. Riemann (1827). English translation: Van Nostrand Co., 1891; Kraus (reprint), 1969
17. Penin, A.: Analysis of Electrical Circuits with Variable Load Regime Parameters. Springer, Cham (2015). https://doi.org/10.1007/978-3-030-35366-7
18. Reis, T.: Circuit synthesis of passive descriptor systems. A modified nodal approach. Int. J. Cir. Theory Appl. **38**, 44–68 (2010)
19. Riaza, R.: Differential-Algebraic Systems. World Scientific, Singapore (2008)
20. Riaza, R.: Circuit theory in projective space and homogeneous circuit models. IEEE Trans. Circuits Syst. I **66**, 463–476 (2019)
21. Riaza, R.: Homogeneous models of nonlinear circuits. IEEE Trans. Circuits Syst. I **67**, 2002–2015 (2020)
22. Riaza, R.: Associate submersions and qualitative properties of nonlinear circuits with implicit characteristics. Intl. J. Bif. Chaos **30**, 2050033 (2020)
23. Riaza, R.: A comprehensive framework for the Thévenin-Norton theorem using homogeneous circuit models. IEEE Trans. Circuits Syst. I **70**, 1671–1684 (2023)
24. Shashkov, V., Cortés García, I., Egger, H.: MONA: a magnetic oriented nodal analysis for electric circuits. Int. J. Cir. Theory Appl. **50**, 2997–3012 (2022)
25. Tischendorf, C.: Coupled systems of differential algebraic and partial differential equations in circuit and device simulation. Modeling and numerical analysis, Habilitationsschrift, Inst. Math., Humboltd University of Berlin (2003)

A Port-Hamiltonian, Index ≤ 1, Structurally Amenable Electrical Circuit Formulation

Lena Scholz[1], John Pryce[2(✉)], and Nedialko Nedialkov[3]

[1] Institute for Mathematics, Technische Universität Berlin, Berlin, Germany
lena.scholz@tu-berlin.de
[2] School of Mathematics, Cardiff University, Cardiff, UK
prycejd1@cardiff.ac.uk
[3] Computing and Software, McMaster University, Hamilton, ON, Canada
nedialk@mcmaster.ca

Abstract. We present a recently developed electrical circuit formulation that has port-Hamiltonian (pH) structure and results in a structurally amenable differential-algebraic equation (DAE) system of index ≤ 1. Being pH assures energy stability—the total energy of the system cannot increase. It also provides compositionality—larger pH models can be assembled from smaller ones in a standard way that facilitates building pH models in software. Structurally amenable and index ≤ 1 eliminate the phases of DAE index analysis and reduction, which are commonly used in circuit simulation software. Thus, standard numerical solvers can be applied directly to integrate the DAE. In addition, it has a known *a priori* block-triangular form that can be exploited for efficient numerical solution. A prototype MATLAB code shows high potential for development of this "compact port-Hamiltonian" (CpH) methodology.

1 Summary

Computer simulation of electrical circuits entails integrating systems of *differential-algebraic equations* (DAEs). Our work forms a DAE of remarkably simple structure promising faster numerics. It is a synergy of three oldish themes

- Energy-based *port-Hamiltonian* modelling, from ∼2000, [1–3]
- Structural analysis (SA) of *circuit topology*, from ∼1960, [4–7] but using ideas of Kron dating to the 1940s, [8]
- Structural analysis of *general DAEs*, from 1988, [9,10]

Port-Hamiltonian (pH) is a philosophy/technology for multi-physical system modelling, making *energy flow* central; while SA deploys combinatorial algorithms, inexpensive compared with the numerical ones, to *reveal structure* and thus speed up the numerics—cf. reordering algorithms in solving sparse linear systems.

2 Constituents

We take the three parts of the title in turn.

"Port-Hamiltonian". The universe runs on energy conservation across all physical domains. It is perilous for modelling to ignore this. The port-Hamiltonian (pH)

M. van Beurden et al. (Eds.): SCEE 2022, MI 43, pp. 23–31, 2024.
https://doi.org/10.1007/978-3-031-54517-7_3

approach [3] splits a system into an *energy-storing* part S, a *resistive* or energy-dissipating part R, a *control or input-output part* P, and lossless *energy-distributing* elements D connecting them, Fig. 1. Stored energy in S is described by a *Hamiltonian* function $\mathscr{H}$. The e, f lines are *ports* where energy flows; $e \cdot f$ has dimension of power. For a circuit they are voltages and currents and D represents Kirchhoff's laws. Multi-physics models and numerics respecting this structure are energy-stable, e.g. perpetual motion is excluded.

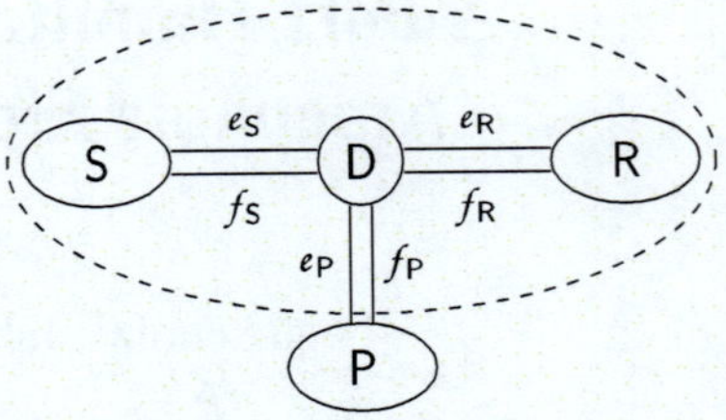

Fig. 1. pH system.

"Index ≤ 1". Unlike ODEs, DAEs typically have *hidden constraints* found by differentiating one or more equations. As an example take the simple DAE

$$x_1 - g(t) = 0,\ \dot{x}_1 - x_2 = 0.$$

To find the solution $x_1 = g(t)$, $x_2 = \dot{g}(t)$, one *must* differentiate the first equation.

An *index* of the DAE measures how much difficulty this causes. There are several; we use the differentiation index [11], the largest number of times some equation must be differentiated so the resulting equations can be solved to give an ODE.

Index > 1 DAEs typically give numerical difficulties. In practice, some index-reduction procedure is applied to arrive at an index-1 DAE, e.g. Pantelides's algorithm [9].

Those of index ≤ 1 are solvable by standard codes e.g. DASSL [12], SUNDIALS [13], MATLAB's `ode15i` etc. Popular circuit models such as the modified nodal analysis (MNA) [14] can give index 2. That our CpH method is of index ≤ 1 was a surprise bonus, and as a consequence, no index reduction is necessary.

"Structurally Amenable". A DAE is structurally amenable (S-amenable) if analysing the sparsity pattern of its equations reveals exactly what differentiations of them are needed. This analysis is inexpensive, only needs be done once, and when successful, allows various efficient numerical methods to be used (e.g., Mattsson–Söderlind Dummy Derivatives [15] for reducing the DAE to index ≤ 1).

The original method to find if a DAE is S-amenable is in Pantelides's 1988 paper [9]. We use the 2001 Pryce Σ-method [10], which is equivalent and more direct.

Consider a DAE with N equations $f_i = 0$ in N variables $x_j(t)$ and some of their t-derivatives. In vector form we can write it as

$$f(t;\, x \text{ and derivatives}) = 0.$$

A sketch of the Σ-method's steps follows.

1. Form the $N \times N$ *signature matrix* $\Sigma = (\sigma_{ij})$, where σ_{ij} is the highest derivative order of x_j in f_i, or $-\infty$ if x_j is absent from f_i.
2. Find suitable *offsets* $c_i \geq 0$, $d_j \geq 0$ with $d_j - c_i \geq \sigma_{ij}$ $(i, j = 1, \ldots, N)$, and equality on some *transversal*, a set of N positions (i, j) with one in each row and each column. This is a linear assignment problem, an efficiently solvable kind of linear programming problem [16].

3. Form the $N \times N$ *system Jacobian* $\mathbf{J}$ with $\mathbf{J}_{ij} = \partial f_i / \partial x_j^{(d_j - c_i)}$, or 0 if $d_j - c_i < 0$.
4. If $\mathbf{J}$ is nonsingular at some arbitrary point, the DAE is S-amenable. Then the offsets say what differentiations of equations are needed, and how to reduce the DAE to an implicit ODE.

3 Circuit Equations

Circuits. We consider circuits made of *2-pin elements* on graph *edges*, joined at *nodes*, see e.g. [7]. Key variables are *voltage drop* v over, and *current* ι in, an edge.

As an example, Fig. 3 shows an RLC circuit schematic as commonly drawn, with elements V = voltage source, I = current source, R = resistor, G = conductor[1], C = capacitor, L = inductor. Figure 3 is the corresponding mathematical graph, showing e.g. that the top ends of R, G, L_2 come together at a single node (Fig. 2(a) and Fig. 2(b)).

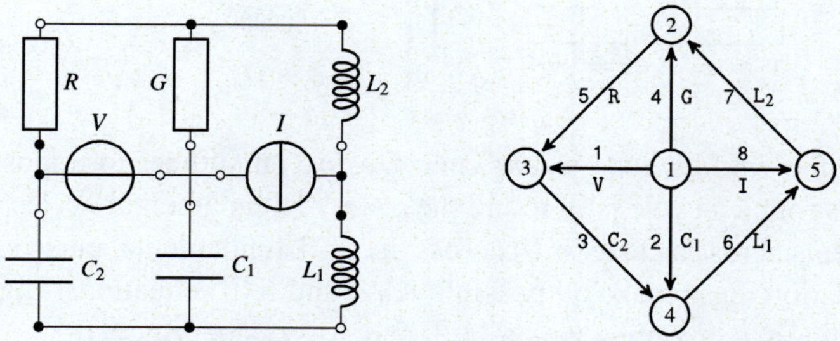

(a) Schematic. (b) Circuit graph with edge orientations.

Fig. 2. RLC circuit example. On the right the edges have been numbered 1–8, with orientations shown, and the nodes 1–5.

In a circuit graph G, we allow multiple edges between two given nodes, but not edges from a node to itself (which have no electrical use). G is undirected but each edge has an *orientation* to say which direction of v and ι counts as positive (switching it doesn't change the physics). Henceforth, we assume G is connected.

Trees. Spanning trees (just called trees in circuit literature) in the graph are key to finding the right set of equations for the DAE. A tree T is a minimal subset of the m edges containing a path from any of the n nodes to any other. Necessarily it contains no cycles and has $n-1$ edges. The $m-n+1$ edges not in T are the cotree T^*.

---

[1] A resistor with Ohm's law written as $\iota = \mathrm{v}G$ instead of $\mathrm{v} = \iota R$.

Each cotree edge specifies a *fundamental cycle*—that edge, plus the unique path between its ends via the tree. Each tree edge specifies a *fundamental cutset*—removing it splits the nodes into two nonempty subsets, the cutset is all edges between these subsets.

In Fig. 3, with tree $\{5,1,2,6\}$, cotree edge 4 specifies fundamental cycle $\{1,4,5\}$; tree edge 2 when removed splits the nodes into $\{1,2,3\}$, $\{4,5\}$, hence the fundamental cutset of edges between them is $\{2,3,7,8\}$.

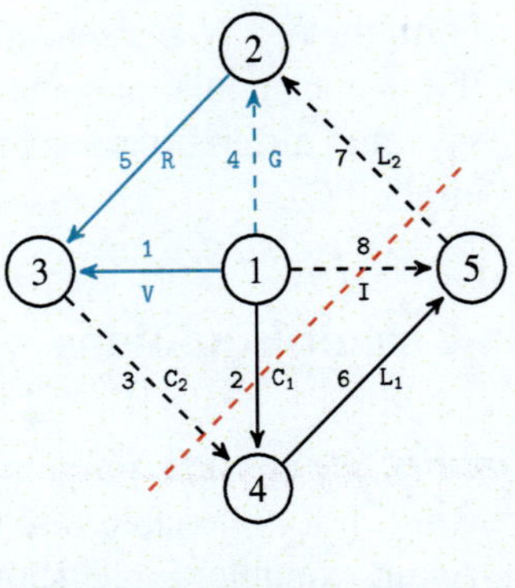

Fig. 3. Cycle and cutset example.

Physics. The physical assumptions on which the DAE is constructed are:

(a) *Constitutive relations*. If we assume standard linear elements, these are:

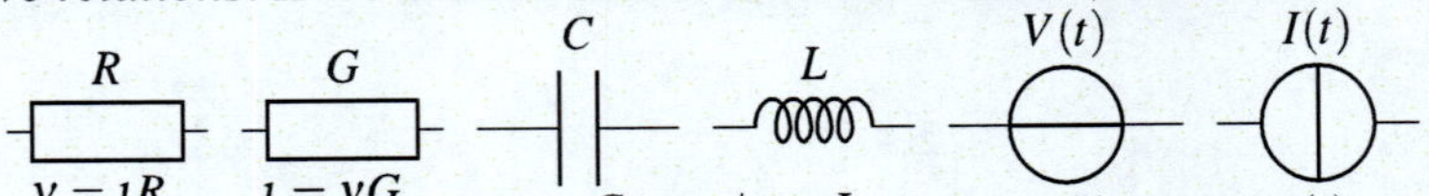

$v = \iota R$ $\iota = vG$ $q = vC$, $\dot{q} = \iota$ $\phi = \iota L$, $\dot{\phi} = v$ $v = V(t)$ $\iota = I(t)$

(b) *Kirchhoff's voltage and current laws*. Given T, it suffices to impose KVL (sum of voltages round a cycle is 0) round the $m-n+1$ fundamental cycles, and KCL (sum of currents across a cutset is 0) across the $n-1$ fundamental cutsets, since by linear combination these make all possible KVL and KCL equations. E.g., the cycle and cutset in Fig. 3 give $v_4 + v_5 - v_1 = 0$ and $\iota_3 + \iota_2 + \iota_8 - \iota_7 = 0$.

Each of the many circuit formulations combines the constitutive relations with selected Kirchhoff equations, to get a DAE $f(t,x,\dot{x}) = 0$ in some variables $x = x(t)$. Methods differ in what variables are in vector x—they can be voltages, currents, capacitor charges and inductor fluxes—and which KVL/KCL equations are used.

4 Graph Linear Algebra

Definition 1. G's $n\times m$ *incidence matrix* A has

$$a_{pj} = 1,\ a_{qj} = -1 \text{ if edge } j \text{ is from node } p \text{ to node } q$$

and zero elsewhere.

Assuming the graph is connected we have the well known facts:

Theorem 1. *A's column space (the linear span of its columns) is the hyperplane* $x_1 + \cdots + x_n = 0$ *in* $\mathbb{R}^n$. *A set of edges is a tree if and only if the corresponding columns of A are a basis of this column space.*

Consider such a graph, its incidence matrix A, a tree T and its cotree T^*.

Definition 2. The $(m-n+1) \times (n-1)$ *Kron matrix* $F = (f_{ij})$ of $\mathtt{T}$ holds the unique representation of cotree columns of A as linear combinations of tree columns[2]:

$$a_i = -\sum_{j \in \mathtt{T}} f_{ij} a_j, \qquad \text{columns indexed by } \mathtt{T}\text{, rows by } \mathtt{T}^*.$$

Theorem 2. *Kirchhoff's laws for the graph can be written as* $\imath_\mathtt{T} = F^\top \imath_{\mathtt{T}^*}$, $v_{\mathtt{T}^*} = -F v_\mathtt{T}$ *or equivalently*

$$\begin{bmatrix} \imath_\mathtt{T} \\ v_{\mathtt{T}^*} \end{bmatrix} = \begin{bmatrix} 0 & F^\top \\ -F & 0 \end{bmatrix} \begin{bmatrix} v_\mathtt{T} \\ \imath_{\mathtt{T}^*} \end{bmatrix} \tag{1}$$

where $\imath_\mathtt{T}, \imath_{\mathtt{T}^*}$ *denote the vectors of currents on tree and cotree edges respectively, and similarly* $v_\mathtt{T}, v_{\mathtt{T}^*}$.

A non-obvious fact [17] is that all nonzeros of the Kron matrix are -1 or 1, and that these nonzeros encode the fundamental cycles and cutsets, with the orientation of each edge thereon. Our example circuit graph has the following 5×8 incidence matrix and 4×4 Kron matrix. In F for instance, the row labeled 4 encodes cycle $\{1,4,5\}$; the column labeled 2 encodes cutset $\{2,3,7,8\}$.

$$A = \begin{bmatrix} 1 & 1 & 0 & 1 & 0 & 0 & 0 & 1 \\ 0 & 0 & 0 & -1 & 1 & 0 & -1 & 0 \\ -1 & 0 & 1 & 0 & -1 & 0 & 0 & 0 \\ 0 & -1 & -1 & 0 & 0 & 1 & 0 & 0 \\ 0 & 0 & 0 & 0 & 0 & -1 & 1 & -1 \end{bmatrix}, \qquad F = \begin{array}{c} \\ 3 \\ 4 \\ 7 \\ 8 \\ \\ \end{array}\!\!\begin{array}{c} \begin{array}{cccc} 1 & 2 & 5 & 6 \end{array} \\ \begin{bmatrix} 1 & -1 & 0 & 0 \\ -1 & 0 & 1 & 0 \\ -1 & 1 & 1 & 1 \\ 0 & -1 & 0 & -1 \end{bmatrix} \\ \begin{array}{cccc} \mathtt{V} & \mathtt{C_1} & \mathtt{R} & \mathtt{L_1} \end{array} \end{array}\!\!\begin{array}{c} \\ \mathtt{C_2} \\ \mathtt{G} \\ \mathtt{L_2} \\ \mathtt{I} \\ \\ \end{array} \tag{2}$$

5 The Compact port-Hamiltonian DAE

Circuit formulations differ in what vector x of *DAE variables*, and which *Kirchhoff equations* they select. The CpH method in its simplest form chooses x with one x_j for each non-source edge, thus:

$$\begin{array}{c} \text{charge } q \text{ for capacitors,} \qquad \text{flux linkage } \phi \text{ for inductors,} \\ \text{voltage } v \text{ or current } \imath \text{ at will for resistors,} \end{array} \tag{3}$$

and applies Kirchhoff's laws in the form in Theorem 2. A circuit must be *well posed*, meaning it has no cycles composed only of voltage sources and no cutsets composed only of current sources ("no V-cycles or I-cutsets"). Otherwise, since we have ideal circuit elements, Kirchhoff's equations are either underdetermined or contradictory.

Circuit SA says that a well-posed circuit has a *normal tree* containing all voltage sources, no current sources, and in a well defined sense the most possible capacitors and the fewest possible inductors. For a simple proof see [18, §2.3].

Constructing the CpH DAE then comprises the following steps.

[2] The $-$ is chosen to match the notation in [7].

1. **Input**: t and the DAE vector x defined in (3).
2. Evaluate the constitutive relations. Given the choice of variables (3), these make v and ι on each edge an explicit function of x or $\dot{x}$ and (for source edges) t.
3. Substitute these v and ι into Kirchhoff's equations in the Kron form (1). Each such equation "belongs" to a unique edge: e.g. each KVL equation is round a fundamental cycle, and belongs to the cotree edge that generates this cycle.
4. Separate out equations belonging to voltage and current source edges. They give control-output to be handled after the DAE is solved.
5. **Output**: the remaining equations as the DAE $f(t,x,\dot{x}) = 0$, of size N equal to the number of non-source edges.

We have assumed simple linear RLC circuit elements above. However this construction and the next theorem work more generally, for nonlinear elements and various kinds of coupling. Thus diodes, transistors, transformers, etc. are supported.

Theorem 3. *Subject to suitable* passivity *assumptions on the circuit elements, the CpH DAE is port-Hamiltonian, S-amenable and index* ≤ 1.

Sketch proof, showing the synergy of our three themes: see [18] for details.
Tree T being normal puts some blocks of zeros in the Kron matrix F ... *[circuit-SA]*
... which improve the sparsity of the system Jacobian $\mathbf{J}$, making it block-triangular with three diagonal blocks *[DAE-SA]*

$$\mathbf{J} = \begin{bmatrix} \text{capacitor data} & & \\ & \text{inductor data} & \\ \times & \times & \text{resistor data} \end{bmatrix}.$$

The passivity assumptions mean *physically* that no circuit element except voltage and current sources can create energy in the system. They have the *mathematical* form that certain Jacobian matrices must be positive definite. *[pH]*

These Jacobians enter into $\mathbf{J}$'s diagonal blocks. Their positive definiteness implies each such block is nonsingular. So $\mathbf{J}$ is nonsingular, proving S-amenability.

Index ≤ 1 is a by-product, and pH structure is immediate from construction. □

For the example circuit's F in (2), the tree that produces it is normal; the blocks of zeros mentioned in the sketch proof are the three zeros in F's top right corner.

6 Conclusion

For details of what is said here, and proofs, see [18]. Other circuit formulations presented at the SCEE 2022 conference are in [20,21], of which the first is port-Hamiltonian and the second is always index ≤ 1; but neither discusses SA-amenability.

CpH Advantages. CpH essentials are to be port-Hamiltonian and structurally amenable. Being pH is desirable *physics*: pH assures *energy-stability* of the mathematical DAE, and also of its numerical solution when suitable methods are used.

Being S-amenable is good for *numerics*. It makes possible, or inexpensive to implement, methods that in its absence are unavailable or expensive. Essentially equivalent for a DAE are that (a) it is S-amenable; (b) the Pantelides method (1988) works on it—e.g. to find consistent initial values; and (c) the Mattsson-Söderlind dummy derivatives method (1993) works on it—e.g. to reduce it to an implicit ODE.

Being pH is good *software engineering*, since pH models are compositional, i.e. can be assembled in a standard way to make larger pH models. This suits them to languages like Modelica, whose essence is to build systems from basic components.

Code Generation. We have implemented our theory in MATLAB, in principle supporting nonlinear dependent elements of the full generality in [18]. An object of a class `pHcircuit`, a "part", is specified by an incidence matrix, and type and parameters for each edge. We build larger circuits by combining existing parts; e.g. for the circuit in Fig. 4, the statement `P = [P0, BJT]/["a4b1", "a3b2", "a1b3"];` was used to join transistor `BJT` to the rest of the circuit `P0` by "soldering" pins 4, 3, 1 of `P0` to pins 1, 2, 3 of `BJT` respectively. Then we generate MATLAB code for the DAE function $f(t,x,\dot{x})$ and use `ode15i` to integrate the DAE. The generated code is readable and efficient. Generation is easily customised to make `C++` or `Fortran`.

Acknowledgements. We thank the anonymous referees for useful comments that led us to improve the paper. LS acknowledges support of the Berlin Mathematics Research Center MATH+, Project AA4-5. NN acknowledges support of the Natural Sciences and Engineering Research Council of Canada (NSERC), FRN RGPIN-2019-07054.

Appendix: Application, and Example CpH Code

We modelled the bipolar junction transistor (BJT) amplifier circuit example from Falaize & Hélie [19]. This shows the CpH method going beyond the linear elements assumed in Sect. 3.

The upper part of Fig. 4 is the circuit schematic. From the graph viewpoint, edges IN, 9V, OUT join to the "ground" node. We number the edges and assign orientations as marked by red arrows.

The MATLAB code was generated automatically, and the lower part shows the output from solving by `odei5i` with absolute and relative tolerances 10^{-3}, with input $V_{\text{in}}(t)$ as in [19], namely zero for 0.3 s to let a steady state be reached, followed by a sinusoidal oscillation of linearly increasing amplitude for 0.01 s. It agrees to graphical accuracy with [19, Fig. 6(b)].

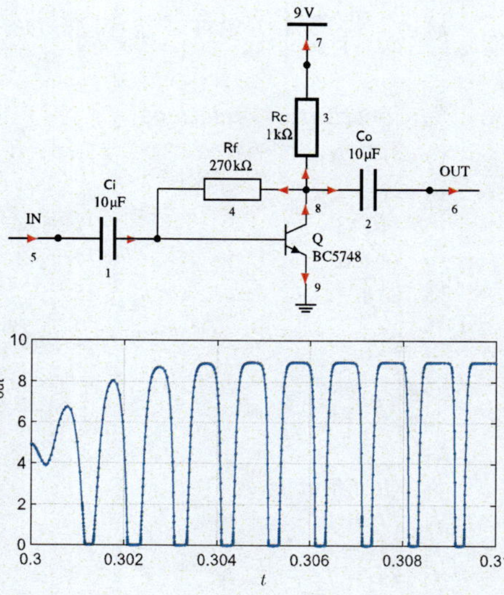

Fig. 4. BJT amplifier circuit, and output.

One could use tolerances down to around 10^{-12}, beyond which step size failure occurred. An upper step size limit was needed (0.005 worked), else the large steps built up in the initial 0.3 s were liable to make the solver go from $t = 0.3$ to 0.31 in one step, not noticing the changed behaviour at 0.3.

Below is the central part of the code—constitutive relations and Kirchhoff equations, expressed in mathematical notation. The physical parameters are in lines 5–11. The transistor is constructed on the Ebers–Moll model, of two Shockley diodes (nonlinear voltage-controlled resistors) with a linear dependence between them. In the code the diode is scalar function D on line 10; note @(...)... is how MATLAB defines anonymous functions. The transistor is modelled by function BJT on line 11, with 2-vector input and output. Values $\alpha_F, \alpha_R > 1$ in matrix M derive from the Ebers–Moll forward and reverse current gains β_F, β_R.

Constitutive relations for independent edges are in lines 14–20; for the dependent transistor edges, in line 21. The Kirchhoff equations derive from the tree of edges $\{1,2,3,5,7\}$. Setting the currents of voltage sources and voltages of current sources to zero (lines 18–20 is a trick to simplify code generation. It makes the y's in lines 28–30 equal minus their correct output values, hence the sign reversal at line 33.

1 **function** $[f,y] = \text{fcnBJTamplifier}(t,x,\dot{x})$
2 % *DAE vector* $x = (x_1,\ldots,x_6)^T = (q_{C1},\ q_{C2},\ i_{R3},\ i_{R4},\ v_{G8},\ v_{G9})^T$

4 % *Physical parameters*
5 $C_i = 10^{-6},\ C_o = 10^{-6},\ R_c = 270\times 10^3,\ R_f = 10^3,\ V_{cc} = -9,\ I_{out} = 0$
6 $t_d = 0.3,\ t_{max} = t_d + 0.01, V_{max} = 0.2,\ \omega = 2\pi 10^3$
7 $V_{in} = @(t) - (t - t_d \geq 0) V_{max} \frac{t-t_d}{t_{max}-t_d} \sin(\omega(t-t_d))$
8 $I_s = 10^{-13},\ V_T = 0.025,\ \beta_F = 250,\ \beta_R = 10,\ \alpha_F = 1 + 1/\beta_F,\ \alpha_R = 1 + 1/\beta_R$
9 % *Shockley diode D and Ebers–Moll transistor BJT*
10 $D = @(v)\, I_s \cdot (e^{v/V_T} - 1)$ %*It accepts vector input*
11 $M = \begin{bmatrix} \alpha_F & -1 \\ -1 & \alpha_R \end{bmatrix},\quad \text{BJT} = @(v)\, M \cdot D(v)\ \%\ v = \begin{bmatrix} v_1 \\ v_2 \end{bmatrix}$

13 % *Constitutive relations*
14 $v_1 = C_i^{-1} x_1 \qquad i_1 = \dot{x}_1$
15 $v_2 = C_o^{-1} x_2 \qquad i_2 = \dot{x}_2$
16 $v_3 = R_c x_4 \qquad i_3 = x_4$
17 $v_4 = R_f x_3 \qquad i_4 = x_3$
18 $v_5 = V_{in}(t) \qquad i_5 = 0$
19 $v_6 = 0 \qquad i_6 = I_{out}$
20 $v_7 = V_{cc} \qquad i_7 = 0$
21 $\begin{bmatrix} v_8 \\ v_9 \end{bmatrix} = \begin{bmatrix} x_5 \\ x_6 \end{bmatrix} \qquad \begin{bmatrix} i_8 \\ i_9 \end{bmatrix} = \text{BJT}\left(\begin{bmatrix} x_5 \\ x_6 \end{bmatrix}\right)$

23 % *Kirchhoff laws*
24 $f_1 = i_1 + i_4 - i_8 - i_9$
25 $f_2 = i_2 - i_6$
26 $f_3 = i_3 + i_4 + i_6 - i_8$
27 $f_4 = v_4 - v_1 - v_3 - v_5 + v_7$
28 $y_1 = i_5 + i_4 - i_8 - i_9$
29 $y_2 = v_6 + v_2 - v_3 + v_7$
30 $y_3 = i_7 - i_4 - i_6 + i_8$
31 $f_5 = v_8 + v_1 + v_3 + v_5 - v_7$
32 $f_6 = v_9 + v_1 + v_5$
33 $y = -y$

35 % *Return:* $f = (f_1,\ldots,f_6)^T$ *and control–output* $y = (y_1,y_2,y_3)^T = (i_{V5}, v_{I6}, i_{V7})^T$

References

1. van der Schaft, A.J.: Port-Hamiltonian systems: network modeling and control of nonlinear physical systems. In: Irschik, H., Schlacher, K. (eds.) Advanced Dynamics and Control of Structures and Machines. ICMS, vol. 444, pp. 127–167. Springer, Vienna (2004). https://doi.org/10.1007/978-3-7091-2774-2_9
2. van der Schaft, A.J.: Port-Hamiltonian differential-algebraic systems. In: Ilchmann, A., Reis, T. (eds.) Surveys in Differential-Algebraic Equations I, pp. 173–226. Springer, Heidelberg (2013). https://doi.org/10.1007/978-3-642-34928-7_5
3. van der Schaft, A.J., Jeltsema, D.: Port-Hamiltonian systems theory: an introductory overview. Found. Trends Syst. Control **1**(2–3), 173–378 (2014)
4. Brown, D.P.: Derivative-explicit differential equations for RLC graphs. J. Franklin Inst. **275**, 503–514 (1963)
5. Bartel, A., Baumanns, S., Schöps, S.: Structural analysis of electrical circuits including magnetoquasistatic devices. Appl. Numer. Math. **61**(12), 1257–1270 (2011)
6. Estévez Schwarz, D., Tischendorf, C.: Structural analysis of electric circuits and consequences for MNA. Int. J. Circuit Theory Appl. **28**(2), 131–162 (2000)
7. Riaza, R.: Differential-Algebraic Systems: Analytical Aspects and Circuit Applications. World Scientific, Singapore (2008)
8. Kron, G.: Tensor Analysis of Networks. Wiley, New York (1939)
9. Pantelides, C.C.: The consistent initialization of differential-algebraic systems. SIAM J. Sci. Statist. Comput. **9**, 213–231 (1988)
10. Pryce, J.D.: A simple structural analysis method for DAEs. BIT Numer. Math. **41**, 364–394 (2001)
11. Campbell, S.L., Gear, C.W.: The index of general nonlinear DAEs. Numer. Math. **72**(2), 173–196 (1995)
12. Petzold, L.R.: Description of DASSL: a differential/algebraic system solver. Technical report, Sandia National Labs, Livermore, CA (USA) (1982)
13. Hindmarsh, A., et al.: SUNDIALS: suite of nonlinear and differential/algebraic equation solvers. ACM TOMS **31**(3), 363–396 (2005)
14. Vlach, J., Singhal, K.: Computer Methods for Circuit Analysis and Design. Van Nostrand Reinhold, New York (1994)
15. Mattsson, S., Söderlind, G.: Index reduction in differential-algebraic equations using dummy derivatives. SIAM J. Sci. Statist. Comput. **14**(3), 677–692 (1993)
16. Jonker, R., Volgenant, A.: A shortest augmenting path algorithm for dense and sparse linear assignment problems. Computing **38**(4), 325–340 (1987)
17. Biggs, N.L.: Algebraic Graph Theory, 2nd edn. No. 67 in Cambridge Tracts in Mathematics. Cambridge University Press, Cambridge (1993)
18. Nedialkov, N., Pryce, J.D., Scholz, L.: An energy-based, always index ≤ 1 and structurally amenable electrical circuit model. SIAM J. Sci. Comput. **44**(4), B1122–B1147 (2022)
19. Falaize, A., Hélie, T.: Passive guaranteed simulation of analog audio circuits: a port-Hamiltonian approach. Appl. Sci. **6**(10) (2016)
20. Günther, M., Bartel, A., Jacob, B., Reis, T.: Dynamic iteration schemes and port-Hamiltonian formulation in coupled differential-algebraic equation circuit simulation. In: Proceedings of the SCEE (2022)
21. Shashkov, V., Cortes Garcia, I., Egger, H.: MONA, A magnetic oriented nodal analysis for electric circuits. In: Proceedings of the SCEE (2022)

Device Simulation

Simulation of a GNR-FET

Giovanni Nastasi(✉) and Vittorio Romano

Department of Mathematics and Computer Science, University of Catania, Viale Andrea Doria 6, 95125 Catania, Italy
giovanni.nastasi@unict.it, romano@dmi.unict.it

Abstract. A field effect transistor is simulated in the case the active area is made by a single graphene nanoribbon. At variance with large area graphene, an energy gap is present and this should improve the performance of the device as transistor. A drift-diffusion model which includes the degenerate effects, coupled to the Poisson equation for the electrostatic potential, is used. The mobility models are obtained, by a fitting procedure, solving numerically the semiclassical Boltzmann equation for the graphene nanoribbon, including also the edges scattering besides the electron-phonon interactions.

1 Introduction

Device engineers devote considerable effort for developing transistor designs in which short-channel effects are suppressed and series resistances are minimized. Scaling theory predicts that a field effect transistor (FET) with a thin barrier and a thin gate-controlled region will be robust against short-channel effects down to very short gate lengths. The possibility of having channels that are just one atomic layer thick is perhaps the most attractive feature of graphene for its use in transistors [1]. Main drawbacks of a large-area monolayer graphene are the zero gap and, for graphene on substrate, the degradation of the mobility. A possible way to overcome this problem consists to adopt narrow strips of graphene, called nanoribbons (GNR), because the spatial confinement of carriers induces a band gap [2,4], even if the mobility reduces with respect to the large area graphene sheet.

The drift-diffusion model represents a common tool for the current prediction and many studies are devoted to GFETs [5,13,17]. In [12] a new geometry with a good switch on/off ratio is presented, in [18] the drift-diffusion model for GFET is presented in detail together with the numerical approach. The adoption of more sophisticated models is under investigation, e.g. hydrodynamic [6], and kinetic [7].

The current behavior in GNRs has been simulated at kinetic level [14] where the charge confinement is taken into account in the dispersion relation [2] and in the collision term [3]. Moreover, hydrodynamic models have been deduced [8,9].

In this paper, we simulate a GNR-FET with the geometry presented in [12] and the drift-diffusion model of [18]. Here, the mobility model is deduced by a fitting procedure of the data obtained with the numerical solution of the semiclassical Boltzmann equation for GNRs [14].

M. van Beurden et al. (Eds.): SCEE 2022, MI 43, pp. 35–42, 2024.
https://doi.org/10.1007/978-3-031-54517-7_4

2 Mathematical Model

We simulate the device depicted in Fig. 1. The active area consists of a layer made of GNR.

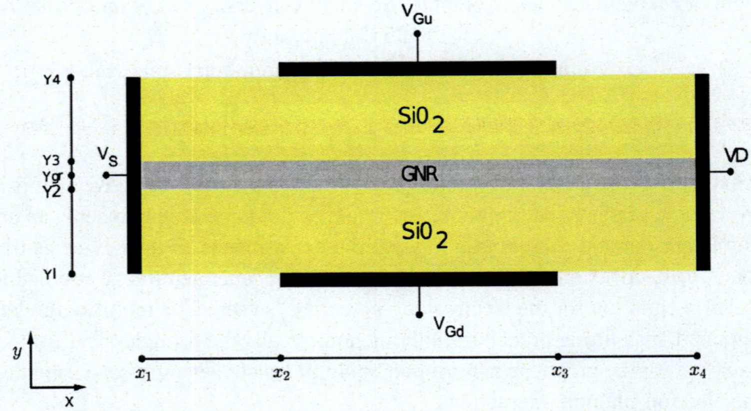

Fig. 1. Schematic representation of the simulated device.

We adopt the 1D bipolar stationary drift-diffusion model

$$\frac{\partial J_n}{\partial x} = eR, \quad \frac{\partial J_p}{\partial x} = -eR, \tag{1}$$

e being the (positive) elementary charge. $J_n(x)$ and $J_p(x)$ are the electron and hole current densities respectively. R denotes the generation-recombination term. The previous equations are coupled with the 2D Poisson equation for the electrostatic potential in the whole section.

The electron and hole current densities are expressed in terms of the quasi-Fermi energies [5]

$$J_n = \mu_n n \frac{\partial \varepsilon_F^{(n)}}{\partial x}, \quad J_p = \mu_p p \frac{\partial \varepsilon_F^{(p)}}{\partial x}, \tag{2}$$

where n and p, $\varepsilon_F^{(n)}$ and $\varepsilon_F^{(p)}$, μ_n and μ_p are the densities, quasi-Fermi energies and mobilities of electrons and holes, respectively.

The electron and hole densities, $n(x)$ and $p(x)$ respectively, are evaluated as

$$\begin{aligned} n(x) &= \frac{g_s g_v}{(2\pi)^2} \int_{\mathbb{R}^2} f_{FD}(\mathbf{k}, \varepsilon_F^{(n)})\, d\mathbf{k}, \qquad x \in [x_1, x_4], \\ p(x) &= \frac{g_s g_v}{(2\pi)^2} \int_{\mathbb{R}^2} f_{FD}(\mathbf{k}, -\varepsilon_F^{(p)})\, d\mathbf{k}, \qquad x \in [x_1, x_4], \end{aligned} \tag{3}$$

with $g_s = 2$ and $g_v = 2$ the spin and valley degeneracy, and the crystal momentum of electrons and holes is assumed to vary over $\mathbb{R}^2$. f_{FD} indicates the Fermi-Dirac distribution

$$f_{FD}(\mathbf{k}, \varepsilon_F) = \left[1 + \exp\left(\frac{(\varepsilon(\mathbf{k}) - \varepsilon_D) - (\varepsilon_F - \varepsilon_D)}{k_B T}\right)\right]^{-1},$$

where k_B is the Boltzmann constant, T is the lattice temperature, kept at 300 K (room temperature), $\varepsilon_D = -e\phi(x, y_{gr})$ is the Dirac energy and $\phi(x, y)$ is the electrical potential, here evaluated on $y = y_{gr}$, y_{gr} being the average y-coordinate of the graphene sheet (see Fig. 1). ε_F denotes the Fermi level (in pristine graphene $\varepsilon_F = \varepsilon_D$). Following [2], the GNR dispersion relation reads

$$\varepsilon(\mathbf{k}) - \varepsilon_D = \hbar v_F \sqrt{k_x^2 + k_y^2 + \left(\frac{\pi}{W}\right)^2},$$

strictly valid around the Dirac points, where $\hbar$ is the reduced Planck constant, v_F the Fermi velocity and W the GNR width. The dispersion relation is the same for electrons and holes (see [10, 11]).

We impose Dirichlet boundary conditions on the system (1), (2) as in [12]

$$\varepsilon_F^{(n)}(x_1) - \varepsilon_D(x_1) = \varepsilon_F^{(p)}(x_1) - \varepsilon_D(x_1) = \Delta\varepsilon_F,$$
$$\varepsilon_F^{(n)}(x_4) - \varepsilon_D(x_4) = \varepsilon_F^{(p)}(x_4) - \varepsilon_D(x_4) = \Delta\varepsilon_F.$$

The quantity $\Delta\varepsilon_F$ is the difference of the work functions between metal and graphene. Its value depends on the material the contact is made of. In this paper, we use the value $\Delta\varepsilon_F = 0.25$ eV, appropriate for copper.

Generally, for the generation-recombination term R the Shockley-Read-Hall model is adopted, as suggested in [13] by analogy with standard semiconductors. In the case of GNR, if $W \lesssim 10$ nm a gap high enough to prevent the generation-recombination effect is induced. In this case $R = 0$ with high accuracy.

The system (1) is coupled with the 2D Poisson equation for the electrostatic potential, solved in the section of the device [17],

$$\nabla \cdot (\varepsilon \nabla \phi) = h(x, y), \tag{4}$$

where

$$h(x,y) = \begin{cases} e(n(x) - p(x) - n_{imp})/t_{gr} & \text{if } (x,y) \in [x_1, x_4] \times [y_2, y_3], \\ 0 & \text{otherwise.} \end{cases}$$

The dielectric function ε is given by

$$\varepsilon = \begin{cases} \varepsilon_{gr} & \text{if } y \in [y_2, y_3], \\ \varepsilon_{ox} & \text{otherwise,} \end{cases}$$

with ε_{gr} and ε_{ox} dielectric constants in graphene and oxide respectively. n_{imp} is the areal density of the impurity charges at the graphene/oxide interface. Since n and p are areal densities, the charge in the graphene layer is considered as distributed in the volume enclosed by the parallelepiped of base the area of the graphene and height t_{gr}. Equation (4) is augmented with the following boundary conditions [12]

$$\begin{aligned} \phi &= 0 && \text{at } y \in [y_1, y_4],\, x = x_1, \\ \phi &= V_b && \text{at } y \in [y_1, y_4],\, x = x_4, \\ \phi &= V_{Gd} && \text{at } y = y_1,\, x \in [x_2, x_3], \\ \phi &= V_{Gu} && \text{at } y = y_4,\, x \in [x_2, x_3], \\ \nabla_\nu \phi &= 0 && \text{at the remaining part of the boundary,} \end{aligned}$$

where V_b is the bias voltage, V_{Gu} is the upper gate-source potential, V_{Gd} is the down gate-source potential. We have denoted by $\nabla_\nu \phi$ the external normal derivative.

The mobility models $\mu_n = \mu_n(n, E)$ and $\mu_p = \mu_p(p, E)$ represent a crucial point for an accurate determination of currents. We adopt a model deduced by a fitting procedure on extensive simulations of the homogeneous Boltzmann equation for charge transport in GNRs solved by a discontinuous Galerkin deterministic numerical method [14] as already done in [15,16] for large area graphene. For several values of the electron density n the low field mobility is extrapolated by a linear regression model

$$\mu_0(n) = \frac{\tilde{\mu}}{1 + \left(\frac{n}{n_{ref}}\right)^{\alpha}}, \tag{5}$$

where $\tilde{\mu}$, n_{ref} and α are fitting parameters estimated by means of the least squares method. For the electric field dependence we propose the following model [15,16]

$$\mu(n,E) = \frac{\mu_0(n) + \mu_1 \left(\frac{E}{E_{ref}}\right)^{\beta_1}}{1 + \left(\frac{E}{E_{ref}}\right)^{\beta_2} + \gamma \left(\frac{E}{E_{ref}}\right)^{\beta_3}}, \tag{6}$$

where E_{ref}, μ_1, β_1, β_2, β_3 and γ are fitting parameters. Since the behaviour is the same for holes on account of the symmetry between the hole and electron distributions we set $\mu = \mu_n = \mu_p$.

3 Numerical Results

Here some preliminary numerical results are presented. We set the width of the GNR $W = 5$nm and estimate the fitting parameters consistently. In Table 1 the fitting parameters for the low field mobility (5) are reported while in Fig. 2 the corresponding plot is shown. In the same way, in Table 2 the fitting parameters for the high field mobility (6) are reported while in Fig. 3 a plot is shown together with the corresponding currents. For each value of the electron density, in a range from $\varepsilon_F = 0.1$ eV to $\varepsilon_F = 0.5$ eV, we calculate the coefficients E_{ref} μ_1, β_1, β_2, β_3 and γ by means of least square method. Then in each interval $[n_i, n_{i+1}]$ a third degree polynomial interpolation has been adopted for the parameters mentioned above. The drift-diffusion equations coupled with the

Table 1. Estimated parameters for the low field mobility.

Parameter	Value
$\tilde{\mu}$	1.493 μm^2/V ps
n_{ref}	4.236·10^4 μm^{-2}
α	1.128

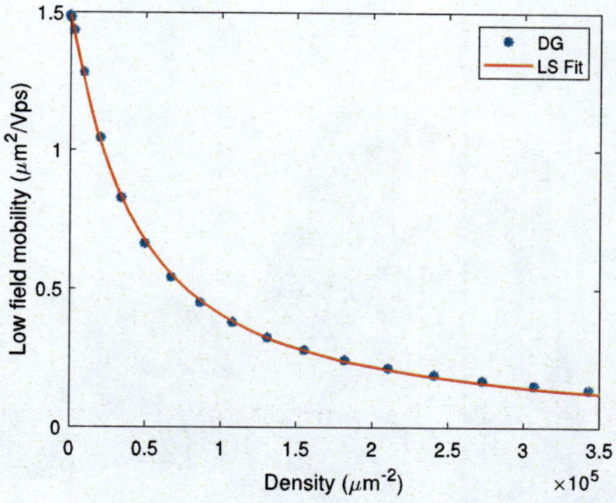

Fig. 2. Comparison between the low field mobility simulated with the DG method and the fitted one.

Table 2. Estimated parameters for the high field mobility.

n (μm^{-2})	E_{ref} (V/μm)	μ_1 (μm^2/V ps)	β_1	β_2	β_3	γ
9.904	$1.197 \cdot 10^{-1}$	7.450	$2.865 \cdot 10^{-4}$	1.363	$1.810 \cdot 10^{-1}$	7.291
$4.560 \cdot 10^{2}$	$1.343 \cdot 10^{-1}$	5.961	$8.865 \cdot 10^{-9}$	1.353	$1.824 \cdot 10^{-1}$	6.010
$9.374 \cdot 10^{3}$	$7.184 \cdot 10^{-2}$	4.476	2.807	3.776	2.669	3.498
$3.414 \cdot 10^{4}$	$5.773 \cdot 10^{-2}$	5.717	2.394	3.387	2.336	6.467
$6.718 \cdot 10^{4}$	$2.232 \cdot 10^{-2}$	8.696	1.804	2.696	1.750	1.520

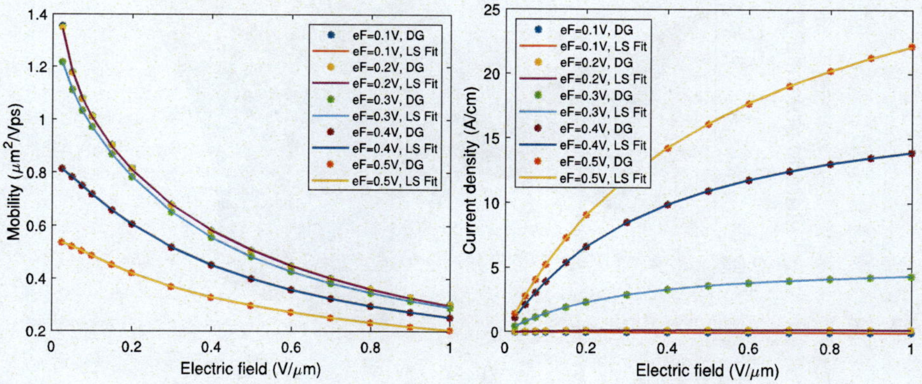

Fig. 3. On the left, comparison between the high field mobility simulated numerically solving the semiclassical Boltzmann equation by the DG method and the fitted one. On the right, the corresponding current densities are plotted.

Poisson equation are solved by the finite difference scheme presented in [12,18] where an iterative procedure is adopted to uncouple the system. The parameters for the geometry of the device are: length 100 nm, height 21 nm, gates length 50 nm, contacts height 21 nm. t_{gr} is set 1 nm. Moreover, we have set $\varepsilon_{gr} = 3.3\varepsilon_0$ and $\varepsilon_{ox} = 3.6\varepsilon_0$, where ε_0 is the dielectric constant in the vacuum, and $n_{imp} = 2.5 \cdot 10^3$ μm^{-2}.

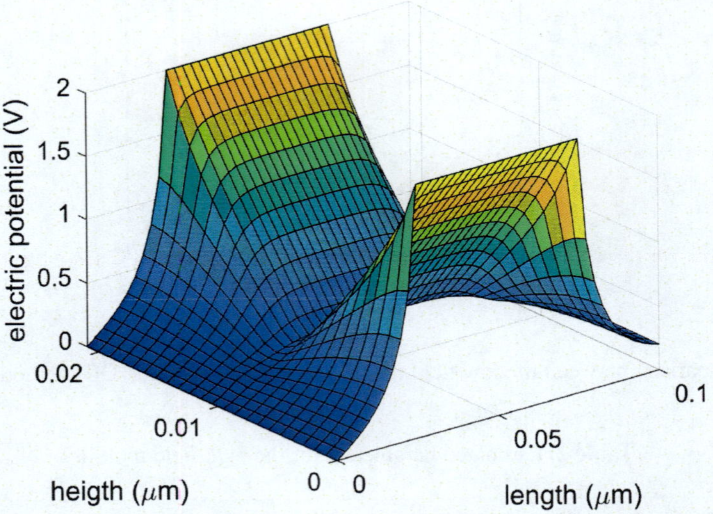

Fig. 4. Electrostatic potential at $V_{sd} = 0.2\text{V}$ and $V_G = 2\text{V}$.

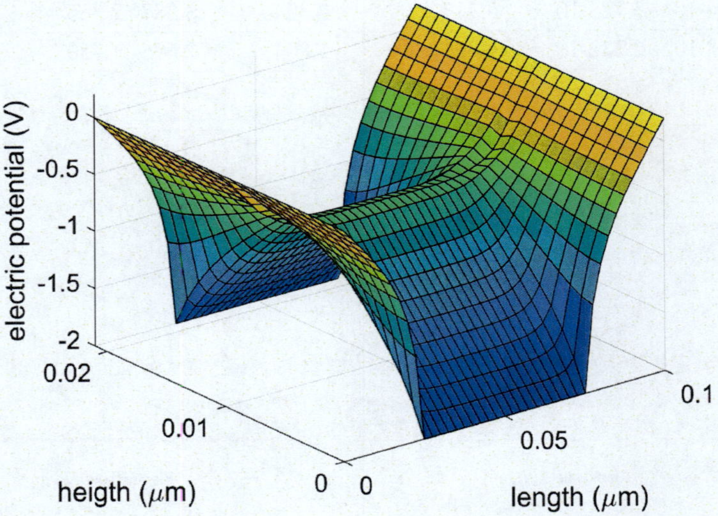

Fig. 5. Electrostatic potential at $V_{sd} = 0.2\text{V}$ and $V_G = -2\text{V}$.

In Figs. 4, 5 the electrostatic potential is shown in a case of current on and in a case of current off. In Fig. 6 the characteristic curves are plotted both in linear and logarithmic scale. Note that in the on region the current has not a monotone behaviour. The switch on/off ratio is about eight orders of magnitude, assuring a good FET behavior.

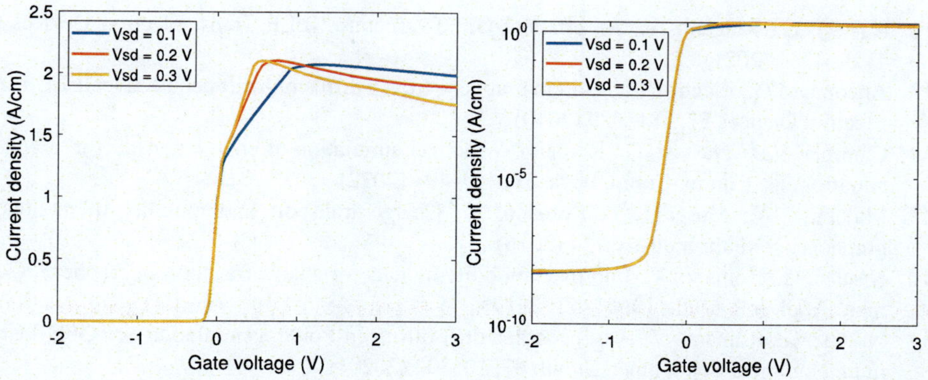

Fig. 6. Current density versus gate voltage for several values of bias in linear scale (left) and logarithmic scale (right).

Acknowledgments. The authors acknowledge the support from INdAM (GNFM) and from Università degli Studi di Catania, *Piano della Ricerca 2020/2022 Linea di intervento 2* "QICT". G. Nastasi acknowledges the financial support from the project PON R&I 2014-2020 "Asse IV - Istruzione e ricerca per il recupero - REACT-EU, Azione IV.4 - Dottorati e contratti di ricerca su tematiche dell'innovazione", project: "Modellizzazione, simulazione e design di transistori innovativi".

References

1. Schwierz, F.: Graphene transistors. Nat. Nanotechnol. **5**, 487–496 (2010)
2. Bresciani, M., Palestri, P., Esseni, D., Selmi, L.: Simple and efficient modeling of the E-k relationship and low-field mobility in Graphene Nano-Ribbons. Solid-State Electron. **54**, 1015–1021 (2010)
3. Dugaev, V.K., Katsnelson, M.I.: Edge scattering of electrons in graphene: Boltzmann equation approach to the transport in graphene nanoribbons and nanodisks. Phys. Rev. B **88**, 235432 (2013)
4. Han, M.Y., Özyilmaz, B., Zhang, Y., Kim, P.: Energy band-gap engineering of graphene nanoribbons. Phys. Rev. Lett. **98**, 206805 (2007)
5. Champlain, J.G.: A first principles theoretical examination of graphene-based field effect transistors. J. Appl. Phys. **109**, 084515 (2011)
6. Luca, L., Romano, V.: Comparing linear and nonlinear hydrodynamical models for charge transport in graphene based on the maximum entropy principle. Int. J. Non-Lin. Mech. **104**, 39–58 (2018)
7. Nastasi, G., Romano, V.: Discontinuous Galerkin approach for the simulation of charge transport in graphene. Ricerche mat. **70**, 149–165 (2021)
8. Camiola, V.D., Nastasi, G.: Hydrodynamical model for charge transport in graphene nanoribbons. Confinement and edge scattering effects. J. Stat. Phys. **184**, 23 (2021)
9. Camiola, V.D., Nastasi, G.: Bipolar hydrodynamical model for charge transport in graphene nanoribbons. J. Comput. Theor. Transp. **51**, 80–100 (2022)
10. Jacoboni, C.: Theory of Electron Transport in Semiconductors. Springer, Heidelberg (2013)
11. Castro Neto, A.H., Guinea, F., Peres, N.M.R., Novoselov, K.S., Geim, A.K.: The electronic properties of graphene. Rev. Mod. Phys. **81**, 109–162 (2009)

12. Nastasi, G., Romano, V.: An efficient GFET structure. IEEE Trans. Electron Devices **68**, 4729–4734 (2021)
13. Ancona, M.G.: Electron transport in graphene from a diffusion-drift perspective. IEEE Trans. Electron Devices **57**, 681–689 (2010)
14. Camiola, V.D., Nastasi, G., Romano, V.: Direct simulation of charge transport in graphene nanoribbons. Comm. Comp. Phys. **31**, 449–494 (2022)
15. Majorana, A., Mascali, G., Romano, V.: Charge transport and mobility in monolayer graphene. J. Math. Industry **7**, 4 (2016)
16. Nastasi, G., Romano, V.: Improved mobility models for charge transport in graphene. Commun. Appl. Ind. Math. **10**, 41–52 (2019)
17. Nastasi, G., Romano, V.: A full coupled drift-diffusion-Poisson simulation of a GFET. Commun. Nonlinear Sci. Numer. Simul. **87**, 105300 (2020)
18. Nastasi, G., Romano, V.: Drift-diffusion models for the simulation of a graphene field effect transistor. J. Math. Ind. **12**, 4 (2022)

Computational Electromagnetics

Solution of Time-Harmonic Maxwell's Equations by a Domain Decomposition Method Based on PML Transmission Conditions

Sahar Borzooei[1(✉)], Victorita Dolean[1], Pierre-Henri Tournier[2], and Claire Migliaccio[3]

[1] Côte d'Azur University, CNRS, LJAD, Nice, France
Sahar.Borzooei@univ-cotedazur.fr, work@victoritadolean.com
[2] Sorbonne University, CNRS, LJLL, Paris, France
tournier@ljll.math.upmc.fr
[3] Côte d'Azur University, CNRS, LEAT, Nice, France
Claire.Migliaccio@univ-cotedazur.fr

Abstract. Numerical discretization of the large-scale Maxwell's equations leads to an ill-conditioned linear system that is challenging to solve. The key requirement for successive solutions of this linear system is to choose an efficient solver. In this work we use Perfectly Matched Layers (PML) to increase this efficiency. PML have been widely used to truncate numerical simulations of wave equations due to improving the accuracy of the solution instead of using absorbing boundary conditions (ABCs). Here, we will develop an efficient solver by providing an alternative use of PML as transmission conditions at the interfaces between subdomains in our domain decomposition method. We solve Maxwell's equations and assess the convergence rate of our solutions compared to the situation where absorbing boundary conditions are chosen as transmission conditions.

1 Introduction

Maxwell's equations need to be solved in many applications, such as medical imaging or electromagnetic compatibility. The Finite Element Method (FEM) is widely used for numerical modeling of these problems due to its ability to handle complex geometrical configurations. Finite element discretization of these frequency-domain wave problems leads to an ill-conditioned linear system with large number of unknowns. To solve this system, the efficiency of direct solvers is limited at larger scales due to scalability problems in memory and computing time. Besides, Krylov subspace iterative solvers have shown slow convergence. An alternative method that tackles the convergence problem of iterative solvers is the domain decomposition method (DDM). The method relies on a division of the computational domain into smaller subdomains, leading to subproblems of smaller sizes manageable by direct solvers. One perfect candidate introduced in [1] and then improved in [2] is the domain decomposition preconditioner which proved to be very robust in large scale computations. However, designing an efficient domain decomposition preconditioner is still challenging for such a system.

M. van Beurden et al. (Eds.): SCEE 2022, MI 43, pp. 45–52, 2024.
https://doi.org/10.1007/978-3-031-54517-7_5

In this paper, we present an efficient PML-based Schwarz-type domain decomposition preconditioner with overlapping subdomains. The convergence rate of Schwarz methods highly depends on the transmission condition on the interfaces between subdomains. Thus, carefully designed transmission conditions play a critical role in the efficiency of the solver. To decrease undesired numerical reflections, one usually adds a PML layer along the boundaries that extend a definite area to the infinity, representing an unbounded volume, and absorbs almost all incident waves, regardless of angle of incidence, so that the waves decay exponentially in magnitude into the PML medium [5,6]. Specifically, using PML is essential for simulating unbounded systems such as infinitely long waveguides or an isolated structure in an infinite vacuum region. While the use of PMLs as boundary conditions when solving a problem in open space is quite common, less things are known about their use as transmission conditions within a domain decomposition algorithm. We propose to assess the performance of a one-level domain decomposition algorithm where the transmission conditions at the boundaries between subdomains are PML conditions, providing a better approximation to the transparent boundary operator. We will investigate the convergence properties and compare them with the more common impedance transmission conditions. In a previous work, PML have been used successfully as transmission conditions in domain decomposition methods in geophysical applications modeled by the Helmholtz equation in [3]. The paper is organized as follows. In Sect. 2, we present mathematical model including Maxwell's equations, PML formulation with different stretching functions as well as its implementation. Then, the DDM with PML-based transmission operators is introduced. In Sect. 3, some numerical examples are presented to analyze the performance of the proposed domain decomposition algorithm. Finally, conclusion is written in Sect. 4.

2 Mathematical Model

Let us consider the computational domain $\Omega \subset \mathbb{R}^3$ to be a homogeneous dielectric medium of complex-valued electric permittivity ε_σ and electrical conductivity $\sigma > 0$. Let μ_0 be the permeability of free space and $\mathbf{n}$ be the unit outward normal to the boundaries $\partial\Omega$. ω is the angular frequency and c is the wave speed. In the frequency domain, the electric field $\xi(\mathbf{x},\mathbf{t}) = \Re(\mathbf{E}(\mathbf{x})e^{i\omega\mathbf{t}})$ has harmonic dependence on time of angular frequency ω, where $\mathbf{E}(\mathbf{x})$ is its complex amplitude depending on the space variable $\mathbf{x}$. Hence $\mathbf{E}(\mathbf{x})$ is a solution to the following second order time-harmonic Maxwell's equation

$$\nabla\times(\nabla\times\mathbf{E}) - \omega^2\varepsilon_\sigma\mu_0\mathbf{E} = f \qquad \text{in } \Omega. \tag{1}$$

Let us denote the boundary of the global domain by $\partial\Omega$ where Robin condition $(\nabla\times\mathbf{E})\times\mathbf{n} + i\frac{\omega}{c}\mathbf{n}\times(\mathbf{E}\times\mathbf{n}) = 0$ is imposed [4]. The Robin or impedance boundary condition (Imp BCs) is a standard first order approximation to the far field Sommerfeld radiation condition enabling the description of the wave behavior in a bounded domain, while the physical domain is not bounded. The finite element discretization of Eq. (1) is written as the following linear system

$$A\mathbf{u} = \mathbf{b}. \tag{2}$$

2.1 PML Formulation

To solve a partial differential equation (PDE) numerically, the computational domain has to be truncated without introducing reflections. The first attempt in this regard is absorbing boundary conditions (ABCs). The first order ABC as regular choice is Robin condition that was mentioned earlier. Due to the limited accuracy of this method, PML were introduced by Berenger [5] as a better alternative. PML provide non-reflecting boundaries so that the numerical solution converges exponentially to the exact solution in the computational domain as the thickness of the layer increases.

PML implementation is done by stretching cartesian coordinates so that stretching is defined in a layer surrounding the computational domain [5] and Dirichlet boundary condition can be imposed at the end of the PML layer. In this regard, we assume the boundaries of the computational domain to be aligned with the coordinate axes. For simplicity we will focus on truncating the problem in the x direction. Let us suppose that the PML layer extends from the boundary of our domain of interest $x = a$ until $x = a^*$. The coordinate mapping in x direction is:

$$\frac{\partial}{\partial x_{pml}} \rightarrow \frac{1}{1 - \frac{i}{\omega}\sigma(x)} \frac{\partial}{\partial x}, \sigma(x) = \begin{cases} 0 & \text{if } x < a \\ > 0 & \text{if } a < x < a^*. \end{cases} \tag{3}$$

In the PML region where $\sigma(x) > 0$, The oscillating solutions turn into exponentially decaying ones. In the rest of the region where $\sigma(x) = 0$, the wave equation is unchanged and the solution is unchanged. In this paper we have studied two different stretching functions $\sigma(x)$ as following

$$\sigma_{-1}(x) = \frac{1}{a^* - x} \quad (a) \quad , \quad \sigma_{-2}(x) = \frac{2}{(a^* - x)^2} \quad (b) \tag{4}$$

To truncate our computational region with a PML layer in other directions, we just need to do the same transformations to get $\frac{\partial}{\partial y_{pml}}$ and $\frac{\partial}{\partial z_{pml}}$. At the corners of the computational cell, we will have PML regions along two or three directions simultaneously, but it will not generate any problem.

Implementing this mapping in the three dimensional domain requires a slight further generalization of Eq. (1), resulting in the following definition of the curl operator to be used in the variational formulation:

$$\nabla_{pml} \times \mathbf{E} = \begin{bmatrix} \frac{\partial \mathbf{E}_z}{\partial y_{pml}} - \frac{\partial \mathbf{E}_y}{\partial z_{pml}} \\ \frac{\partial \mathbf{E}_x}{\partial z_{pml}} - \frac{\partial \mathbf{E}_z}{\partial x_{pml}} \\ \frac{\partial \mathbf{E}_y}{\partial x_{pml}} - \frac{\partial \mathbf{E}_x}{\partial y_{pml}} \end{bmatrix} \tag{5}$$

2.2 Domain Decomposition Preconditioner

To solve our large and ill conditioned linear system (2), the use of a robust and efficient preconditioner is necessary in a Krylov iterative solver (GMRES) [6]. A preconditioner M^{-1} is a linear operator that approximates the inverse of matrix $\mathbf{A}$ whose cost of the associated matrix-vector product is much cheaper than solving the original linear system. In this regard, we employ right preconditioning to solve (2) that will give us:

$$AM^{-1}\mathbf{y} = \mathbf{f}, \quad where \quad \mathbf{u} = M^{-1}\mathbf{y} \tag{6}$$

This right preconditioned system benefits from a residual that is preconditioner independent compared to the left-preconditioned variant.

As an overlapping Schwarz method, the optimized restricted additive Schwarz (ORAS) domain decomposition preconditioner is chosen here

$$M_{ORAS}^{-1} = \sum_{s=1}^{N_{sub}} R_s^T D_s A_s^{-1} R_s \tag{7}$$

where N_{sub} is the number of overlapping subdomains Ω_s into which the domain Ω is decomposed. Here, matrices A_s stem from the discretisation of local boundary value problems on Ω_s with transmission conditions at the subdomain interfaces. Let N be an ordered set of the unknowns of the whole domain and let $N = \bigcup_{s=1}^{N_{sub}} N_s$ be its decomposition into the (nondisjoint) ordered subsets corresponding to the different (overlapping) subdomains Ω_s. Matrix R_s is the restriction matrix from Ω to subdomain Ω_s; it is a $N_s \times N$ Boolean matrix. R_s^T is then the extension matrix from subdomain Ω_s to Ω. D_s is a $N_s \times N_s$ diagonal matrix that gives a discrete partition of unity, i.e., $\sum_{s=1}^{N_{sub}} R_s^T D_s R_s = I$.

The convergence rate of this method highly depends on the choice of transmission conditions between the subdomains [7]. The optimal convergence is obtained by imposing the Dirichlet-to Neumann (DtN) map related to the complementary of each subdomain [8,9]. However, since the cost of computing the exact DtN is prohibitive, low-order absorbing boundary conditions (ABCs) to approximate the DtN have been developed. Nonetheless, these methods have limited accuracy, which led to developing domain decomposition strategies with high order transmission conditions [10]. But the problem with high order transmission conditions is the difficulty of their implementation. A good approximation of ABCs in terms of providing better convergence rate and easy implementation would be to use PML on the interface boundaries of the cuboid-shaped subdomains [11,12], that is what we consider here. In this purpose a PML layer is added in each direction in the overlap region. Note that the width of the overlap has to be larger than the PML layer for a good transmission of the data between subdomains.

3 Numerical Results

The performance of the proposed PML-based preconditioner for Maxwell's equations is studied in a 3D homogeneous domain Ω, while length of the domain in each direction is 10 m. We have excited the $z = 0$ surface with plane wave incident term $e^{(-ikz)}$, where $k = \frac{2\pi}{\lambda}$, with propagation in $+z$ direction shown in the Fig. 1. The convergence rate is studied while there is PML or Impedance as global boundary conditions (BCs) or interface conditions (ICs), which leads to four different situations reported in Table 1. The finite element discretization is done for the first order edge elements for two different frequencies. Let #DoF represents the number of degrees of freedom. For $f = 0.5$ Hz, we have #DoF =511775 and for $f = 1$ Hz, we have #DoF = 2098100. The global domain is

Fig. 1. Plane wave propagation.

decomposed into N = 100 number of cuboid-shaped subdomains, that PMl layer is set along their interfaces with the length L_{pmli}.

In Table 1, simulations are done for the σ_{-2} stretching function and PML length on the interfaces is shown with $L_{pmli} = 8h$ where $h = \frac{\lambda}{n_\lambda}$ is the mesh size and n_λ is the number of points per wavelength. In all simulations, PML length on the global boundary is $L_{pml} = 2\lambda$ and the overlapping layers between subdomains is changed from 2 to 8 layers in four steps. Looking at the Table 1, for $f = 0.5$ and considering 8 number of overlapping layers to be larger than L_{pmli}, we can see while we have PML BCs and Imp ICs, number of iterations is 21, but with PML BCs and PML ICs, this number decrease to 16. It is while with Imp on BCs and ICs, number of iterations is 26. In this table, • means that, solution has not converged for 200 number of iterations.

Table 1. Function σ_{-2}. $n_\lambda = 5$, $L_{pml} = 2\lambda$, $L_{pmli} = 8h$, $c = 1$, $N = 100$ is number of subdomains

BCs	ICs	f=0.5				f=1			
		2	4	6	8	2	4	6	8
Imp	Imp	29	24	23	26	34	27	25	25
Imp	PML	75	30	22	28	•	50	29	23
PML	Imp	23	21	20	21	28	23	21	20
PML	PML	65	25	18	16	•	43	25	19

To see the influence of the L_{pmli}, we did the simulation with smaller PML layer on the subdomains, mentioned in Table 2, only for the case with PML as BCs and ICs. Comparing with the equivalent row in the Table 1 for 6 and 8 overlaping layers, we see that with larger length of PML on subdomains we have better convergence. Although the rate of convergence has become better for lower overlapping layers, with smaller PML length, due to the better data transmission between subdomains. Comparing the results for the use of stretching function σ_{-1} instead of σ_{-2} is mentioned in the Table 3.

For 2 number of overlapping layers, better number of iterations is seen with σ_{-1}, however for higher number of overlapping layers, σ_{-2} results in better convergence.

Table 2. Function σ_{-2}. $n_\lambda = 5$, $L_{pml} = 2\lambda$, $L_{pmli} = 4h$, $c = 1$, $N = 100$

BCs	ICs	f = 0.5				f = 1			
		2	4	6	8	2	4	6	8
PML	PML	59	22	18	17	183	35	30	29

Table 3. Function σ_{-1}. $n_\lambda = 5$, $L_{pml} = 2\lambda$, $L_{pmli} = 8h$, $c = 1$, $N = 100$

BCs	ICs	f = 0.5				f = 1			
		2	4	6	8	2	4	6	8
PML	Imp	24	20	18	21	29	24	22	21
PML	PML	46	28	19	17	66	35	26	21

The performance of the proposed preconditioner in a heterogeneous domain is studied in Tables 4 and 5. Here, we have defined a medium with two values of ε_r with the dimension of 6.3 m in X and Y directions and 2.5 m in Z direction inside the free space computational domain. In this experiment, rhs is chosen as a random value, $L_{pmli} = 4h$, $f = 1$ Hz, stretching function is chosen as σ_{-2} in Table 4 and σ_{-1} in Table 5. Results show, increasing value of ε_r increase number of iterations, but with PML interface conditions we can have faster convergence. Comparing two tables, better performance is obtained by σ_{-2} stretching function. Here we have considered maximum number of iterations as 600. In the results, - means problem is not solved due to memory limitation.

Table 4. Function σ_{-2}. $n_\lambda = 5$, $L_{pml} = 2\lambda$, $f = 1.0$ Hz, $c = 1$, $N = 100$.

BCs	ICs	$\varepsilon_r = 4$				$\varepsilon_r = 5$			
		2	4	6	8	2	4	6	8
PML	Imp	262	207	181	168	586	425	375	326
PML	PML	•	249	199	160	•	514	440	311

Results are obtained on the Université Côte d'Azur's High-Performance Computing (HPC) center. In this HPC center, cluster is composed of 48 CPU computing nodes, including 32 nodes with Dual Intel Xeon Gold processor, providing 40 cores per node and 192 GB of memory and 16 nodes with 2 AMD Epyc processors, providing 32 cores per node and 256 GB of memory.

Table 5. Function σ_{-1}. $n_\lambda = 5, L_{pml} = 2\lambda,\ f = 1.0\,\text{Hz}, c = 1, N = 100.$

BCs	ICs	$\varepsilon_r = 4$				$\varepsilon_r = 5$			
		2	4	6	8	2	4	6	8
PML	Imp	264	207	183	-	587	427	374	-
PML	PML	411	221	201	-	•	402	353	-

4 Conclusions

In this work, we have developed a numerical model for an accurate and fast simulation of Maxwell's equations. To achieve this goal, the PML layer is implemented as physical boundaries and as transmission conditions in domain decomposition preconditioner for a three dimensional domain. A better convergence rate is achieved with PML layer, compared to Impedance interface conditions. Numerical results shows that the performance of the PML depends on a well chosen stretching function and length of the PML. This work is a preliminary study that was inspired by a similar work done for Helmholtz equations [13] where the results were very encouraging. More investigations can be done in next works, like evaluating performance of the PML as interface conditions for higher order edge elements or in a heterogeneous domain. Note that PML have some limitations, for the time being it has been applied only along the straight interfaces but variants for circular boundaries exist that can be further explored in the context of other applications.

Acknowledgements. This project has received funding from the European Union's Horizon 2020 research and innovation programme under grant agreement No 847581 and is co-funded by the Région Provence-Alpes-Côte d'Azur and IDEX UCA^{JEDI}. This work was supported by the French government, through the UCA^{JEDI} Investments in the Future project managed by the National Research Agency (ANR) under reference number ANR-15-IDEX-01. The authors are grateful to the OPAL infrastructure from Université Côte d'Azur and the Université Côte d'Azur's Center for High-Performance Computing for providing resources and support.

References

1. Tournier, P.-H., et al.: Numerical modeling and high-speed parallel computing: new perspectives on tomographic microwave imaging for brain stroke detection and monitoring. IEEE Antennas Propag. Mag. **59**(5), 98–110 (2017)
2. Bonazzoli, M., Dolean, V., Graham, I., Spence, E., Tournier, P.-H.: Domain decomposition preconditioning for the high-frequency time-harmonic Maxwell equations with absorption. Math. Comput. **88**(320), 2559–2604 (2019)
3. Tournier, P.-H., Jolivet, P., Dolean, V., Aghamiry, H.S., Operto, S., Riffo, S.: Three-dimensional finite-difference and finite-element frequency-domain wave simulation with multi-level optimized additive Schwarz domain-decomposition preconditioner: a tool for FWI of sparse node datasets, arXiv:2110.15113 (2021)
4. Borzooei, S., Dolean, V., Migliaccio, C., Tournier, P.H.: An efficient, high order finite element method for the time-harmonic Maxwell's equations. In: 2021 IEEE Conference on Antenna Measurements and Applications (CAMA), pp. 340–344. IEEE (2021)

5. Berenger, J.-P.: A perfectly matched layer for the absorption of electromagnetic waves. J. Comput. Phys. **114**(2), 185–200 (1994)
6. Bao, G., Haijun, W.: Convergence analysis of the perfectly matched layer problems for time-harmonic Maxwell's equations. SIAM J. Numer. Anal. **43**(5), 2121–2143 (2005)
7. Dolean, V., Gander, M.J., Gerardo-Giorda, L.: Optimized Schwarz methods for Maxwell's equations. SIAM J. Sci. Comput. **31**(3), 2193–2213 (2009)
8. Nataf, F.: Interface connections in domain decomposition methods. In: Bourlioux, A., Gander, M.J., Sabidussi, G. (eds.) Modern Methods in Scientific Computing and Applications, pp. 323–364. Springer, Dordrecht (2002). https://doi.org/10.1007/978-94-010-0510-4_9
9. Bootland, N., Dolean, V., Jolivet, P., Tournier, P.-H.: A comparison of coarse spaces for Helmholtz problems in the high frequency regime. Comput. Math. Appl. **98**, 239–253 (2021)
10. Marsic, N., De Gersem, H.: Convergence of classical optimized non-overlapping Schwarz method for Helmholtz problems in closed domains. arXiv preprint arXiv:2001.01502 (2020)
11. Royer, A., Geuzaine, C., Béchet, E., Modave, A.: A non-overlapping domain decomposition method with perfectly matched layer transmission conditions for the Helmholtz equation. Comput. Methods Appl. Mech. Eng. **395**, 115006 (2022)
12. Thierry, B., et al.: GetDDM: an open framework for testing optimized Schwarz methods for time-harmonic wave problems. Comput. Phys. Commun. **203**, 309–330 (2016)
13. Bootland, N., Borzooei, S., Dolean, V., Tournier, P.-H.: Numerical assessment of PML transmission conditions in a domain decomposition method for the Helmholtz equation. arXiv preprint arXiv:2211.06859 (2022)

Validation-Oriented Modelling of Electrical Stimulation Chambers for Cartilage Tissue Engineering

Lam Vien Che[1(✉)], Julius Zimmermann[1], Henning Bathel[1], Alina Weizel[2], Hermann Seitz[2,3], and Ursula van Rienen[1,3,4]

[1] Institute of General Electrical Engineering, University of Rostock, Rostock, Germany
{lam.che,julius.zimmermann,henning.bathel, ursula.van-rienen}@uni-rostock.de

[2] Chair of Microfluidics, Faculty of Mechanical Engineering and Marine Technology, University of Rostock, Rostock, Germany
{alina.weizel,hermann.seitz}@uni-rostock.de

[3] Department Life, Light and Matter, University of Rostock, Rostock, Germany

[4] Department of Ageing of Individuals and Society, Interdisciplinary Faculty, University of Rostock, Rostock, Germany

Abstract. The capability of electrical stimulation to enhance cell activity, proliferation, and differentiation makes it an attractive method in cell-based therapies. Due to its biocompatibility, capacitive coupling has emerged as a favourable method to deliver electric fields to cartilaginous cells. Unfortunately, there exists no means to measure the electric field directly. It can solely be inferred from other measurement results. Nonetheless, numerical simulations by the finite element method provide a possibility to estimate the electric field distribution and magnitude. The experimental validation of numerical models, however, receives insufficient attention. This study aims to bridge the gap between theory and experiment by applying validation-oriented modelling. The impact of different uncertain input parameters on relevant observables was assessed to suggest validation experiments using uncertainty quantification. The estimated capacitance was found to be in excellent agreement with the experimental result, indicating that the model is accurate. However, the electric field remains uncertain since the electric field and capacitance are dependent upon different input parameters. The electric field is primarily determined by the conductivity of the medium. Hence, a more precise conductivity measurement will allow for more accurate computation of the electric field magnitude.

1 Introduction

Recent years have seen increased interest in a variety of biophysical stimuli, such as mechanical, electrical, and (electro)magnetic fields as a potential tool in tissue engineering and regenerative medicine [11,15]. The clinical market for therapeutic devices for electrical stimulation is currently expanding, but only a limited number of studies addresses the electrical stimulation of cartilaginous cells [10,13]. Besides the common direct contact electrical stimulation, it is feasible to use electrodes electrically isolated

M. van Beurden et al. (Eds.): SCEE 2022, MI 43, pp. 53–60, 2024.
https://doi.org/10.1007/978-3-031-54517-7_6

from the sample. They are referred to as capacitive coupling (CC) approaches. CC has been favoured since it has been shown to enhance chondrogenesis (i.e., production of cartilage) [11] and (re-)differentiation of stem cells into cartilage cells [9]. Furthermore, CC can overcome drawbacks of direct contact stimulation, such as the formation of cytotoxic compounds and electrochemical reactions [11], because a thin layer insulates the electrodes. The goal of the device design is to control the electric field induced by the displacement current in the area where the cells are placed. However, electric field strength is rarely reported in most CC studies [16]. As it is not straightforward to adjust the electric field in an experimental setting, numerical simulations are required to determine the recommended voltage and/or current to be applied. The local electric field is also obtained from numerical simulations. This contribution focuses on the computation of measurable quantities to design appropriate validation approaches. For that purpose, we present a model of a device for cartilage regeneration that has already been employed in practice [9]. In many cases, numerical models are approached with suspicion, so uncertainty quantification (UQ) becomes increasingly crucial. We prepared UQ by establishing a parametrised geometrical model with an automated, adaptive modelling approach to assess the impact of various uncertain parameters (both geometrical and physical) and propose validation experiments.

2 Materials and Methods

A stimulation device for the application of capacitively coupled electrical fields in vitro was designed by Krueger et al. [9]. Figure 1 shows a photograph of a sample carrier comprising a polycarbonate plate with 12 cavities (wells) inserted by drilling. A flexible printed circuit board (Multi Circuit Boards, Poole, UK) made from polyimide was fixed at the bottom of the plate. An electrode array with 12 pairs of two annular ring-shaped electrodes and one small circular electrode in between was embedded in the circuit board. The medium is usually a scaffold positioned in the centre of the cavity to support cell cultivation. The schematic illustration with the dimensions of one well and its geometry are shown in Fig. 1.

The stimulations are often performed in the kHz range. In this study, the simulation was conducted at 60 kHz because this frequency has been found to be the most suitable for the differentiation of chondrocytes and cartilage tissue by electrical stimulation [9, 10]. The applied voltage was 1 V. The electric fields for other input voltages can be determined by scaling the result due to the linearity of the system. We used 3 ml potassium chloride (KCl) as the medium to ease the experimental validation of our numerical results because its conductivity at 25 °C is reported by the manufacturer (Hanna Instruments) (0.1413 S/m). Its relative permittivity is 78.57 [18]. The wall (polycarbonate) and plastic layer (polyimide) were assumed to have a conductivity of 10^{-8} S/m and 10^{-9} S/m, respectively. Both were assumed to possess the same relative permittivity of 3.4.

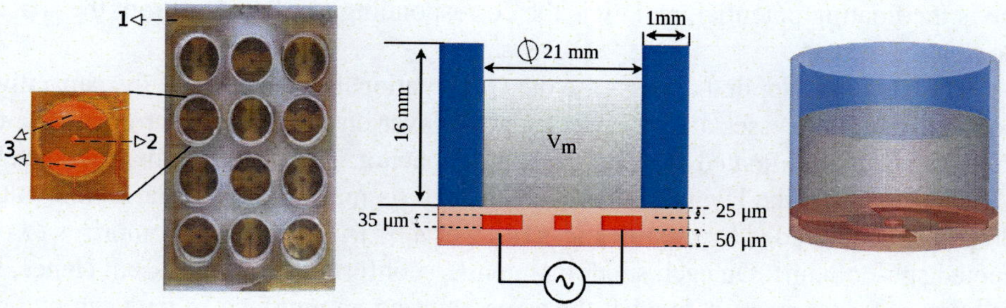

Fig. 1. Left: The sample carrier consists of (1) a polycarbonate plate with 12 cavities, (2) a central electrode and (3) annular ring-shaped electrodes and the enlarged image from the bottom view of one cavity. Central: Schematic view of a single stimulation device with the geometrical parameters, including the wall (blue), the insulation layer (red), the electrodes (dark red) and the medium (grey). Right: The geometry of the single stimulation device. V_m indicates the volume of the medium.

Electromagnetic fields can be considered slowly varying in several therapeutic procedures involving the electrical stimulation of biological samples [4]. Because magnetic field and eddy currents are assumed to be negligibly small, the electro-quasistatic (EQS) potential ϕ can be computed by solving

$$\nabla \cdot [(\sigma + i\omega\varepsilon)\nabla\phi] = 0 \tag{1}$$

where σ is the conductivity, ε is the permittivity, and ω is the angular frequency. The solution of (1) can be obtained using either a full discretization, here called full-fidelity (FF), or an approximate discretization based on a thin layer approximation (TLA). The open-source finite element solvers (*NGSolve*[1] [3] with the mesh generator *Netgen* [2]) and the proprietary software code (*COMSOL Multiphysics*®v5.5[2]) were employed to perform the simulation of both FF and TLA models.

In the FF approach, Dirichlet BCs were imposed on the annular ring-shaped electrodes, i.e., a voltage-controlled stimulation was used. The central electrode was modelled with a floating potential BC. This electrode has the potential to cause a distortion of the electric field in the medium, even if it is not connected to a power source. To examine its influence, we considered two cases: the FF model with the central electrode included in the final geometry and the FF model without the central electrode. The weak form of (1) [14] for the FF model can be formulated as

$$\int_{\Omega} (\sigma + i\omega\varepsilon)\nabla\phi\nabla v \mathrm{d}\Omega + \int_{\Gamma_F} (\lambda(v - v_F) + (\phi - \phi_F)\mu)\mathrm{d}S = 0 \tag{2}$$

where v is the test function defined on the entire domain Ω. λ and μ are the solution and the test function of the Lagrange multipliers belonging to an $H^{-\frac{1}{2}}$-conforming space.

1 https://www.ngsolve.org/.
2 https://www.comsol.com/.

ϕ_F is the floating potential, and v_F is the corresponding test function from the space of complex numbers.

The TLA was applied to decrease the computational effort because the capacitively coupled setup possesses a very thin insulation layer on top of the electrodes. The thin layer can be approximated as a parallel-plate capacitor. The TLA assumes that the electric field **E** inside the layer is homogeneous, and its magnitude is equal to the voltage drop across U divided by the thickness of layer d_l [5]. The problem comprises M connected sub-domains. On each subdomain an H^1-conforming space is used. Hence, two degrees of freedom are defined at the same point on an interface Γ_I between two subdomains, demonstrating the potential jump across an interface. Consequently, the weak form of the TLA model [17] is defined as

$$\sum_{m=1}^{M}\int_{\Omega_m}(\sigma+\mathrm{i}\omega\varepsilon_m)\nabla\phi_m\nabla v_m \mathrm{d}\Omega+\sum_{I=1}^{M-1}\int_{\Gamma_I}(\frac{\sigma_\mathrm{I}+\mathrm{i}\omega\varepsilon_\mathrm{I}}{d_\mathrm{l}})[\phi][v]\mathrm{d}S=0 \tag{3}$$

Here, [•] indicates a jump of the potential across the interface Γ_I.

The impedance Z of the considered chamber can be determined from the solution of (1) using the instantaneous power dissipation P [1]. We applied adaptive mesh refinement (AMR) for the base geometry described in the previous section using the Zienkiewicz-Zhu error estimator [6]. Moreover, scalable iterative solvers (GMRes, BiCGSTAB) with algebraic multigrid preconditioners were employed to solve (1). Regarding uncertainty quantification (UQ), polynomial chaos expansion was used to obtain the model output's statistical metrics. We utilized a modified version[3] of the Python library *Uncertainpy* [8]. A point collocation method was utilized to estimate the expansion coefficients. The polynomial order was set to four, and 10^4 samples were drawn from the surrogate model. The simulations and UQ computations were carried out using parallel computing on the HAUMEA high-performance computing cluster of the University of Rostock (each computing node supplied with 2 Intel Xeon Gold 6248 CPUs with a total of 40 cores and 192 GB RAM). We assessed the impact of different geometrical and physical uncertain parameters to suggest validation experiments. All possible error sources and presumed hypotheses in the UQ analysis are summarised in Table 1. Note that we did not consider the influence of the plastic conductivity and the medium permittivity. The plastic acts as an insulator, while the medium acts as a conductor; hence those above-mentioned dielectric properties do not play an essential role in the model output. Further, we analyzed only uniform distributions to reflect the current state of our knowledge about the uncertainty of individual parameters. The large ε_p variation was used to investigate how different plastic materials can affect electrical stimulation as the material properties are not disclosed by the manufacturer. The considerable range of σ_m covers all possible values from standard measurement solutions to biological tissue at various temperatures. The goal was to understand if a change in medium conductivity can be detected by impedance/capacitance measurements.

[3] https://github.com/j-zimmermann/uncertainpy.

Table 1. Model parameters for UQ given in terms of the uniform distribution $\mathcal{U}$. σ_m: conductivity of the medium, ε_p: relative permittivity of the plastic, V_m: volume of the medium, t_e: thickness of the electrode, t_{tl}: thickness of the top plastic layer, t_{bl}: thickness of the bottom plastic layer

Parameter	Distribution	Reasoning
σ_m	$\mathcal{U}[0.1,1.8]$ (S/m)	Assumptions from previous works [9]
ε_p	$\mathcal{U}[1.5,5]$	Assumptions from previous works [9]
V_m	$\mathcal{U}[3,4]$ (ml)	Pipetting inaccuracies
t_e	$\mathcal{U}[20,50]$ (μm)	Manufacturer error
t_{tl}	$\mathcal{U}[20,30]$ (μm)	Manufacturer error
t_{bl}	$\mathcal{U}[0,2]$ (mm)	Variation in the thickness of the sample holder

Impedance spectroscopy was employed to validate the numerical outputs because CC results in a frequency-dependent impedance. The device was characterized over a wide bandwidth using a frequency response analyzer (Rohde&Schwarz RTB2004 Digital Oscilloscope). A frequency sweep was conducted from 10 Hz to 25 MHz with an input voltage of 2.5 Vpp. To calculate the unknown impedance, the applied voltage was divided by the resulting current, calculated from the voltage drop across a shunt resistor. This method is commonly known as the I-V method.

3 Results and Discussion

Both the FF models with and without the central electrode included in the final geometry exhibit an insignificant discrepancy in the electric field in the medium. The negligible influence of the floating potential conductor is comprehensible as it is located in an extremely low electric field. In the vicinity of this electrode, there was a distortion of the electric field distribution. However, this change can be neglected because the field strength is approximately 20,000 times smaller than the one around the annular ring-shaped electrodes. Therefore, the vast majority of the electric field can be traced to the field adjacent to the annular ring-shaped electrodes. The central electrode, thus, was eliminated from the final geometry when applying the TLA approach and the UQ analysis to save time and computational effort. Concretely, the computational time of the FF model with and without the central electrode is 728 and 397 s, respectively. TLA further reduces the runtime by 20 s, which is significant as nearly 500 simulations are required to perform meaningful UQ. The electric field distribution and magnitude in the medium are shown in Fig. 2. The field strength ranges from 0.01 V/m to 3.6 V/m. The impedance is $(888 - 55\,250\text{i})\,\Omega$. This result indicates, as expected, a primarily capacitive behaviour. The numerically estimated capacitance of the device is about 48 pF. The convergence of the impedance was proved by using AMR techniques. After the fifth refinement, the variation in the impedance and the estimated error became negligible. Regarding the electric field distribution and magnitude in the medium as well as the global impedance, no significant difference between *COMSOL Multiphysics®* and *NGSolve*, irrespective of the chosen approach (FF or TLA model), could be found. The experimental result of the mean capacitance was calculated to be 49 ± 2 pF, using the ImpedanceFitter package[4]. The numerical and experimental capacitance values conform well, indicating that we have a solid and accurate model (the relative difference is roughly 2%).

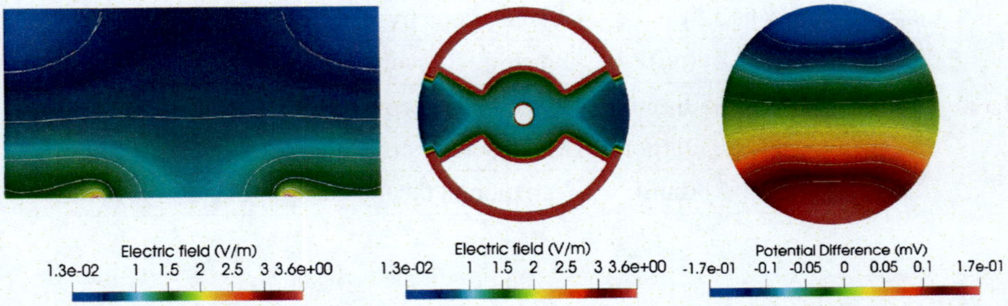

Fig. 2. Left: Electric field distribution in the medium. Cut through the central YZ plane. Central: Electric field distribution in the insulator layer. For visualisation, the electric field strength in the insulator layer was fixed to a maximum of 3.6 V/m; however, the actual field strength maximum is around 20,000 V/m near the annular ring-shaped electrodes. Cut through the central YX plane. Right: The voltage difference distribution in the medium just above the insulation is shown. The voltage difference was evaluated with respect to the mean value of the voltage in the medium.

Subsequently, UQ of the input parameters was conducted to examine the accuracy and reliability of the numerical simulation as well as to suggest validation experiments. Aside from the mean and standard deviation, the Sobol indices are computed in the UQ approach. We studied the first-order Sobol indices to assess the sensitivity of the model outcome to variations of individual parameters. If a parameter has a low Sobol index, variations in this parameter result in comparatively slight changes in the final model output and vice versa [8]. As shown in Fig. 3, the thickness of the electrode and the bottom plastic layer, and the volume of the medium do not influence the model output. In contrast, all features are sensitive to the permittivity of plastic. The capacitance depends on the thickness of the top plastic layer and the permittivity of the plastic. However, the impact of the plastic permittivity is more dominant than the top plastic layer thickness. The variations in the electric field strength are mainly attributed to variations in the medium conductivity, followed by the permittivity of plastic. Based on UQ results, the computed capacitance is 47.15 ± 15.67 pF and the field strength at the fixed point, located in the centre of the YX plane and 100 μm above the bottom of the medium, is 0.27 ± 0.3 V/m. Moreover, the 90% prediction interval of the capacitance in pF is [23.73, 73.99]. The field strength at the fixed point in V/m is [0.04, 0.94]. The high standard deviation and the wide range of prediction intervals suggest that more information about the model inputs is required to obtain a robust and reliable numerical prediction.

The most interesting property of electrically active implants is the electric field distribution and amplitude. They can be mapped by measuring the voltage gradient or the current through the sample, which should match a predicted value [12]. However, it

[4] https://github.com/j-zimmermann/ImpedanceFitter.

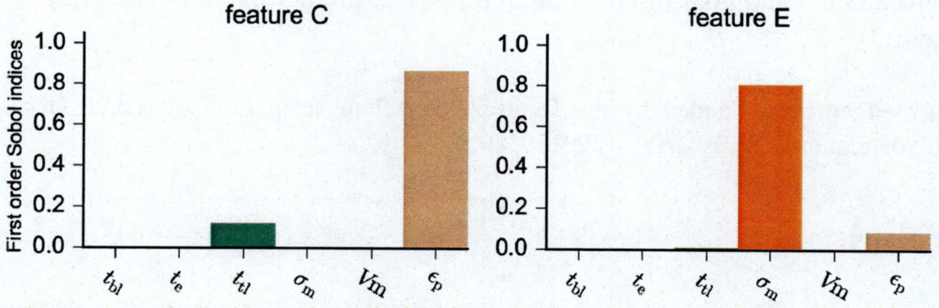

Fig. 3. The first order Sobol indices for the capacitance (C) and the electric field at a fixed point (E) 100 µm above the bottom of the medium, computed for the parameters from Table 1.

is impossible to measure the voltage gradient in our case due to significantly minor changes in voltage (0.34 mV) across the CC chamber (Fig. 2). A similar situation exists for the impedance phase; the phase change is approximately 0.2°, whereas the smallest experimentally measurable difference is about one degree. Measuring the capacitance has emerged as the feasible option to validate the numerical model considering all previous aspects. Unfortunately, based on the UQ results, the uncertainty of the capacitance and the electric field rely upon different model parameters. Therefore, knowing the experimental result of the capacitance is not fully informative about the electric field in electrical stimulation. Because the conductivity of the medium largely determines the electric field, knowing the conductivity well allows for fair confidence in the computed electric field. To investigate the case when measured conductivity and permittivity data is available, which is rare in a biological laboratory, we utilized the normal distributions for σ_m and ε_p. The uncertainty of the model outputs still mainly depends on t_{tl}, σ_m and ε_p. If the interval of three standard deviations from the mean interval is small for this parameter, the 90% prediction range will also be small and vice versa.

4 Conclusion

Using different simulation software and models of different complexity yielded consistent results for the capacitance and electric field. Eventually, a simplified geometric model could be identified, which permitted an accelerated UQ analysis. The estimated capacitance from the simulation is in excellent agreement with the experimental result of approximately 48 pF, indicating a reliable numerical model. Nevertheless, the electric field is still uncertain because, according to UQ analysis, the electric field and capacitance depend upon distinct input parameters. The conductivity of the medium largely determines the electric field distribution and magnitude. Consequently, knowing the conductivity well allows for fair confidence in the computed electric field. In future work, we suggest the development of a digital twin (DT) of CC devices because a DT for direct contact devices coupling numerical simulations and impedance spectroscopy has been realised to make local electric field strengths accessible [12]. Otherwise, the lack of electric field measurement might hamper translation to clinics [16]. In sum, a DT will combine both validation-oriented modelling (e.g., computing and measuring the capacitance) and goal-oriented modelling (i.e., predicting and optimising the electric field).

Acknowledgements. Funded by the Deutsche Forschungsgemeinschaft (DFG, German Research Foundation) - SFB 1270/2 - 299150580.

References

1. Bondeson, A., Rylander, T., Ingelström, P.: Computational Electromagnetics. Springer, New York (2012). https://doi.org/10.1007/978-1-4614-5351-2
2. Schöberl, J.: NETGEN an advancing front 2D/3D-mesh generator based on abstract rules. Comput. Vis. Sci. **1**(1), 41–52 (1997)
3. Schöberl, J.: C++11 Implementation of Finite Elements in NGSolve, ASC Report 30/2014, Institute for Analysis and Scientific Computing, Vienna University of Technology (2014)
4. van Rienen, U., et al.: Electro-quasistatic simulations in bio-systems engineering and medical engineering. Adv. Radio Sci. **3**, 39–49 (2005)
5. Pucihar, G., et al.: Numerical determination of transmembrane voltage induced on irregularly shaped cells. Ann. Biomed. Eng. **34**(4), 642–652 (2006)
6. Zienkiewicz, O.C., Zhu, J.Z.: The superconvergent patch recovery and a posteriori error estimates. Part 2: error estimates and adaptivity. Int. J. Numer. Methods Eng. **33**(7), 1365–1382 (1992)
7. Xiu, D.: Numerical Methods for Stochastic Computations. Princeton University Press, Princeton (2010)
8. Tennøe, S., Geir, H.: Uncertainpy: a Python toolbox for uncertainty quantification and sensitivity analysis in computational neuroscience. Front. Neuroinform. **12**, 49 (2018)
9. Krueger, S., et al.: Establishment of a new device for electrical stimulation of non-degenerative cartilage cells in vitro. Int. J. Mol. Sci. **22**(1), 394 (2021)
10. Brighton, C.T., et al.: The effect of electrical fields on gene and protein expression in human osteoarthritic cartilage explants. J. Bone Joint Surg. **90**(4), 833–848 (2008)
11. Thrivikraman, G., Boda, S.K., Basu, B.: Unraveling the mechanistic effects of electric field stimulation towards directing stem cell fate and function: a tissue engineering perspective. Biomaterials **150**, 60–86 (2018)
12. Zimmermann, J., et al.: Using a digital twin of an electrical stimulation device to monitor and control the electrical stimulation of cells in vitro. Front. Bioeng. Biotechnol. **9**, 765516 (2021)
13. Vaca-González, J.J., et al.: Biophysical stimuli: a review of electrical and mechanical stimulation in hyaline cartilage. Cartilage **10**(2), 157–172 (2019)
14. Brandstetter, G., Govindjee, S.: A high-order immersed boundary discontinuous-Galerkin method for Poisson's equation with discontinuous coefficients and singular sources. Int. J. Numer. Meth. Eng. **101**(11), 847–869 (2015)
15. da Silva, L.P., et al.: Electric phenomenon: a disregarded tool in tissue engineering and regenerative medicine. Trends Biotechnol. **38**(1), 24–49 (2020)
16. Nicksic, P.J., et al.: Electronic bone growth stimulators for augmentation of osteogenesis in in vitro and in vivo models: a narrative review of electrical stimulation mechanisms and device specifications. Front. Bioeng. Biotechnol. **10**, 793945 (2022)
17. Ben Belgacem, F., et al.: Finite element methods for the temperature in composite media with contact resistance. J. Sci. Comput. **63**(2), 478–501 (2015)
18. Drake, F.H., et al.: Measurement of the dielectric constant and index of refraction of water and aqueous solutions of KCl at high frequencies. Phys. Rev. **35**(6), 613 (1930)

Matrix-Free Parallel Preconditioned Iterative Solvers for the 2D Helmholtz Equation Discretized with Finite Differences

Jinqiang Chen(✉), Vandana Dwarka, and Cornelis Vuik

Delft Institute of Applied Mathematics, Delft University of Technology, Mekelweg 4, 2628 CD Delft, The Netherlands
{j.chen-11,v.n.s.r.dwarka,c.vuik}@tudelft.nl

Abstract. We present a matrix-free parallel iterative solver for the Helmholtz equation related to applications in seismic problems and study its parallel performance. We apply Krylov subspace methods, GMRES, Bi-CGSTAB and IDR(s), to solve the linear system obtained from a second-order finite difference discretization. The Complex Shifted Laplace Preconditioner (CSLP) is employed to improve the convergence of Krylov solvers. The preconditioner is approximately inverted by multigrid iterations. For parallel computing, the global domain is partitioned blockwise. The standard MPI library is employed for data communication. The matrix-vector multiplication and preconditioning operator are implemented in a matrix-free way instead of constructing large, memory-consuming coefficient matrices. These adjustments lead to direct improvements in terms of memory consumption. Numerical experiments of model problems show that the matrix-free parallel solution method has satisfactory parallel performance and weak scalability. It allows us to solve larger problems in parallel to obtain more accurate numerical solutions.

1 Introduction

The Helmholtz equation describes the phenomena of time-harmonic wave scattering in the frequency domain. It is widely studied in computational electromagnetics, with applications in seismic exploration, sonar, antennas, and medical imaging. To solve the Helmholtz equation numerically, we discretize it and obtain a linear system $Ax = b$. The linear system matrix is sparse, symmetric, non-Hermitian, and indefinite [1]. Iterative methods and parallel computing are commonly considered for a large-scale linear system resulting from a practical problem. However, the indefiniteness of the linear system brings a great challenge to the numerical solution method, especially for large wavenumbers. The convergence rate of many iterative solvers is affected significantly for increasing wavenumber. Therefore, the research problem of how to solve the systems efficiently and economically, while at the same time maintaining a high accuracy by minimizing pollution error arises in this field.

Many efforts have been made to solve the problem accurately with high performance. Originally derived from [2], the industry standard, also known as the Complex Shifted Laplace Preconditioner (CSLP) [3,4] does show good properties for medium

M. van Beurden et al. (Eds.): SCEE 2022, MI 43, pp. 61–68, 2024.
https://doi.org/10.1007/978-3-031-54517-7_7

wavenumbers. Nevertheless, the eigenvalues shift to the origin as the wavenumber increases. These near-zero eigenvalues have an unfavorable effect on the convergence speed of Krylov-based iterative solvers. Recently, a higher-order approximation scheme to construct the deflation vectors was proposed to reach close to wavenumber-independent convergence [5].

The development of scalable parallel Helmholtz solvers is also ongoing. One approach is to parallelize existing advanced algorithms. Kononov and Riyanti [6,7] first developed a parallel version of Bi-CGSTAB preconditioned by multigrid-based CSLP. Gordon and Gordon [8] parallelized their so-called CARP-CG algorithm (Conjugate Gradient acceleration of CARP) blockwise. The block-parallel CARP-CG algorithm shows improved scalability as the wavenumber increases. Calandra et al. [9] proposed a geometric two-grid preconditioner for 3D Helmholtz problems, which shows strong scaling properties in a massively parallel setup. Another approach is the Domain Decomposition Method (DDM), which originates from the early Schwarz Methods. DDM, as a preconditioner mostly, has been widely used to develop parallel solution methods for the Helmholtz problems. For comprehensive surveys, we refer the reader to [10–14] and references therein.

This work describes parallel versions of Krylov subspace methods, such as the Generalized minimal residual method (GMRES), Bi-CGSTAB, and IDR(s), preconditioned by the multigrid-based CSLP for the Helmholtz equation. We consider the CSLP preconditioner because it is the first and most popular method where the number of iterations scales linearly within medium wavenumbers. Based on a block-wise domain decomposition and a matrix-free implementation, our parallel framework contributes to robust parallel CSLP-preconditioned Krylov solvers for Helmholtz problems. It is the basis for scalable parallel computing. Numerical experiments show that, compared to [5,6] that assemble matrices, the matrix-free framework allows us to solve the Helmholtz problem with a larger grid size to reduce pollution errors related to grid resolution. The parallel efficiency is up to 70%. Its weak scaling performance means that a larger problem can be solved in about the same amount of time as a smaller problem as long as the number of tasks increases proportionally.

The rest of this paper is organized as follows. Section 2 describes the mathematical model that we will discuss. All numerical methods we use are given in Sect. 3. The numerical performance is explored in Sect. 4. Finally, Sect. 5 contains our conclusions.

2 Mathematical Model

We will consider the following 2D Helmholtz equation on a rectangular domain Ω with boundary $\Gamma = \partial\Omega$. The Helmholtz equation reads

$$-\Delta u(x,y) - k^2 u(x,y) = b(x,y), \text{ on } \Omega \tag{1}$$

supplied with Dirichlet boundary conditions $u(x,y) = g(x,y)$ or first-order Sommerfeld boundary conditions $\frac{\partial u(x,y)}{\partial n} - \mathrm{i}ku(x,y) = 0$, on $\partial\Omega$. i is the imaginary unit. n and $g(x,y)$ represent the outward normal and the given data of the boundary respectively. $b(x,y)$ is the source function. k is the wavenumber. The frequency is f, the speed of propagation is c, which are related by $k = \frac{2\pi f}{c}$.

3 Numerical Methods

3.1 Discretization

Structural vertex-centered grids are used to discretize the computational domain. Suppose the mesh width in x and y direction are both h. A second-order finite difference scheme is used. The discrete Helmholtz operator A_h can be obtained by adding the diagonal matrix $-k^2 I_h$ to the Laplacian operator $-\Delta_h$, *i.e.* $A_h = -\Delta_h - k^2 I_h$. Therefore, the stencil of the discrete Helmholtz operator is

$$[A_h] = \frac{1}{h^2}\begin{bmatrix} 0 & -1 & 0 \\ -1 & 4-k^2h^2 & -1 \\ 0 & -1 & 0 \end{bmatrix} \tag{2}$$

In the case of the Sommerfeld radiation condition, ghost points located outside the boundary points can be introduced for the boundary points. For instance, suppose $u_{0,j}$ is a ghost point on the left of $u_{1,j}$, the normal derivative can be approximated by $\frac{\partial u}{\partial n} - \mathrm{i}ku = \frac{u_{0,j}-u_{2,j}}{2h} - \mathrm{i}ku_{1,j} = 0$. We can rewrite it as $u_{0,j} = u_{2,j} + 2hiku_{1,j}$, which can be used in the above computational stencil. The discretization of first-order Sommerfeld boundary conditions will result in a complex-valued linear system.

3.2 Preconditioned Krylov Subspace Methods

Among the representative Krylov subspace methods, GMRES and Bi-CGSTAB are suitable choices for the Helmholtz equation, as they are designed for non-singular problems. Also, the IDR(s) developed by Sonneveld and van Gijzen [17] is an efficient alternative to Bi-CGSTAB for Helmholtz problems. Compared with full GMRES, Bi-CGSTAB and IDR(s) have short recurrences and are easily parallelizable.

As for the preconditioner, we will focus on the CSLP due to its satisfactory performance and easy setup. The CSLP is defined by

$$M_h = -\Delta_h - (\beta_1 + \beta_2 \mathrm{i})\, k^2 I_h \tag{3}$$

We need to compute the inverse of preconditioner M_h in the preconditioned Krylov-based algorithms. It is usually too costly to invert a preconditioner like CSLP directly. One idea is to approximately solve the preconditioner by using the multigrid method [4]. It is necessary to choose a proper complex shift [18], since a small complex shift may affect the convergence of the multigrid method. In the numerical experiments of this paper, $\beta_1 = 1, \beta_2 = 0.5$ will be used.

A multigrid method involves several components that need a careful design to achieve excellent convergence. In this paper, damped Jacobi smoother with relaxation $\omega = 0.8$ is used. The so-called full weighting restriction operator and the bilinear interpolation operator are employed for the inter-grid transfer operations. The coarse grid operator M_{2h} is constructed by re-discretizing on the coarse mesh in the same way that the operator M_h is obtained on the fine mesh. This is known as the discretization coarse grid operator (DCG). The classical multigrid V-cycle is performed. Instead of solving the coarse-mesh problem directly, we will solve it by full GMRES.

Suppose a problem with N unknowns is solved by the CSLP-preconditioned Krylov subspace method by assembling the matrices. According to the complexity analysis in [5,19], except the variables vectors, we need extra memory to store the sparse matrix A_h with $5N$ non-zero elements, M_h with $5N$ non-zero elements, M_{2h} with $\frac{9N}{4}$, inter-grid transfer operator Z with $\frac{9N}{4}$, etc. To minimize memory limitations and solve real-world large-scale problems, we implement the preconditioned Krylov subspace methods in a matrix-free way instead of constructing the coefficient matrices explicitly.

3.3 Matrix-Free Parallel Implementation

For the matrix-vector multiplication, like $A_h\mathbf{u}_h$ in outer iterations, $M_h\mathbf{u}_h$ in the smoother and $M_{2h}\mathbf{u}_{2h}$ in the preconditioner solver, can be replaced by stencil computations. For example, results of $A_h\mathbf{u}_h$ can be obtained by Algorithm 1. The inter-grid transfer operations can also be performed in a matrix-free way according to the linear interpolation/restriction polynomials.

Algorithm 1. Matrix-free $\mathbf{v_h} = A_h\mathbf{u_h}$.

1: Input array $\mathbf{u_h}$;
2: According to Eq. 2: $ap = \frac{4-k^2h^2}{h^2}$, $aw = ae = as = an = -\frac{1}{h^2}$;
3: Internal grid points ($i = 2\cdots nx-1$, $j = 2\cdots ny-1$):
4: $\quad v_h(i,j) = a_p u(i,j) + a_w u(i-1,j) + a_e u(i+1,j) + a_s u(i,j-1) + a_n u(i,j+1)$;
5: Boundary grid points ($i = 1, nx$, $j = 1, ny$): update a_p, a_w, a_e, a_s and a_n and compute $v_h(i,j)$;
6: Return $\mathbf{v_h}$.

To implement parallel computing, the standard MPI library is employed for data communications among the processors. Based on the MPI Cartesian topology, we can partition the computational domain blockwise. The partition is carried out between two grid points. One layer of overlapping grid points is introduced outward at each interface boundary to represent the adjacent grid points. In our method, the grid unknowns are stored as an array based on the grid ordering (i,j) instead of a column vector based on x-line lexicographic ordering.

We implement the parallel multigrid iteration based on the original global grid. According to the relationship between the fine grid and the coarse grid, the parameters of the coarse grid are determined by the grid parameters of the fine one. For example, point (i_c, j_c) in the coarse grid corresponds to point $(2i_c-1, 2j_c-1)$ in the fine grid. For a V-cycle, after reaching a manually predefined coarsest grid size, the coarsening operation will stop and solve the coarsest problem by GMRES in parallel, which may incur some efficiency loss. In this paper, the predefined coarsest global grid size is $nc_x \times nc_y = 9\times 9$ as the maximum number of processors we use is 4×4.

4 Numerical Experiments

The solver is developed in Fortran and compiled by GNU Fortran and runs on a Linux compute node with Intel(R) Xeon(R) Gold 6152 (2.10GHz) CPUs. For outer iterations, the L_2-norm of preconditioned relative residuals are reduced to 10^{-6}. According to pre-experiments, the stopping criterion for the coarse grid preconditioner solver should be 2–3 orders of magnitude smaller than the stopping criterion for the outer iteration. We use 10^{-8} as the stopping criterion for the coarse grid preconditioner solver. The WallClockTime for the preconditioned Krylov-based solver to reach the stopping criterion is denoted by t_w. The speedup S_p is defined by $S_p = \frac{t_1}{t_p}$, where t_1 and t_p are the WallClockTime for sequential and parallel computation, respectively. The parallel efficiency E_p is given by $E_P = \frac{S_P}{np} \times 100\%$, where np is the number of processors.

First, we consider a model problem (MP-1) with a point source described by the Dirac delta function $\delta(x,y)$, imposed at the center $(x_0,y_0) = (0.5,0.5)$. The wave propagates outward from the center of the domain. Dirichlet boundary conditions are imposed. The analytical solution for this problem is given in [15].

Compared to the analytical solutions given by [15], our parallel preconditioned GMRES gives a fair approximation of the exact solution with relative errors (RErr.) less than 5×10^{-6}. Parallel partitioning also has no effect on the results. The main differences in the amplitude of the waves are caused by the finite-difference approximation of the Dirac function. As shown in Table 1, if we simultaneously and proportionally scale the problem size and the number of processors (in bold), the WallClockTime almost stays constant. It means our parallel framework has weak scalability. It indicates that a parallel efficiency of up to 75% is satisfactory.

Table 1. Parallel performance of CSLP preconditioned GMRES for MP1 with wavenumber $k = 100$.

grid size	np	#Iter	RErr	t_w	Speedup	E_p
161 × 161	1	350	1.638E−06	**7.07**	1.00	–
	4	350	1.596E−06	2.06	3.43	85.68
321 × 321	1	348	1.476E−06	32.18	1.00	–
	4	348	1.592E−06	**8.10**	3.97	99.34
481 × 481	1	358	4.319E−06	78.22	1.00	–
	9	359	4.444E−06	**9.02**	8.68	96.41
641 × 641	1	339	2.000E−06	121.44	1.00	–
	4	339	1.657E−06	33.75	3.60	89.97
	16	339	2.158E−06	**9.88**	12.29	76.79

Most physical problems of geophysical seismic imaging describe a heterogeneous medium. The so-called Wedge problem (MP-2) is a typical problem with simple heterogeneity. It mimics three layers with different velocities hence, different wavenumbers. The rectangular domain $\Omega = [0,600] \times [-1000,0]$ is split into three layers, where the wave velocity c is constant within each layer but different from each other. A point source is located at $(x_0,y_0) = (300,0)$. The wavenumber is $k(x,y) = \frac{2\pi f}{c(x,y)}$, where f is the frequency. The distribution of wave velocity $c(x,y)$ refers to [6]. First-order Sommerfeld boundary conditions are imposed on all boundaries.

The 2D wedge problem is used to evaluate the performance of our parallel solution method for a simple heterogeneous medium. Besides, the matrix-free parallel framework is not limited to the GMRES algorithm. All the ingredients can be directly generalized to other Krylov methods like Bi-CGSTAB and IDR(s). In Table 2 we give the WallClockTime and the required number of matrix-vector multiplications of different CSLP-preconditioned Krylov methods for the wedge problem. It illustrates that our matrix-free parallel CSLP-preconditioned method is still suitable for heterogeneous Helmholtz problems. It still leads to satisfactory scalability if we increase the number of processors correspondingly while refining the grid.

Table 2. MP-2: CPU time consumed by different parallel CSLP-preconditioned Krylov methods while refining the grid, $f = 40\,\text{Hz}$. The number of matrix-vector multiplications is in parentheses.

Grid size	np	GMRES	Bi-CGSTAB	IDR(4)
385×641	2	25.34 (278)	6.20 (321)	6.10 (282)
769×1281	8	31.92 (283)	7.98 (301)	7.41 (251)

The so-called Marmousi problem [16] is a well-known benchmark problem (MP-3). It contains 158 horizontal layers in the depth direction, making it highly heterogeneous, see [6] for an illustration. In our numerical experiments, first-order Sommerfeld boundary conditions are imposed on all boundaries. The source frequency $f = 40\,\text{Hz}$, grid size 2945×961, which indicates $kh \leq 0.54$ and guarantees more than 10 grids per wavelength. In [6], the Marmousi problem with grid size 2501×751 has to be solved on at least two cores due to memory limitations. The matrix-free framework allows us to solve larger problems within even a single core.

Table 3 presents the required number of matrix-vector multiplications (denoted by #Matvec), CPU-time, and relative speedup of different CSLP-preconditioned Krylov methods for the Marmousi problem. One can find that a huge number of iterations are required. GMRES has the least number of matrix-vector multiplications but requires the most CPU time reducing the parallel efficiency to 50%. This is due to the Arnoldi process in GMRES requiring a lot of dot product operations, which need global communication in parallel computing. IDR(4) and Bi-CGSTAB exhibit higher parallel efficiency than GMRES. The results illustrate that the matrix-free parallel CSLP-preconditioned method also works for the highly heterogeneous Helmholtz problems.

Table 3. MP-3: parallel performance of different parallel CSLP-preconditioned Krylov methods, $f = 40\,\text{Hz}$, grid size 2945×961.

GMRES					Bi-CGSTAB					IDR(4)				
np	#Matvec	Time(s)	Sp	Ep	np	#Matvec	Time(s)	Sp	Ep	np	#Matvec	Time(s)	Sp	Ep
1	2872	61705.58	–	–	1	4124	2431.94	–	–	1	4438	2881.58	–	–
3	2872	21892.21	2.82	93.95	3	4435	712.34	3.41	113.80	3	4688	854.72	3.37	112.38
12	2872	10309.02	5.99	49.88	12	4513	279.09	8.71	72.61	12	4484	334.59	8.61	71.77

5 Conclusions

In this paper, we studied a matrix-free parallel solution method using preconditioned Krylov methods for Helmholtz problems. The Complex Shifted Laplace Preconditioner is used, which is approximately inverted by multigrid iterations. The matrix-free parallel framework is suitable for different Krylov methods. Numerical experiments of model problems demonstrate the robustness, satisfactory parallel performance, and weak scalability of our matrix-free parallel solution method. It allows us to solve larger problems in parallel to obtain more accurate numerical solutions.

Acknowledgements. Funding by China Scholarship Council is acknowledged. The authors are grateful to the referees for their helpful comments and suggestions that improved the quality of the paper considerably.

References

1. Freund, R., Golub, G., Nachtigal, N.: Iterative solution of linear systems. Acta Numer. **1**, 57–100 (1992)
2. Bayliss, A., Goldstein, C.I., Turkel, E.: An iterative method for the Helmholtz equation. J. Comput. Phys. **49**(3), 443–457 (1983)
3. Erlangga, Y.A., Vuik, C., Oosterlee, C.W.: On a class of preconditioners for solving the Helmholtz equation. Appl. Numer. Math. **50**(3–4), 409–425 (2004)
4. Erlangga, Y.A., Oosterlee, C.W., Vuik, C.: A novel multigrid based preconditioner for heterogeneous Helmholtz problems. SIAM J. Sci. Comput. **27**(4), 1471–1492 (2006)
5. Dwarka, V., Vuik, C.: Scalable convergence using two-level deflation preconditioning for the Helmholtz equation. SIAM J. Sci. Comput. **42**(2), A901–A928 (2020)
6. Kononov, A.V., Riyanti, C.D., de Leeuw, S.W., Oosterlee, C.W., Vuik, C.: Numerical performance of a parallel solution method for a heterogeneous 2D Helmholtz equation. Comput. Vis. Sci. **11**(3), 139–146 (2008)
7. Riyanti, C.D., et al.: A parallel multigrid-based preconditioner for the 3D heterogeneous high-frequency Helmholtz equation. J. Comput. Phys. **224**(1), 431–448 (2007)
8. Gordon, D., Gordon, R.: Robust and highly scalable parallel solution of the Helmholtz equation with large wave numbers. J. Comput. Appl. Math. **237**(1), 182–196 (2013)
9. Calandra, H., Gratton, S., Vasseur, X.: A geometric multigrid preconditioner for the solution of the Helmholtz equation in three-dimensional heterogeneous media on massively parallel computers. In: Lahaye, D., Tang, J., Vuik, K. (eds.) Modern Solvers for Helmholtz Problems. GM, pp. 141–155. Springer, Cham (2017). https://doi.org/10.1007/978-3-319-28832-1_6
10. Engquist, B., Ying, L.: Sweeping preconditioner for the Helmholtz equation: moving perfectly matched layers. Multisc. Model. Simul. **9**(2), 686–710 (2011)
11. Chen, Z., Xiang, X.: A source transfer domain decomposition method for Helmholtz equations in unbounded domain. SIAM J. Numer. Anal. **51**(4), 2331–2356 (2013)
12. Stolk, C.C.: A rapidly converging domain decomposition method for the Helmholtz equation. J. Comput. Phys. **241**, 240–252 (2013)
13. Gander, M.J., Zhang, H.: A class of iterative solvers for the Helmholtz equation: factorizations, sweeping preconditioners, source transfer, single layer potentials, polarized traces, and optimized Schwarz methods. SIAM Rev. **61**(1), 3–76 (2019)
14. Taus, M., Zepeda-Núñez, L., Hewett, R.J., Demanet, L.: L-sweeps: a scalable, parallel preconditioner for the high-frequency Helmholtz equation. J. Comput. Phys. **420**, 109706 (2020)

15. Dwarka, V., Vuik, C.: Pollution and accuracy of solutions of the Helmholtz equation: a novel perspective from the eigenvalues. J. Comput. Appl. Math. **395**, 113549 (2021)
16. Versteeg, R.: The Marmousi experience: velocity model determination on a synthetic complex data set. Lead. Edge **13**(9), 927–936 (1994)
17. Sonneveld, P., Van Gijzen, M.B.: IDR (s): a family of simple and fast algorithms for solving large nonsymmetric systems of linear equations. SIAM J. Sci. Comput. **31**(2), 1035–1062 (2009)
18. Gander, M.J., Graham, I.G., Spence, E.A.: Applying GMRES to the Helmholtz equation with shifted Laplacian preconditioning: what is the largest shift for which wavenumber-independent convergence is guaranteed? Numer. Math. **131**(3), 567–614 (2015)
19. Dwarka, V., Vuik, C.: Scalable multi-level deflation preconditioning for highly indefinite time-harmonic waves. J. Comput. Phys. **469**, 111327 (2022)

Implementation and Validation of the Dual Full-Wave E and H Formulations with Electric Circuit Element Boundary Conditions

Gabriela Ciuprina[1(✉)], Daniel Ioan[1], and Ruth V. Sabariego[2]

[1] Politehnica University of Bucharest, Spl. Independentei 313, 060042 Bucharest, Romania
{gabriela,daniel}@lmn.pub.ro

[2] KU Leuven, Campus Energyville, Thor Park 8310, 3600 Genk, Belgium
ruth.sabariego@kuleuven.be

Abstract. Dual full-wave (FW) frequency-domain **E** and **H** formulations, with scalar potentials on the boundary, and with electric circuit element boundary conditions are discussed and details about their implementation in the finite element method are given. For some magneto-quasi-static devices this duality frames the exact solution thus allowing the accuracy control. In such cases the geometric mean of the dual solutions exhibits a better accuracy and higher convergence rate than the individual numerical solutions. For FW devices the dual formulations allow a compromise between model accuracy and computational effort, especially if the models are not 3D. Implementation is available for free in `onelab`. Validation for test cases with analytic solution are provided: a conducting cylinder and a coaxial cable.

1 Introduction

Electric Circuit Element (ECE) boundary conditions allow a natural coupling between electromagnetic devices and electric circuits. This paper shows how ECE can be implemented in the 3D-finite element method, for a full-wave (FW) field regime in dual formulations, in terms of the electric **E** and magnetic **H** fields. This is an extension of our previous results referring to duality published for electrostatics (ES) and magnetostatics (MS) [1], and referring to the recent use of ECE boundary conditions for FW **E**-based formulation [2], implemented in `onelab` [3]. When providing bounds, the duality concept is important for the assessment of the numerical results accuracy, with implications to optimal mesh refinement and efficient parameter extraction. In general, bilateral bounds cannot be ensured [4]. However, dual formulations have been investigated in the literature. e.g. in [5,6].

Given the non perfect duality between the **E** and **H** fields, we investigate the behaviour of the numerical implementations of the **E** and **H** formulations when used for FW models and ECE BC. In this paper we aim to implement and use the frequency domain FEM formulation in [2], with an electric field strictly inside the domain and electric scalar potential on the boundary, and a dual counterpart, with a magnetic field inside the domain and magnetic scalar potential on boundary. A sound and general theory for weak **E**-based and **H**-based formulations, with electric and magnetic ports, for

M. van Beurden et al. (Eds.): SCEE 2022, MI 43, pp. 69–77, 2024.
https://doi.org/10.1007/978-3-031-54517-7_8

multiply connected domains can be found in [7]. We consider the **H** formulation proposed here different from the one in [7] because its time domain correspondent involves convolution integrals. A detailed numerical implementation of the **H** formulation is provided as well. Since our final goal is the extraction of lumped circuits equivalent to electromagnetic devices with distributed parameters, we adopt here only frequency domain formulations. This eases the **H**-based approach, for which the elimination of **E** in a time domain formulation is not straightforward. The implementation is done in `onelab` and the files are available for download from https://gitlab.onelab.info/doc/models. We provide a strong validation by using two simple examples for which analytical solutions exist. More complicated structures are currently under investigation.

2 H-Based Formulation for ECE and Frequency Domain FW Field

Herein we provide only the **H**-based formulation. The **E**-based counterpart can be found in [2]. For simplicity, we assume a simply connected domain Ω, with a Lipschitz connected boundary $\partial\Omega$, linear and isotropic materials.

1. Second order strong formulation in **H**, with $\alpha = j\omega/(\sigma + j\omega\varepsilon)$ is:

$$\nabla \times (\alpha \nabla \times \underline{\mathbf{H}}) + (j\omega)^2 \mu \underline{\mathbf{H}} = \mathbf{0}, \tag{1}$$

where ε, μ, σ are the material permittivity, permeability and conductivity, underlined symbols are complex vectors, and ω is the angular frequency. Classical BC means that $\underline{\mathbf{E}}_t = \mathbf{n} \times (\underline{\mathbf{E}} \times \mathbf{n})$ is given on $S_E \subset \partial\Omega$, and $\underline{\mathbf{H}}_t = \mathbf{n} \times (\underline{\mathbf{H}} \times \mathbf{n})$ on $S_H = \partial\Omega - S_E$, where $\mathbf{n}$ is the outer normal on $\partial\Omega$ and $\underline{\mathbf{E}} = (\nabla \times \underline{\mathbf{H}})/(\sigma + j\omega\varepsilon)$.

2. Strong formulation in **H** and magnetic scalar potential φ with ECE BC.

Equation (1) holds. ECE BC are: the boundary includes m disjoint parts S_k, $k = 1, m$ (device terminals), so that:

(ece1) there is no magnetic coupling with the exterior: $\mathbf{n} \cdot \underline{\mathbf{H}}(\mathbf{r}) = 0, \forall \mathbf{r} \in \partial\Omega$;

(ece2) the electric coupling is carried out only through the terminals:
$\mathbf{n} \cdot (\nabla \times \underline{\mathbf{H}}(\mathbf{r})) = 0, \forall \mathbf{r} \in \Sigma \stackrel{\text{def}}{=} \partial\Omega - \cup_{k=1}^{m} S_k$;

(ece3) each terminal is equipotential: $\mathbf{n} \times \underline{\mathbf{E}}(\mathbf{r}) = \mathbf{0}, \forall \mathbf{r} \in S_k$, $k = 1, m$.

Note that here, contrary to the classical BC (without ECE) where S_E and S_H are the usual notations for the parts of the boundary where the tangential components of **E** and **H** are given, there is no S_E and/or S_H parts. On Σ, the tangential component of **H** is not known, but the normal component and the normal component of its curl are zero. To have (ece2), a magnetic scalar potential φ may be defined on Σ. However, since Σ is not simply connected, cuts are needed to transform it into a simply connected one, for a proper definition of φ. Currents and voltages associated at first to the terminals will be transferred to these cuts. Each terminal is controlled in voltage $V_k = \int_{C_k \subset \partial\Omega} \mathbf{E} \cdot \mathrm{d}\mathbf{r}$, C_k linking S_k to S_m, or current $I_k = \oint_{\partial S_k} \mathbf{H} \cdot \mathrm{d}\mathbf{r}$, with $\{1 : m\} = \mathscr{I}_v \cup \mathscr{I}_c$. Here $\mathscr{I}_v$, $\mathscr{I}_c$ are the sets of indices of voltage and current excited terminals, respectively.

3. Weak formulation in **H** with classical BC:

Find $\underline{\mathbf{H}}$ *in* $\mathscr{H}$ *such that* $a(\underline{\mathbf{H}}, \underline{\mathbf{H}}') = b(\underline{\mathbf{H}}')\ \forall \underline{\mathbf{H}}' \in \mathscr{H}_0$, *where*

$$\mathscr{H} = \{\mathbf{u} \in \mathscr{H}(\text{curl}, \Omega) | \mathbf{n} \times (\mathbf{u} \times \mathbf{n}) = \underline{\mathbf{H}}_t \text{ on } S_H\}, \quad \mathscr{H}_0 \text{ as } \mathscr{H} \text{ with } \underline{\mathbf{H}}_t = \mathbf{0}, \tag{2}$$

$\underline{\mathbf{H}}$ *and* $\underline{\mathbf{H}}'$ *are curl-conform with essential BC (zero for* $\underline{\mathbf{H}}'$*), where*

$$a(\underline{\mathbf{H}},\underline{\mathbf{H}}') = \int_{\Omega} (\alpha \nabla \times \underline{\mathbf{H}}) \cdot (\nabla \times \underline{\mathbf{H}}')\, \mathrm{d}x + \int_{\Omega} j\omega(j\omega\mu)\underline{\mathbf{H}} \cdot \underline{\mathbf{H}}'\, \mathrm{d}x, \tag{3}$$

$$b(\underline{\mathbf{H}}') = j\omega \int_{S_E} (\mathbf{n} \times \underline{\mathbf{E}}_t) \cdot \underline{\mathbf{H}}'\, \mathrm{d}A. \tag{4}$$

$\underline{\mathbf{E}}_t$ is a natural BC, whereas $\underline{\mathbf{H}}_t$ is essential.

4. Weak formulation in $\mathbf{H}, \varphi$ with ECE BC.

Find $\underline{\mathbf{H}} \in \mathscr{H}_H$ *and* $\underline{\varphi} \in \mathscr{H}_\varphi$*, so that* $a(\underline{\mathbf{H}},\underline{\mathbf{H}}') = b(\underline{\mathbf{H}}')$*,* $\forall \underline{\mathbf{H}}' \in \mathscr{H}_{H,0}$*; and*

$$\int_{C_k} \mathbf{E} \cdot \mathbf{dl} = \underline{V}_k, k \in \mathscr{I}_v; \quad \underline{\mathbf{H}}_t = \nabla_2 \varphi \text{ on } \partial\Omega - \cup_{k=1}^m S_k, \tag{5}$$

$\underline{\mathbf{H}}_t' = \nabla_2 \varphi'$, $\underline{\varphi}' \in \mathscr{H}_{\varphi,0}$. Let $T_k = C_k$ *be a cut in* $\partial\Omega - \cup_{k=1}^m S_k$ *assoc. to terminal* k.

$$\left.\begin{aligned} \mathscr{H}_H &= \{\mathbf{u} \in \mathscr{H}(\mathrm{curl}, \Omega) |\ \mathbf{n} \times (\mathbf{u} \times \mathbf{n}) = \nabla_2 \underline{\varphi}' \text{ on } \partial\Omega - \cup_{k=1}^m S_k,\ \underline{\varphi}' \in \mathscr{H}_\varphi \\ \mathscr{H}_\varphi &= \{u \in \mathscr{H}(\mathrm{grad}, \partial\Omega - \cup_{k=1}^m S_k) |\ [u]_k = \underline{I}_k\ k \in \mathscr{I}_c, [u]_k = \mathrm{ct.}\ k \in \mathscr{I}_v\} \end{aligned}\right\} \tag{6}$$

$\mathscr{H}_{H,0}$ is as $\mathscr{H}_H$ but $\underline{\varphi}' \in \mathscr{H}_{\varphi,0}$, $\mathscr{H}_{\varphi,0}$ is as $\mathscr{H}_\varphi$ but the jump on T_k denoted by $[u]_k$, for $k \in \mathscr{I}_c$ is zero. The left side of the functional equality is (3), but $b(\underline{\mathbf{H}}')$ is[1]

$$b(I') = -j\omega \sum_{k \in \mathscr{I}_v} \underline{V}_k I_k'. \tag{7}$$

Note that with this formulation, voltage excitation is a natural BC, current excitation is essential. All test functions have zero essential BC, but the trial functions have imposed values for them.

5. Discrete formulation in $\mathbf{H}, \varphi$ with ECE BC.

To satisfy ECE BC, the numerical solution is searched as

$$\underline{\mathbf{H}} = \sum_{j=1}^{\mathrm{Ne}} \underline{U_m}_j \mathbf{N}_j + \sum_{j=1}^{\mathrm{NnBndNotTerm}} \underline{\varphi}_j \nabla \varphi_j + \sum_{k=1}^{m} \left(\underline{I}_k \sum_{j=1}^{\mathrm{NeCutTermK}} \mathbf{N}_j \right), \tag{8}$$

where Ne are the edges inside the domain and on the surface of terminals. NnBndNotTerm are the nodes on the boundary that does not include the terminals, NeCutTermK are the edges that belong to the cut that corresponds to the terminal k. The cuts are automatically generated with the cohomology solver of `gmsh` [8].

The ECE implementation[2] in `onelab` uses (3) and (7), and the function spaces (6) and (8) and choices for $\mathbf{N}_j$ (edge elements) and φ_j (nodal elements).

Two academic examples with analytical solutions validate the implementations: 1) a cylindrical homogeneous conducting domain, for extracting the inner impedance; 2) a cylindrical conducting domain surrounded by air ended by the return conductor - a coaxial cable. In both models the EM field on the boundary satisfies ECE BC.

---

[1] Classic to ECE: $\oint_{\partial\Omega} (\mathbf{n} \times \underline{\mathbf{E}}_t) \cdot \underline{\mathbf{H}}'\, \mathrm{d}A = \int_\Sigma -(\mathbf{n} \times \nabla_2 \underline{V}) \cdot \underline{\mathbf{H}}'\, \mathrm{d}A = \sum_{k=1}^m \underline{V}_k \oint_{\partial S_k} \mathbf{H}_t' \cdot \mathbf{dl} = \sum_{k \in \mathscr{I}_v} \underline{V}_k I_k'$.

[2] Other resources are available at www.lmn.pub.ro/$\sim$gabriela/ece..

3 MQS Test - Conducting Cylinder

The first 3D test consists of a cylindrical domain with radius $a = 2.5\,\mu\text{m}$, length $l = 10\,\mu\text{m}$, and linear and homogeneous material ($\varepsilon_r = 1, \mu_r = 1, \sigma = 6.6 \cdot 10^7$ S/m). Its ends are two terminals, one grounded and the other excited either in current or voltage. This configuration has the advantage that a formulation with classical boundary conditions is equivalent to the one with ECE BC, the classical having an analytic solution for the current excitation case, and thus a reference. The first numerical tests aimed 2.5D vs. 3D checks[3], current vs voltage excitation, **E** vs. **H** formulation, 1st vs 2nd order FEM, and then *h*-convergence studies were done[4].

Figure 1 depicts the impedance obtained from the FE solution of the two 3D-FW formulations with ECE BC and current excitation. There is no coupling with an external circuit. Note that we can derive an equivalent reduced circuit from the frequency dependent impedance, which can be interconnected with any external circuit [10]. Bilateral bounds are obtained for both frequency characteristics: magnitude and phase of the impedance $\underline{Z}$. Note that the FW Maxwell equations were used for solving, even though this is an MQS problem (indeed $\tau_e < \tau_{em} < \tau < \tau_m$, see [11]).

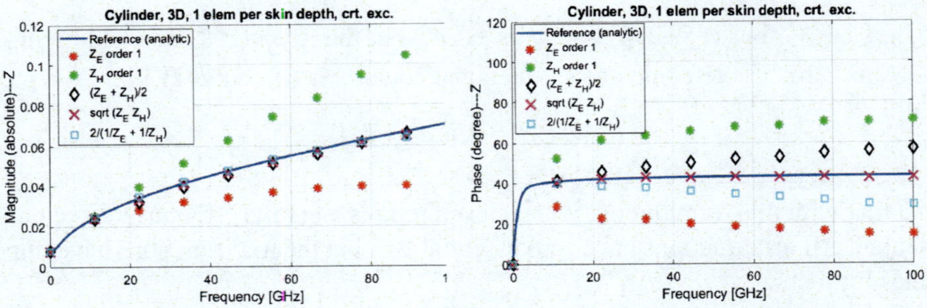

Fig. 1. Cylinder test (3D). The dual formulations provide scissors type bounds for the frequency characteristic in the magnitude (left) and phase (right) representation of the impedance $\underline{Z}$. Their average $\underline{\mathbf{Z}} = \sqrt{\underline{\mathbf{Z}}_E \underline{\mathbf{Z}}_H}$ gives surprisingly better results than the individual formulations or their arithmetic or harmonic averages. A relatively coarse frequency dependent mesh, was used (1 element/skin depth, 10 nodes along the length, a coarse mesh in the middle).

Even if MQS, this test is useful to check the FW **E** and **H** implementations. The following results are obtained for an axisymmetrical (2.5D) model. The same FEM formulation file is used for **E**, whereas the files for **H** in 3D and 2.5D are different. In 2.5D the edge elements for the **H** formulation become nodal elements multiplied by the

[3] Here 2.5D was adopted as an acronym for the axisymmetric problems because from mathematical point of view the model is 2D, but from the physical point of view the model is 3D, no domain truncation is done along the azimuth direction. Numerically, this is encoded in the computation of the Jacobian [9]. This is different from plane-parallel 2D problems where the model is 2D both from mathematical and physical points of view, there is a domain truncation along the Oz axis (physical end effects are neglected).

[4] Error convergence for uniform meshes of increasing fineness; *h* is the characteristic length of the element.

azimuth direction unit vector and φ is no more needed, the boundary consists only of terminals and cuts. In the 2.5D **H** formulation, the unknown has to be set to zero on the axis to avoid numerical instabilities. In the 2.5D **E** formulation, nothing has to be done on the axis, the axis is treated like the domain's interior.

Figure 2 shows an h-convergence study for first order elements, where $h = sa/10$ is the mesh size, with s the mesh factor - a real positive number. Note that the minimum skin depth $\delta_{\min} = 0.2\,\mu\text{m}$ is close to h for $s = 1$, i.e. for $s = 1$ at $f_{\max}$ there is 1 element per skin depth. The **H** formulation is better than **E** at low frequencies, whereas at high frequencies, they are about the same. Figure 3 shows the results for $f_{\max}$, and the effect of computing the square root of the solutions and Richardson extrapolations[5]. The *deviations in* **E** *and* **H** have almost the same magnitude but they frame the exact solution. For a discretization with a mesh having the characteristic length of elements equal $h = \delta_{\min}(s = 1)$, the relative errors are about 10 % (11.8 % for **E** and 9.3 % for **H**), but they are four times smaller if the characteristic length decreases twice (3.3 % for **E** and 2.5 % for **H** when $h = \delta_{\min}/2$). Their *geometric mean* has a relative error which is one order of magnitude less, i.e. 1.4 % for $h = \delta_{\min}$, and 0.4 % for $h = \delta_{\min}/2(s = 0.5)$. The *Richardson extrapolation* of the geometric average is better, it has a relative error of 0.2 %, which is two times smaller than the one corresponding to the geometric average for the finest mesh (0.4 %). It is one order of magnitude less than the Richardson extrapolation carried out separately for each formulation (1.8 % for **E** and 1.4 % for **H**). Both limits are acceptable for most engineering models, and only in exceptional cases recurrent Richardson is needed, with smaller steps, reaching $h = \delta_{\min}/4$, in order to obtain even smaller relative errors. The *convergence rates* for **E** and **H** are in the range 1.7–2, but the Richardson extrapolation applied to the geometric average converges faster, with a convergence rate of about 2.7. By using the geometric mean of the dual solutions, the numerical error obtained for this test is surprisingly much smaller than the error in **E** or **H** formulations, the arithmetic or harmonic averages, or their difference (which gives thus pessimistic information). However, the difference can be used as an error estimator for low and medium frequencies.

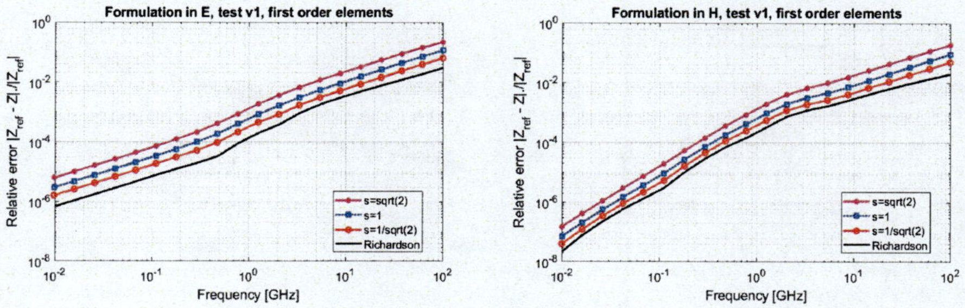

Fig. 2. Cylinder test (2.5D, uniform mesh). Relative errors for the whole frequency range for 3 meshes and the Richardson extrapolation. **E** formulation (left), **H** formulation (right).

[5] Three different computations are needed for the Richardson extrapolation, based on the assumption $\underline{Z}(h) = \underline{Z}_0 + Ah^p$, with $\underline{Z}_0$, A and h unknowns. If the mesh sizes are such that $h_1/h_2 = h_2/h_3$, the extrapolated value is computed as $\underline{Z}_0 = (\underline{Z}_1\underline{Z}_3 - \underline{Z}_2^2)/(\underline{Z}_1 - 2\underline{Z}_2 + \underline{Z}_3)$, with $\underline{Z}_k = \underline{Z}(h_k)$, $k = 1,2,3$.

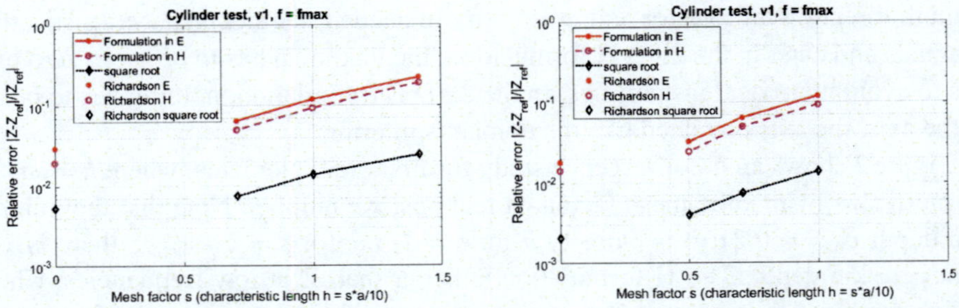

Fig. 3. Cylinder test (2.5D, uniform mesh). Two different Richardson extrapolations of the square root solution.

4 FW Test - Coaxial Cable

This test case is adapted from [12]. It consists of an air-filled 1 m long coaxial cable, with radius of the inner circular cylindrical electrode of 4 mm. The inner radius of the outer circular cylindrical electrode is 8 mm. The cable is ended by a lumped circuit that includes a 100 pF capacitor in series with the parallel combination of a 5 Ω resistor with a 10 nH inductor. On the driving end there is a 5 Ω shunt resistor connecting the two electrodes. The problem admits an analytic result, based on TLs theory. We use a 2.5D model (Fig. 4-left). This problem is truly FW at high frequencies (Fig. 4-right). A difference with respect to the formulation in [12] is that we added the inner conductor in the model, so that not to have holes in it. Adding the inner conductor in the model will not have a visible effect at high frequencies, particularly when looking at the terminal impedance, but it makes the model closer to reality by including the propagation in the dielectric and the skin effect. Thus, the model can be used for frequencies in the MQS range as well, where the skin effect is relevant and not the propagation.

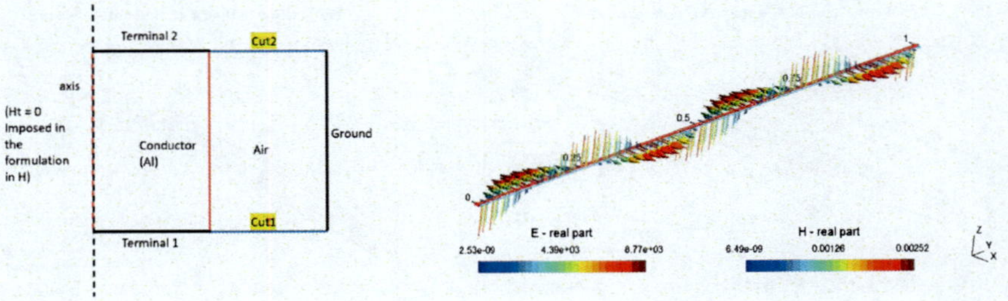

Fig. 4. Coax test (2.5D). Domain - the figure is not at scale (left); Fields along the full length of the cable at $f_{\max}$, $\underline{\mathbf{E}}$ - arrows, $\underline{\mathbf{H}}$ - lines (right). Propagation is obvious.

We simulated the coaxial cable alone, as a MIMO (multiple input multiple output) system with 2 floating terminals (ends of the inner electrode) and a ground - the outside electrode. The results obtained for the dual formulations are shown in Fig. 5. Results for

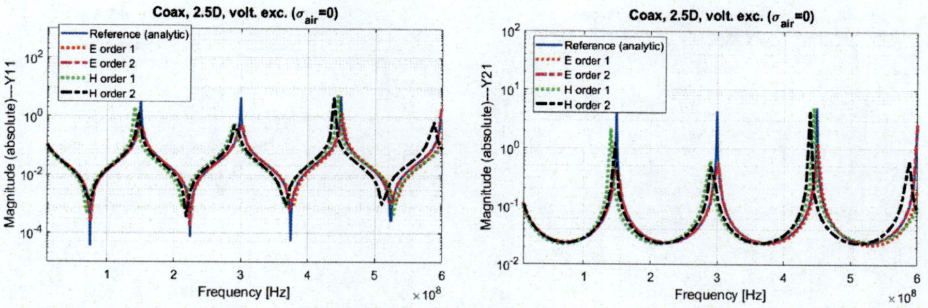

Fig. 5. Coax test (2.5 D). Frequency characteristics of the cable simulated alone.

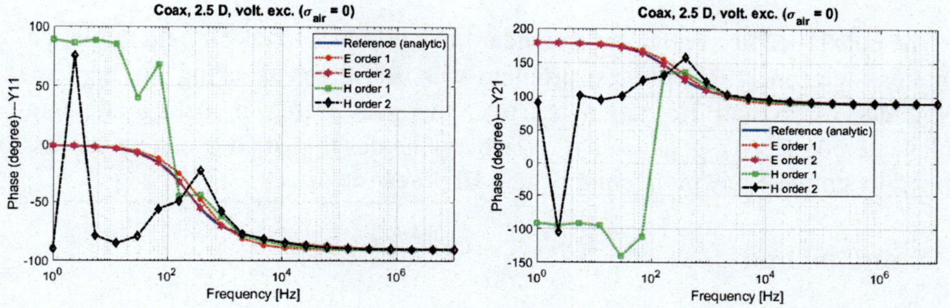

Fig. 6. At low frequencies, **H** formulation is unstable if $\sigma_{\text{air}} = 0$, which does not happen for the **E** formulation. The **H** formulation can be stabilized by choosing $\sigma_{\text{air}} = 10^{-5}$ S/m.

1st and 2nd order elements for the **E**-based formulation are very good for the considered frequency range (60–600 MHz). Second order results for the **H** formulation are also good. However, first order FEM results for the **H** formulation are not so good at high frequencies (HF). This time, no bounds are obtained at HF and also **H** formulation is not as accurate as the **E**-based one of the same FEM order.

Figure 6 shows a zoom-in the low frequency (LF) range. Stability at LF is a known concern for FW numerical models [13]. At LF **H**-based formulation suffers from instabilities (under 10 kHz) if $\sigma_{\text{air}} = 0$, which is not the case for the **E**-based formulation which proved robust for all frequencies. Further analysis of the stability issues is currently ongoing.

The coupling with the circuit can be done in several ways: with a code implementing the simple circuit equations (as it was done here - Fig. 7); in the field simulator (e.g. in onelab); or reduce the model, realize a circuit and use a circuit simulator [2]. The easiness of field-circuit coupling is the strongest point of our approach. Figure 7 validates the correctness of the **H** formulation for FW with ECE. It seems to indicate that the **E** formulation works better at HF. However, this test problem has a very particular field distribution, and therefore, in order to draw fair conclusions about **E** vs. **H**, 1 vs. 2 order, problems with less particular configurations should be investigated. This is beyond the scope of this paper.

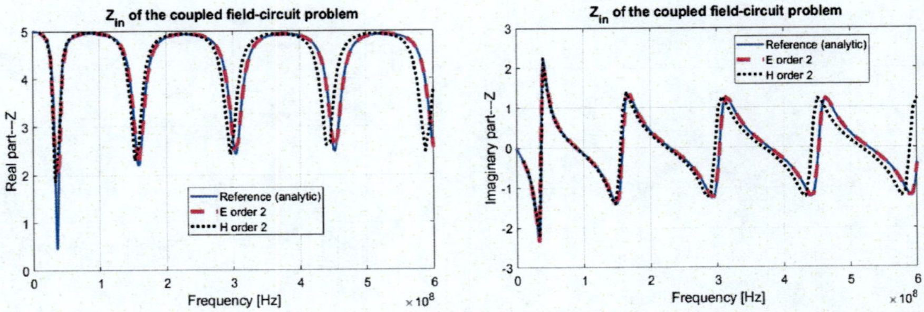

Fig. 7. The input impedance of the cable model connected to its feeding and load circuits.

The coaxial cable is however a standard test problem to check field numerical formulations, e.g. in [13]. This test problem was useful for checking the implemented FW codes for **E** and **H**, but in practice, there are better modeling strategies for cables/interconnects, e.g. as in [14]. However, the developed code can be successfully applied to simulate any more complicated RF linear devices.

5 Conclusions

The main contribution of this paper is the detailed derivation of the full-wave **H** formulation in the frequency domain: strong, weak (continuous and discrete), so that its numerical implementation becomes clear. The implementation is validated by examples with analytical solution. The benefits of using dual formulations are investigated as well. It is known that dual formulations for elliptic PDEs give scissors type bounds for extracted lumped quantities. The tests shown here indicate that it is possible that, under certain conditions, such a framing can also be obtained in MQS, allowing a reliable accuracy control. This statement is solely based on numerical observations. In this case, the *square root* of the solutions obtained from the **E** and **H** formulations had an impressive correction effect at HF. In FW, a possible useful strategy could be to *combine* the **E** and **H** solutions according to a weighted formula inspired from a similar successful CFIE strategy used for integral equations [15]. This strategy was not successful for the coaxial cable, but it works in other tests, e.g. a monopole antenna. Even if bilateral bounds cannot be ensured in general cases, the availability of dual formulations enrich the possibilities of choosing the best compromise between model accuracy and computational effort. This is especially useful for 2D/2.5D models where **E** and **H** solutions of the same order have a very different number of unknowns. ECE BC can be generalized to include parts of the boundary through where radiation is permitted, thus being useful for antenna modeling. Our ongoing work aims at the implementation of radiant ECE.

References

1. Ioan, D., Radulescu, M.-C., Ciuprina, G.: Fast extraction of static electric parameters with accuracy control. In: Schilders, W.H.A., et al. (eds.) Scientific Computing in Electrical Engineering. Mathematics in Industry, vol. 4, pp. 248–256. Springer, Heidelberg (2004). https://doi.org/10.1007/978-3-642-55872-6_26

2. Ciuprina, G., Ioan, D., Sabariego, R.V.: Electric circuit element boundary conditions in the finite element method for full-wave passive electromagnetic devices. J. Math. Ind. **12**(7), 1–13 (2022). http://www.lmn.pub.ro/gabriela/ece/
3. Dular, P., Geuzaine, C.: GetDP reference manual: the documentation for GetDP, a general environment for the treatment of discrete problems. http://getdp.info
4. Bossavit, A.: Complementary formulations in steady-state eddy-current theory. IEE Proc. A (Sci. Meas. Technol.) **139**(6), 265–272 (1992)
5. Dular, P., et al.: Dual magnetodynamic formulations and their source fields associated with massive and stranded inductors. IEEE Trans. Magn. **36**(4), 1293–1299 (2000)
6. Ren, Z., Qu, H.: Investigation of the complementarity of dual eddy current formulations on dual meshes. IEEE Trans. Magn. **46**(8), 3161–3164 (2010)
7. Hiptmair, R., Ostrowski, J.: Electromagnetic port boundary conditions: topological and variational perspective. Int. J. Numer. Modell. Electron. Netw. Devices Fields 1–23 (2021)
8. Pellikka, M., et al.: Homology and cohomology computation in finite element modeling. SIAM J. Sci. Comput. **35**(5), B1195–B1214 (2013)
9. Henrotte, F., Meys, B., Hedia, H., Dular, P., Legros, W.: Finite element modelling with transformation techniues. IEEE Trans. Magn. **35**(3), 1434–1437 (1999)
10. Ciuprina, G., et al.: Parameterized model order reduction. In: Günther, M. (ed.) Coupled Multiscale Simulation and Optimization in Nanoelectronics. MI, vol. 21, pp. 267–359. Springer, Heidelberg (2015). https://doi.org/10.1007/978-3-662-46672-8_5
11. Steinmetz, T., Kurz, S., Clemens, M.: Domains of validity of quasistatic and quasistationary field approximations. COMPEL - Int. J. Comput. Math. Electr. Electron. Eng. **30**(4), 1237–1247 (2011)
12. Wu, H., Cangellaris, A.C.: Model-order reduction of finite-element approximations of passive electromagnetic devices including lumped electrical-circuit models. IEEE Trans. Microwave Theory Tech. **52**(9), 2305–2313 (2004)
13. Jochum, M., et al.: A new low-frequency stable potential formulation for the finite-element simulation of electromagnetic fields. IEEE Trans. Magn. **51**(3), 1–4 (2015)
14. Ioan, D., Ciuprina, G., Kula, S.: Reduced order models for HF interconnect over lossy semiconductor substrate. In: 2007 IEEE Workshop on SPI, pp. 233–236 (2007)
15. Wang, C.-F., Ling, F., Song, J., Jin, J.-M.: Adaptive integral solution of combined field integral equation. Microw. Opt. Technol. Lett. **19**(5), 321–328 (1998)

A Yee-Like Finite Element Scheme for Maxwell's Equations on Hybrid Grids with Mass-Lumping

Herbert Egger[1(✉)] and Bogdan Radu[2]

[1] Institute for Numerical Mathematics, Johannes-Kepler University Linz, Linz, Austria
herbert.egger@jku.at

[2] Johann Radon Institute for Computational and Applied Mathematics, Linz, Austria
bogdan.radu@ricam.oeaw.ac.at

Abstract. A novel finite element method for the approximation of Maxwell's equations over hybrid two-dimensional grids is studied. The choice of appropriate basis functions and numerical quadrature leads to diagonal mass matrices which allow for efficient time integration by explicit methods.On purely rectangular grids, the proposed schemes coincide with well-established FIT and FDTD methods. Additional internal degrees of freedom introduced on triangles allow for mass-lumping without the usual constraints on the shape of these elements. A full error analysis of the method is developed and numerical tests are presented for illustration.

1 Introduction

The propagation of electromagnetic waves through a non-dispersive linear medium can be described by the time-dependent Maxwell's equations

$$\varepsilon \partial_t E + \operatorname{curl} H = -j, \tag{1}$$
$$\mu \partial_t H + \operatorname{curl} E = 0, \tag{2}$$

together with appropriate initial and boundary conditions. Here E, H denote the electric and magnetic field intensities, ε, μ the corresponding material parameters, and j describes the density of source and eddy currents. An efficient discretization of (1)–(2) can be achieved by the finite difference time domain (FDTD) method or the finite integration technique (FIT), see e.g. [14, 15], and for isotropic materials and orthogonal grids, second-order convergence can be obtained in space and time. In order to handle complex geometries, several attempts have been made to generalize these methods to non-orthogonal and unstructured grids; see e.g. [2, 12, 13] and also [3, 4] for more recent results. A rigorous error analysis of a Yee-like scheme on triangles and tetrahedra was given in [6], and first-order convergence in space on general unstructured grids was demonstrated theoretically and numerically.

M. van Beurden et al. (Eds.): SCEE 2022, MI 43, pp. 78–85, 2024.
https://doi.org/10.1007/978-3-031-54517-7_9

Scope. In this paper, we propose a novel Yee-like discretization scheme for hybrid grids in two space dimensions, consisting of triangles and rectangles. The method is based on a finite element approximation with mass-lumping through numerical quadrature, which allows for a rigorous error analysis; see [5,9] for background. On rectangular grid cells, the resulting discretization coincides with that of the FIT or FDTD method. Following [8], additional internal degrees of freedom are introduced on triangular grid cells, which allows us to prove discrete stability without severe restrictions on the mesh. The lowest order approximation on two-dimensional hybrid grids is studied in detail. The main ideas behind the construction of the method and its analysis however carry over to three dimensions and higher-order approximations; see [7,8,11] and the discussion at the end of the paper.

2 Description of the Problem

Let us start with completely specifying the model problem to be considered in the rest of the paper. We choose $\varepsilon = \mu = 1$ and abbreviate $f = -\partial_t j$. Moreover, we consider the second-order form of Maxwell's equations, i.e.,

$$\partial_{tt} E + \operatorname{curl}(\operatorname{curl} E) = f, \qquad \text{in } \Omega, \tag{3}$$

$$n \times \operatorname{curl}(E) = 0, \qquad \text{on } \partial\Omega, \tag{4}$$

with simple boundary conditions. The computational domain $\Omega \subseteq \mathbb{R}^2$ is assumed to be a bounded Lipschitz polygon and $\operatorname{curl} E = \partial_x E_2 - \partial_y E_1$ denotes the curl of a vector field $E = (E_1, E_2)$ in two space dimensions. The above differential equations are considered on a finite time interval $[0,T]$, and complemented by suitable initial conditions $E(0) = E_0$ and $\partial_t E(0) = F_0$. The existence of a unique solution can then be established by semi-group theory or Galerkin approximation. Solutions of (3)–(4) can further be characterized equivalently by the variational identities

$$(\partial_{tt} E(t), v) + (\operatorname{curl} E(t), \operatorname{curl} v) = (f(t), v), \tag{5}$$

for all $v \in H(\operatorname{curl}, \Omega) = \{E \in L^2(\Omega)^2 : \operatorname{curl} E \in L^2(\Omega)\}$ and a.a. $t \in [0,T]$. For abbreviation, we write $(a,b) = \int_\Omega a \cdot b \, dx$ for the scalar product on $L^2(\Omega)$ and $L^2(\Omega)^2$.

3 A Finite Element Method with Mass-Lumping

Let $\mathscr{T}_h = \{K\}$ be a geometrically-conforming quasi-uniform shape-regular partition of Ω into triangular and/or rectangular elements K. By assumption, all edges of the mesh are of similar length and we call the size h of the longest edge the mesh size.

Finite Element Spaces. For the approximation of the field E on individual elements, we consider local polynomial spaces defined by

$$V(K) = \begin{cases} \mathscr{N}_0(K), & \text{if } K \text{ is a square,} \\ \mathscr{N}_0^+(K) = \mathscr{N}_0(K) + \mathscr{B}(K), & \text{if } K \text{ is a triangle.} \end{cases} \tag{6}$$

Here $\mathcal{N}_0(K)$ is the lowest order Nedelec space for triangles or rectangles [1,10], and $\mathcal{B}(K)$ is a space of three quadratic functions with vanishing tangential components. The corresponding degrees of freedom are depicted in Fig. 1, and details on the basis functions are presented in Sect. 5.The global finite element space is finally defined by $\mathcal{V}_h = \{v_h \in H(\mathrm{curl};\Omega) : v_h|_K \in V(K)\ \forall K \in \mathcal{T}_h\}$.

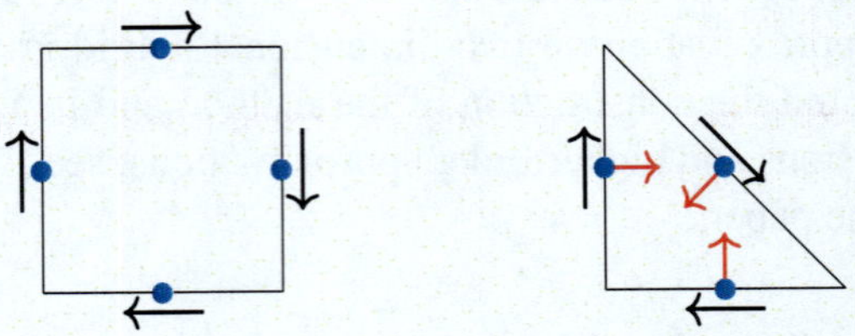

Fig. 1. Degrees of freedom for the space $\mathcal{N}_0(K)$ on the rectangle (left) and the space $\mathcal{N}_0^+(K)$, introduced in [8], on the triangle (right). The three internal degrees of freedom for the bubble functions are displayed in red and the corresponding quadrature points are depicted as blue dots. (Color figure online)

Quadrature. We use an approximation $(u,v)_h := \sum_K (u,v)_{h,K}$ for the L^2-scalar product, with contributions obtained by numerical integration. On the triangle, we set

$$(u,v)_{h,K} = |K| \sum_{i=1}^{3} \tfrac{1}{3} u(m_{K,i}) \cdot v(m_{K,i}), \tag{7}$$

where $m_{K,i}$ is the midpoint of the edge e_i opposite to vertex i; see Fig. 1. For the rectangle, we proceed differently: Here we decompose $(u,v) = (u_1,v_1) + (u_2,v_2)$ into two contributions for the orthogonal directions, and then use different quadrature rules for the two contributions, i.e.

$$(u,v)_{h,K} = |K| \left(\sum_{i=1}^{2} \tfrac{1}{2} u_1(m_{K,h,i}) v_1(m_{K,h,i}) + \sum_{j=1}^{2} \tfrac{1}{2} u_2(m_{K,v,j}) v_2(m_{K,v,j}) \right). \tag{8}$$

Here $m_{K,h,i}$ and $m_{K,h,j}$ are the midpoints of the horizontal and vertical edges, respectively; see again Fig. 1. For the semi-discretization of our model problem in space, we then consider the following inexact Galerkin approximation.

Problem 1. Let $E_{h,0}$, $E_{h,1} \in \mathcal{V}_h$ be given. Find $E_h : [0,T] \to \mathcal{V}_h$ such that

$$(\partial_{tt} E_h(t), v_h)_h + (\mathrm{curl}\, E_h(t), \mathrm{curl}\, v_h) = (f(t), v_h) \tag{9}$$

for all $v_h \in \mathcal{V}_h$ and all $t \in [0,T]$, and such that $E_h(0) = E_{h,0}$ and $\partial_t E_h(0) = E_{h,1}$.

As we will indicate below, the implementation of this method leads to a diagonal mass matrix, which allows using explicit methods for efficient time integration.

4 Main Results

By elementary computations, one can verify the following assertions, which ensure the well-posedness of Problem 1 and yield a starting point for our error analysis.

Lemma 1. *The quadrature rule (8) is exact for polynomials of degree $k \leq 2$ on triangles and for polynomials of degree $k \leq 1$ on squares. Moreover, the inexact scalar product $(\cdot,\cdot)_h$ induces a norm $\|\cdot\|_h$ on $\mathscr{V}_h$, which is equivalent to the L^2-norm on $\mathscr{V}_h$, and consequently Problem 1 has a unique solution.*

As a second ingredient, let us recall some results about polynomial interpolation. We denote by $\Pi_h : H^1(\mathscr{T}_h)^2 \to \mathscr{V}_h$ the projection defined element-wise by

$$(\Pi_h E)|_K := \Pi_K E|_K \tag{10}$$

where $\Pi_K : H^1(K) \to \mathscr{N}_0(K)$ is the standard interpolation operator for the lowest order Nedelec space $\mathscr{N}(K)$ on both triangles and squares; see [1,10] for details. We further denote by $\pi_h^0 : L^2(\Omega) \to P_0(\mathscr{T}_h)$ the L^2-orthogonal projection onto piecewise constants; the same symbol is used for the projection of vector-valued functions.

Lemma 2. *Let $K \in \mathscr{T}_h$ and Π_h defined as in (10). Then*

$$\|E - \Pi_h E\|_{L^2(K)} \leq Ch\|E\|_{H^1(K)}, \tag{11}$$

$$\|\operatorname{curl}(E - \Pi_h E)\|_{L^2(K)} \leq Ch\|\operatorname{curl} E\|_{H^1(K)}, \tag{12}$$

$$\|E - \pi_h^0 E\|_{L^2(K)} \leq Ch\|E\|_{H^1(K)}, \tag{13}$$

whenever E is regular enough, with a constant C independent of h.

Having introduced all the required tools, we can now state and prove our main result.

Theorem 1. *Let E and E_h denote the solutions of (5) and (9) with initial values set by $E_h(0) = \Pi_h E(0)$ and $\partial_t E_h(0) = \Pi_h \partial_t E(0)$. Then*

$$\|\partial_t(E - E_h)\|_{L^\infty(0,T;L^2(\Omega))} + \|\operatorname{curl}(E - E_h)\|_{L^\infty(0,T;L^2(\Omega))} \leq C(E,T)\,h$$

with constant C depending on the norm of E but independent of the mesh size h.

Proof. Apart from some technical details, the analysis follows by standard arguments. For completeness and convenience of the reader, we present all the details.

Step 1. Error Splitting and Estimate for the Projection Error. We first split the discretization error into a projection error and a discrete error component via

$$E - E_h = (E - \Pi_h E) + (\Pi_h E - E_h) =: -\eta + \psi_h. \tag{14}$$

By the estimates of Lemma 2, we immediately obtain

$$\begin{aligned}\|\partial_t \eta\|_{L^\infty(0,T;L^2(\Omega))} &+ \|\operatorname{curl} \eta\|_{L^\infty(0,T;L^2(\Omega))} \\ &\leq Ch\left(\|\partial_t E\|_{L^\infty(0,T;H^1(\mathscr{T}_h))} + \|\operatorname{curl} E\|_{L^\infty(0,T;H^1(\mathscr{T}_h))}\right),\end{aligned}$$

which already covers the first error component.

Step 2. Discrete Error Equation. By subtracting (8) from (5) with $v = v_h$, we can see that the discrete error ψ_h satisfies the identity

$$
\begin{aligned}
(\partial_{tt}\psi_h(t), v_h)_h + (\operatorname{curl}\psi_h(t), \operatorname{curl} v_h) = \\
(\partial_{tt}\eta(t), v_h) + (\operatorname{curl}\eta(t), \operatorname{curl} v_h) + \sigma_h(\Pi_h\partial_{tt}u(t), v_h)
\end{aligned}
$$

for all $v_h \in \mathscr{V}_h$ and $0 \le t \le T$, with quadrature error

$$\sigma_h(E, v) = (E, v)_h - (E, v). \tag{15}$$

We can further split $\sigma_h(E,\phi) = \sum_{K\in\mathscr{T}_h} \sigma_K(E,\phi)$ into element contributions defined by $\sigma_K(E,\phi) = (E,\phi)_{h,K} - (E,\phi)_K$. Moreover, $\psi_h(0) = \partial_t\psi_h(0) = 0$, due to the choice of initial conditions for the discrete problem.

Step 3. Estimates for the Quadrature Error. To further proceed in our analysis, we now quantify the local quadrature error in more detail.

Lemma 3. *Let $E \in L^2(\Omega)^2$ with $E|_K \in H^1(K)^2$ for all $K \in \mathscr{T}_h$. Then*

$$|\sigma_K(\Pi_h E, \phi_h)| \le Ch\|E\|_{H^1(K)}\|\phi_h\|_{L^2(K)}$$

for all $\phi_h \in \mathscr{V}_h$ and all $K \in \mathscr{T}_h$ with constant C independent of the element K.

Proof. Using Lemma 1, we deduce that $(u_h^0, v_h)_K = (u_h^0, v_h)_{h,K}$ for all $u_h^0 \in P_0(K)^2$ and $v_h \in V(K)$. We can then estimate the quadrature error by

$$
\begin{aligned}
|\sigma_K(\Pi_h u, v_h)| = |\sigma_K(\Pi_h u - \pi_h^0 u, v_h)| &\le c\|\Pi_h u - \pi_h^0 u\|_{L^2(K)}\|v_h\|_{L^2(K)} \\
&\le c' h\|u\|_{H^1(K)}\|v_h\|_{L^2(K)},
\end{aligned}
$$

where we used the Cauchy-Schwarz inequality and the norm equivalence of Lemma 1 and the approximation properties of the projections from Lemma 2.

Step 4. Estimate for the Discrete Error. Taking $v_h = \partial_t\psi_h(t)$ as test function in the discrete error equation and integrating from 0 to t leads to

$$
\begin{aligned}
&\frac{1}{2}\left(\|\partial_t\psi_h(t)\|_h^2 + \|\operatorname{curl}\psi_h(t)\|_{L^2(\Omega)}^2\right) \qquad (16) \\
&\quad = \int_0^t (\partial_{tt}\eta(s), \partial_t\psi_h(s)) + (\operatorname{curl}\eta(s), \operatorname{curl}\partial_t\psi_h(s)) + \sigma_h(\Pi_h\partial_{tt}u(s), \partial_t\psi_h(s))\,ds.
\end{aligned}
$$

The three terms can now be estimated separately. Using Cauchy-Schwarz and Young inequalities, the first term may be bounded by

$$(i) \le ch^2\|\partial_{tt}E\|_{L^1(0,t,H^1(\mathscr{T}_h))}^2 + \tfrac{1}{4}\|\partial_t\psi_h\|_{L^\infty(0,t,L^2(\Omega))}^2.$$

For the second term, we utilize that

$$
\begin{aligned}
(ii) &= \int_0^t (\operatorname{curl}(E - \Pi_h E), \operatorname{curl} \partial_t \psi_h))\, ds \\
&= (\operatorname{curl}(E - \Pi_h E)(t), \operatorname{curl} \psi_h(t)) - \int_0^t (\operatorname{curl}(\partial_t E - \Pi_h \partial_t E), \operatorname{curl} \psi_h)\, ds \\
&\leq Ch^2 \big(\|\operatorname{curl} E\|^2_{L^\infty(0,t;H^1(\mathscr{T}_h))} + \|\operatorname{curl} \partial_t E\|^2_{L^1(0,t;L^2(\Omega))} \big) + \tfrac{1}{4} \|\psi_h\|^2_{L^\infty(0,t;L^2(\Omega))}.
\end{aligned}
$$

The third term can finally be estimated using Lemma 3 according to

$$
(iii) \leq ch^2 \|\partial_{tt} E\|^2_{L^1(0,t,H^1(\Omega))} + \tfrac{1}{4} \|\partial_t \psi_h\|^2_{L^\infty(0,t,L^2(\Omega))}
$$

Using these estimates in the inequality (16), absorbing all the terms with the test function into the left side, and taking the supremum over $t \in [0,T]$, after applying the norm equivalence of Lemma 1 to some terms, then leads to the estimate

$$
\begin{aligned}
&\|\partial_t \psi_h\|^2_{L^\infty(0,T;L^2(\Omega))} + \|\operatorname{curl} \psi_h\|^2_{L^\infty(0,T;L^2(\Omega))} \\
&\qquad \leq Ch^2 \big(\|\partial_{tt} E\|^2_{L^1(0,T;H^1(\mathscr{T}_h))} + \|\partial_t E\|^2_{L^1(0,T;H^1(\mathscr{T}_h))} + \|\operatorname{curl} E\|^2_{L^\infty(0,T;H^1(\mathscr{T}_h))} \big)
\end{aligned}
$$

for the discrete error component; one may also take the square root in all terms.

Step 5. The proof of the theorem is completed by applying the triangle inequality to the error splitting in Step 1 and adding up the estimates for the projection error η and the discrete error component ψ_h.

5 Implementation

For completeness of the presentation, let us briefly discuss the choice of basis functions for the local finite element spaces $\mathscr{N}_0(K)$ and $\mathscr{N}_0^+(K)$ which, together with the numerical quadrature leads to diagonal mass matrices.

Rectangle. On quadrilateral elements K, we choose the standard basis for the lowest order Nedelec space $\mathscr{N}_0(K) = \operatorname{span}\{\Phi_{h,i}, \Phi_{v,i} : i = 1, \ldots, 2\}$; see [1, 10]. These functions have the following properties: The function $\Phi_{h,i}$ associated to a horizontal edge $e_{h,i}$ vanishes identically on the opposite horizontal edge, and $\Phi_{v,j}$ associated to for the vertical edge $e_{v,j}$ vanishes on the opposite vertical edge. Hence the local mass matrix produced by the quadrature rule $(u,v)_{K,h}$ for every rectangle is diagonal.

Triangle. Let $\{\lambda_i\}$ be the barycentric coordinates of the element K. For every edge $e_k = e_{ij}$ pointing from vertex i to j, and thus opposite to k, we define the two basis functions

$$
\begin{aligned}
\Phi^B_{ij} &= \lambda_i \lambda_j \nabla \lambda_k \qquad \text{and} \\
\Phi_{ij} &= \lambda_i \nabla \lambda_j - \lambda_j \nabla \lambda_i + \alpha_{ij} \Phi^B_{ij} + \beta_{ij} \Phi^B_{jk} + \gamma_{ij} \Phi^B_{ki}.
\end{aligned}
$$

Then $\mathcal{N}_0^+(K) = \text{span}\{\Phi_{12}, \Phi_{23}, \Phi_{31}, \Phi_{12}^B, \Phi_{23}^B, \Phi_{31}^B\}$. The bubble functions Φ_{ij}^B have vanishing tangential components on the edge are e_k, and they vanish identically on the two remaining edges e_i, e_j. The functions Φ_{ij} are modified Nedelec basis functions. They have vanishing tangential components on the two edges e_i, e_j, and by appropriate choice of the parameters α_{ij}, β_{ij}, γ_{ij}, their normal components on all edge midpoints $m_{K,i}$ can be made zero. As a consequence, the local mass matrix produced by the scalar product $(u,v)_{K,h}$ for the triangle becomes diagonal.

Summary. The global mass matrix is obtained by assembling the local mass matrices, which are diagonal, and hence has inherits this property.

6 Numerical Illustration

We consider the computational domain $\Omega = \Omega_1 \cup \Omega_2$ where $\Omega_1 = (0,2) \times (-1,1)$ and $\Omega_2 = ((2,4) \times (-1,1)) \setminus B_{0.3}(3,0)$, where $B_r(x,y)$ denotes the ball with radius r around midpoint (x,y). The two subdomains are meshed by rectangles and triangles, respectively. For our test problem, we consider the wave Eq. (3). The boundary $\partial\Omega$ is split into several parts and as boundary conditions, we impose

$$\begin{aligned} n \times E &= \sin(10 \cdot t) \cdot e^{-10y^2}, && \text{on } \partial\Omega_{\text{left}}, \\ n \times E &= 0, && \text{on } \partial\Omega_{\text{ball}}, \\ n \times \operatorname{curl} E &= 0, && \text{else}. \end{aligned}$$

The initial conditions are chosen as $E(0) = \partial_t E(0) = 0$. This corresponds to a pulse entering at the left boundary, propagating through the domain, and getting reflected at the walls of the box and the circular inclusion. Some snapshots of the solution are depicted in Fig. 2. Let us remark that no reflections are observed at the interface between the two meshes. In our numerical tests, we observe linear convergence of the error in space. This coincides with the theoretical predictions of Theorem 1, and also demonstrates that the error estimates are sharp. Note that second-order convergence is in general lost for Yee-like approximations on unstructured grids; also see [11].

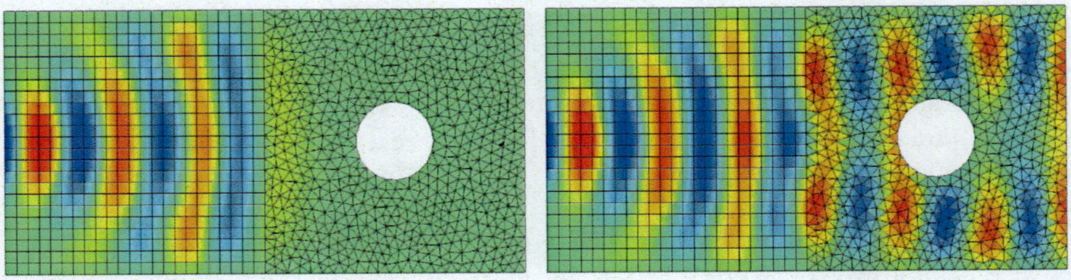

Fig. 2. The first E_1 component of the solution $E = (E_1, E_2)$ at time steps $t = 2.3$ and $t = 5$ showing the scattering at the sphere.

References

1. Boffi, D., Brezzi, F., Fortin, M.: Mixed finite element methods and applications. Springer Series in Computational Mathematics, vol. 44. Springer, Heidelberg (2013). https://doi.org/10.1007/978-3-642-36519-5
2. Bossavit, A., Kettunen, L.: Yee-like schemes on a tetrahedral mesh, with diagonal lumping. Int. J. Numer. Model. Electron. Netw. Devices Fields **12**(1–2), 129–142 (1999)
3. Codecasa, L., Kapidani, B., Specogna, R., Trevisan, F.: Novel FDTD technique over tetrahedral grids for conductive media. IEEE Trans. Antennas Propag. **66**(10), 5387–5396 (2018)
4. Codecasa, L., Politi, M.: Explicit, consistent, and conditionally stable extension of FDTD to tetrahedral grids by FIT. IEEE Trans. Magn. **44**(6), 1258–1261 (2008)
5. Cohen, G.: Higher-Order Numerical Methods for Transient Wave Equations. Springer, Heidelberg (2002). https://doi.org/10.1007/978-3-662-04823-8
6. Egger, H., Radu, B.: A mass-lumped mixed finite element method for maxwell's equations. In: Nicosia, G., Romano, V. (eds.) SCEE 2018. MI, vol. 32, pp. 15–24. Springer, Cham (2020). https://doi.org/10.1007/978-3-030-44101-2_2
7. Egger, H., Radu, B.: A second-order finite element method with mass lumping for Maxwell's equations on tetrahedra. SIAM J. Numer. Anal. **59**(2), 864–885 (2021)
8. Elmkies, A., Joly, P.: Éléments finis d'arête et condensation de masse pour les équations de Maxwell: le cas 2D. Comptes Rendus de l'Académie des Sciences - Series I - Math. **324**(11), 1287–1293 (1997)
9. Monk, P.: Analysis of a finite element methods for Maxwell's equations. SIAM J. Numer. Anal. **29**, 714–729 (1992)
10. Nédélec, J.C.: Mixed finite elements in $\mathbb{R}^3$. Numer. Math. **35**, 315–341 (1980)
11. Radu, B.: Finite element mass lumping for H(div) and H(curl). PhD thesis, Technische Universität Darmstadt, Darmstadt (2022)
12. Rylander, T., Bondeson, A.: Stable FEM-FDTD hybrid method for Maxwell's equations. Comput. Phys. Commun. **125**(1), 75–82 (2000)
13. van Rienen, U.: Triangular grids: a review of resonator and waveguide analysis with classical FIT and some reflections on Yee-like FIT- and FEM-schemes. ACES J. **19**(1b), 73–83 (2004)
14. Weiland, T.: Finite integration method and discrete electromagnetism. In: Monk, P., Carstensen, C., Funken, S., Hackbusch, W., Hoppe, R.H.W. (eds.) Computational Electromagnetics, pp. 183–198. Springer, Heidelberg (2003). https://doi.org/10.1007/978-3-642-55745-3_12
15. Yee, K.: Numerical solution of initial boundary value problems involving Maxwell's equations in isotropic media. IEEE Trans. Antennas Propag. **14**, 302–307 (1966)

Time-Domain Electromagnetic Modeling and Simulation of a Nonlinear Electro-Optical Mixer

Arif Can Gungor(✉), Hande Ibili, Jasmin Smajic, and Juerg Leuthold

Institute of Electromagnetic Fields (IEF), ETH Zurich, 8092 Zurich, Switzerland
arifg@ethz.ch

Abstract. A full-wave electromagnetic solver coupled with a Poisson's solver based on time-domain finite element method (TD-FEM) is developed. This solver aims to simulate the side-band frequency generation on optical signal due to the imposed radio frequency (RF) signal through a nonlinear material. The optical signal propagating within an optical waveguide is simulated in time-domain by solving the electromagnetic wave equation, whereas Poisson's equation is numerically solved to compute the strength of the slowly-varying RF signal. The applied RF signal changes the permittivity of the nonlinear material BTO, and this changing permittivity affects the transient wave behavior of the light. As opposed to the available frequency-domain Maxwell solvers, this proposed time-domain solver is capable of simulating the nonlinear effects introduced by an electro-optical material, and implemented for the modeling of an application where RF signal is mixed into the optical frequencies. As a result of the simulations, nonlinear dielectric constant of electro-optical material is computed, and resulting side-band frequency generation is observed in the spectrum of the time-domain output signal.

1 Introduction

New generation mobile network systems offer faster communication speeds and extensive data transfer rates. New 5G and 6G applications utilize higher frequency bands to satisfy the demand in broader bandwidths and high data transfer rates, and photonic and plasmonic communication systems proved to be vital in this pursuit to replace highly lossy traditional electronic counterparts. In [1–5] authors have successfully demonstrated plasmonic modulation and photodetection reaching up to 500 GHz. Furthermore, a transparent optical-subTHz-optical link with single line rate of 240 Gbit/s over 115 m at subTHz frequencies over 200 GHz is reported [5]. The permanent progress in high frequency communication systems and continuous increase in data transfer rates rely on conversion of electronic and optical signals to each other [6], and ultra-fast, energy-efficient electro-optical switches. Recently, electro-optic modulation schemes via quantum effects [7], 2D materials [8], resonant structures [9], or plasmonic effects [10] have been demonstrated. To avoid the speed limitations that can be introduced by some of these methods, electro-optic devices with plasmonics that enhances the nonlinear effects via light confinement into sub-diffraction limits are reported [11]. Including plasmonics,

M. van Beurden et al. (Eds.): SCEE 2022, MI 43, pp. 86–93, 2024.
https://doi.org/10.1007/978-3-031-54517-7_10

one common way of mixing optical and electrical signals is through phase modulation of an optical signal by an RF signal that produces a series of side-band frequencies in the optical domain by utilizing nonlinear materials [12]. Therefore, advanced nonlinear electro-optical devices offer solutions for the next generation high speed communication systems. The performance of these devices are strongly dependent on the electro-optic coefficients of these nonlinear materials [13]. Moreover, the physical properties of the waveguides and light confinement are affecting the device operation as well [14]. Consequently, this article focuses on time-domain modeling of phase modulation of an optical signal and the mentioned nonlinear electro-optical effect. The objective of this article is to present a nonlinear material integrated optical waveguide in order to observe and model the mixing of the RF frequencies to the optical frequencies.

This article is organized as follows. Section 2 presents the optical waveguide and the material nonlinearity, as well as theoretical backgrounds for the computation. Section 3 discusses the results and Sect. 4 concludes the paper with remarks and future work description.

2 Physical Model and Theory

Considered problem requires electromagnetic analysis of an optical waveguide structure that has a nonlinear material insertion. The defined scheme is depicted in Fig. 1a, which consists of a 2-D optical waveguide, and in its middle section, a nonlinear material, Barium Titanate (BTO), is inserted as a phase modulator. Electric permittivity ε_{rBTO} of BTO depends on the electric field $|E_{local}|$ it encounters, and can provide Pockels coefficients as high as 1600 pm/V [15, 16]. Assuming single crystal BTO, the relation between the permittivity of BTO and the electric field is given as follows [16]:

$$\varepsilon_{rBTO}(|E_{local}|) = \frac{\varepsilon_0}{\left(1 + k\varepsilon_0^3 |E_{local}|^2\right)^{1/3}}, \tag{1}$$

where $\varepsilon_0 = 1000$ is the typical zero-field permittivity for BTO, and $k = 3\beta\varepsilon_0^3 = 10^{-8} m^2 V^{-2}$ is a constant that includes the nonlinear Johnson's parameter β.

The core material of the waveguide has 400 nm width and is assumed to be Silicon (with $\varepsilon_{rSi} = 12.04$). Whereas the cladding is 2 μm wide on each side of the waveguide and is taken as Silicondioxide ($\varepsilon_{rSiO_2} = 2.07$). The electro-optical (EO) modulator section with BTO insertion is chosen such that the effective permittivity of that section varies as $\varepsilon_r = 0.9\varepsilon_{rSi} + 0.1\varepsilon_{rBTO}$, since the fabricated 3-D devices have dominant contributions from the core material in addition to the deposited BTO. In other words, the EO section consists of a layer of BTO on the Silicon waveguide which leads to consideration of an effective electric permittivity that is combination of electric permittivities of Silicon and BTO.

The electromagnetic solver is developed using electromagnetic wave equation in time domain as in (2). The boundary conditions are perfect electric conductor (PEC) boundary condition on the top and the bottom sides of the structure ($\mathbf{n} \times \mathbf{E} = 0$, where $\mathbf{n}$ is unit vector pointing outwards direction), and the port boundary condition (3) is used for both the input and the output waveports (without excitation at the output, $\mathbf{E_0} = 0$). On the other hand, the field distribution due to the applied RF signal is computed by

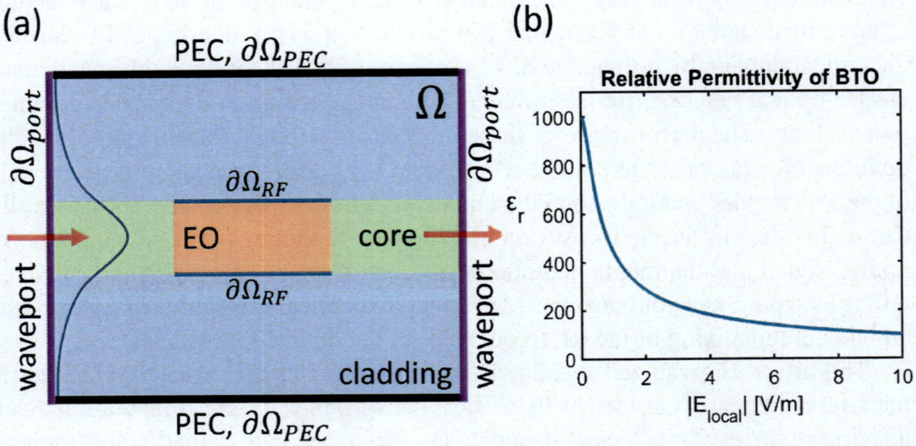

Fig. 1. Figure illustrating the nonlinear electro-optical signal mixer. (a): 2-D optical waveguide is depicted. The input waveport is on the left, whereas the output is on the right. The core section has also an electro-optical (EO) phase modulator (the middle part) in the middle to provide the nonlinearity. (b): Relative permittivity ε_r vs. $|E_{local}|$ is shown for the pure nonlinear material. The RF signal is directly applied on the contacts ($\partial\Omega_{RF}$) near the nonlinear material whereas the light excitation is provided from the waveport ($\partial\Omega_{port}$) on the left, and it is absorbed from the right.

the Poisson's equation as in (4). The RF signals are applied as Dirichlet conditions on the boundaries of the electro-optical (EO) material as in (5) and PEC boundary conditions are implemented for the computation domain boundaries for the RF field analysis ($\varphi = 0$ on outer boundaries).

The computed potential distribution φ is used to obtain the RF field $\mathbf{E}_{RF} = -\nabla\varphi$ and $|E_{local}| = |\mathbf{E}_{RF}|$ since the RF field (in the order of V/m) is much higher than the optical field (in the order of $\mu V/m$). Then, $|E_{local}|$ is used to compute the electric permittivity of BTO and the effective electric permittivity of the phase modulator sections based on (1).

For this coupled analysis to be accurate, the main assumptions are that the wavelength of the RF signal (in the order of millimeters) is very large compared to the computation domain (in the order of micrometers), and RF signal also varies very slowly compared to the time domain optical signal. Therefore, it is safe to solve only Poisson's equation for the RF field and resultant nonlinear contribution. Furthermore, the given waveguide is single-moded and the fundamental mode operation only requires z-component of the electric field for the light propagation. Consequently, Eq. (2) can be further reduced to a scalar equation which is be very convenient for the numerical analysis.

$$\nabla \times \left(\frac{1}{\mu_r}\nabla \times \mathbf{E}\right) + \mu_0\sigma\frac{\partial \mathbf{E}}{\partial t} + \mu_0\varepsilon_0\varepsilon_r\frac{\partial^2 \mathbf{E}}{\partial t^2} = 0 \; in \; \Omega \tag{2}$$

$$\mathbf{n}\times\left(\frac{1}{\mu_r}\nabla\times\mathbf{E}\right)+\frac{\mu_0}{Z_{port}}\mathbf{n}\times\left(\mathbf{n}\times\frac{\partial\mathbf{E}}{\partial t}\right) = \frac{-2\mu_0}{Z_{port}}\mathbf{n}\times\left(\mathbf{n}\times\frac{\partial\mathbf{E_0}}{\partial t}\right)\quad over\ \partial\Omega_{port} \tag{3}$$

$$\nabla(\varepsilon\nabla\varphi)=0\ in\ \Omega \tag{4}$$

$$\varphi=\varphi_{RF}\ over\ \partial\Omega_{RF} \tag{5}$$

$$\frac{\partial\mathbf{E}}{\partial t}\approx\frac{\mathbf{E}_t-\mathbf{E}_{t-1}}{\Delta t} \tag{6}$$

The introduced equations are discretized using Finite Element Method (FEM) and using first order linear shape functions by adopting the formulation in [17] and [18]. This discretization resulted around 14000 triangular elements for the complete geometry. The time discretization is done by Backward Difference Formula (BDF) as in (6) that ensures the stability of the wave equation [18]. Furthermore, the coupling of the wave equation (2) and the Poisson's equation is done through the effective permittivity of the EO material which directly effects the wave propagation. The time-domain approach requires very small time steps in the order of femtoseconds (0.1–1 fs) for the full wave simulation of the propagating light, whereas the RF field is changing very slowly and hence solving the static Poisson's equation at every time step is sufficient. In other words, the static Poisson's equation is solved at every time step to compute the RF field and the instantaneous effective permittivity before solving the wave equation in the respective time step. One complete simulation with sufficient number of periods of the optical signal requires around 200 MB memory, and it takes around 0.3 s CPU time for the computation of a single time step on an Intel i7-7500 CPU. A snapshot of the electric field (E_z) with the propagating fundamental mode is shown in Fig. 2 where the scattering due to nonlinear material is visible.

3 Results

The influence of the 'slowly varying' RF field directly reveals itself on the propagating light due to nonlinear electric permittivity of the EO section. From the input port, optical wave with wavelength of 1550 nm is excited with the spatial shape of the fundamental mode of the single-moded waveguide. Fast Fourier Transform (FFT) of the input signal, (the blue curve in Fig. 3b) also proves the excitation initially has purely one wavelength. After propagating through the waveguide and going through the nonlinear EO material in the presence of an RF excitation of 500 GHz, light reaches the output waveport, where its FFT is again computed. Figure 3b clearly shows that at the output, in addition to the excitation signal, other frequencies emerge. These frequencies occur at $f_{out}=f_0\pm mf_{RF}$, where f_0 denotes the original optical carrier frequency, f_{RF} is the RF frequencies applied to EO section and $m=\pm1,\pm2,...$ is an integer. With the applied RF field in this simulated case, ε_{rBTO} changes between around 25 and 100, which in return provides the mixing of RF signal into the optical carrier signal.

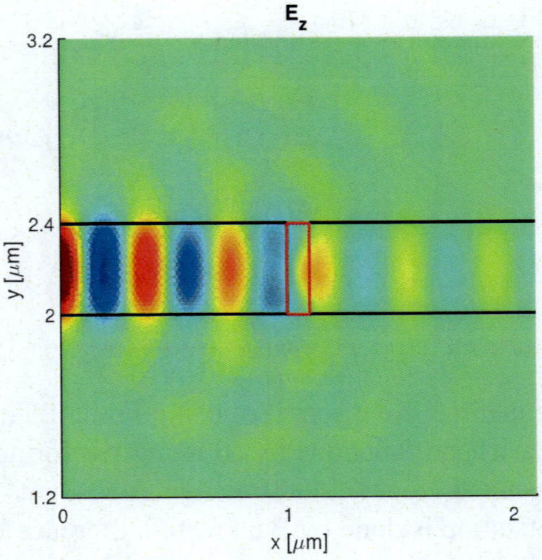

Fig. 2. Snapshot of the normalized field due to propagating light in the given waveguide with the nonlinear insertion. Here the plot depicts the z-component of the propagating electric due to light, and the applied in-plane RF field is not shown. Light enters into the computation domain from the left boundary, where the fundamental mode excitation (having only z-component) is applied, and travels within the Si core between the black lines; and absorbed at the output port on the right. The box, with 100 nm length (in x dimension), indicates where BTO is inserted. Scattering of the light due to discontinuity of permittivity can be seen.

Furthermore, amplitude of the applied RF signal have an effect on how efficient the signal mixing takes place. This also reveals itself as changing side-band ratios in the spectrum of the output signal. In other words, the emerging side-band frequencies have different peaks and different ratios with respect to the central optical frequency f_0. This can be observed in Fig. 4, as the higher RF field amplitude results higher sideband peaks when the spectrum are all normalized with respect to the central peak f_0. Also in Fig. 5, the side-band ratios for the first three side-bands are given with respect to varying RF field amplitude. Since the permittivity dependence of the BTO is nonlinear with respect to E_{local}, the side-band ratios also change nonlinearly and they saturate, which makes the design of electro-optical signal mixers more challenging and interesting.

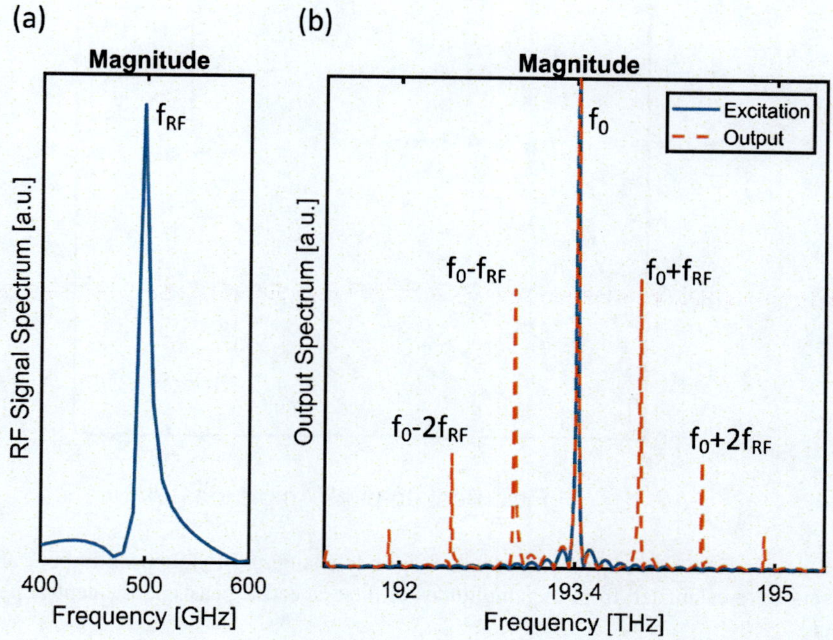

Fig. 3. Figure showing the spectrum of the signals indicating the signal mixing at the output. (a): Spectrum of the RF signal, single peak at 500 GHz. (b): Normalized spectrum of the optical signals: excitation signal, and the signal at the output port. The spectrum of the output optical signal shows that side-band frequencies are generated at $f_{out} = f_0 \pm m f_{RF}$.

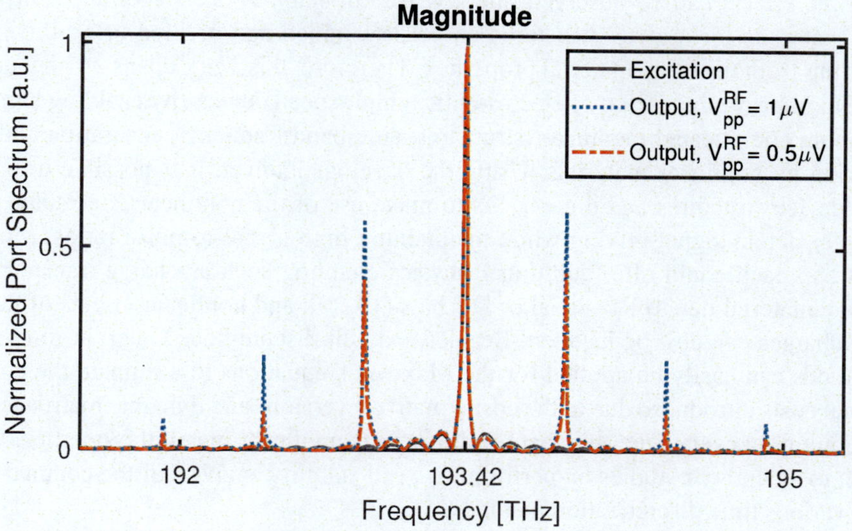

Fig. 4. Normalized (with respect to corresponding f_0 peak) spectrum of the mixed signals when different RF amplitudes are applied. The two curves depict the spectrum when the peak-to-peak potential through the EO section is $0.5\,\mu V$ and $1.0\,\mu V$. As expected, high RF field results side-bands with higher amplitude, but the effect is nonlinear due to material's field dependent electric permittivity.

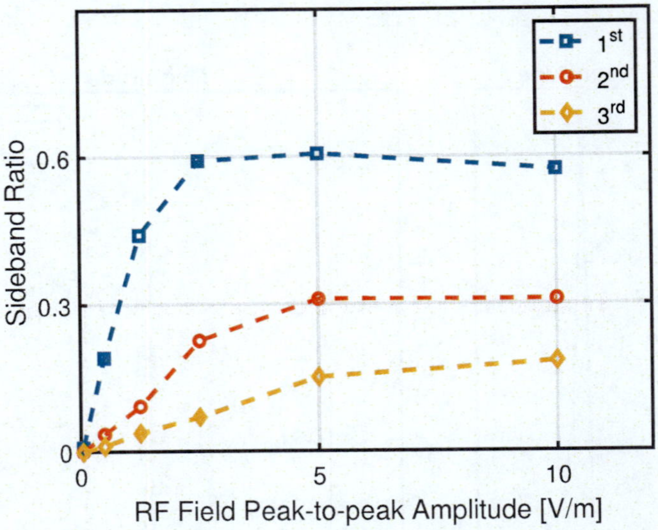

Fig. 5. 1st, 2nd and 3rd sideband ratios with respect to changing RF field peak-to-peak amplitude. The ratios are calculated from the amplitudes of the respective peak and the central peak at the output.

4 Conclusion

Thanks to the developed solver, nonlinear EO materials can be considered accurately and the effects can be observed directly. Additionally, Pockels coefficient of nonlinear materials for a given structure can be determined and verified using this method starting from the basic material properties. Therefore, this work opens up new perspective to improve the design of electro-optical high-speed devices by enabling simulation of more complicated modulator structures, side-band frequency generation and signal mixing by nonlinear materials. Using the developed solver, it is possible to optimize the device structures and dimensions to make use of the nonlinear effects and to efficiently obtain signal mixing while maintaining high signal-to-noise ratios. Moreover, this approach could offer additional physics coupling such as charge transport in the active material due to RF signal or DC bias [19,20], and nonlinear effects of the moving charges can also be inserted. Besides, explicit discontinous Galerkin time-domain methods can easily be applied for the Maxwell's equations to eliminate the computational costs introduced due to the costly matrix inversions and dynamic matrix assembly operations at every time step that result from the nonlinear material properties. We also plan to extend our studies by performing a full stability analysis of the coupled system for various time discretization methods.

References

1. Haffner, C., et al.: All-plasmonic Mach-Zehnder modulator enabling optical high-speed communication at the microscale. Nat. Photonics **9**, 525–528 (2015)
2. Burla, M., et al.: 500 GHz plasmonic Mach-Zehnder modulator enabling sub-THz microwave photonics. APL Photonics **4**, 056106 (2019)
3. Salamin, Y., et al.: 300 GHz plasmonic mixer. In: IEEE 2019 International Topical Meeting on Microwave Photonics (MWP), pp. 1-4 (2019)
4. Koepfli, S., et al.: High-speed graphene photodetection: 300 GHz is not the limit. In: 2021 Conference on Lasers and Electro-Optics Europe & European Quantum Electronics Conference (2021)
5. Horst, Y., et al.: transparent Optical-THz-optical link at 240/192 Gbit/s over 5/115 m enabled by plasmonics. J. Lightwave Technol. (2022)
6. Hu, Y., et al.: On-chip electro-optic frequency shifters and beam splitters. Nature **599**, 587–593 (2021)
7. Segev, A., Sa'ar, A., Oiknine-Schlesinger, J., Ehrenfreund, E.: Quantum interference versus Stark intersubband electro-optical modulation in asymmetrical quantum wells. Superlattices Microstruct. **19**, 47–57 (1996)
8. Sun, Z., Martinez, A., Wang, F.: Optical modulators with 2D layered materials. Nat. Photonics **10**, 227–238 (2016)
9. Timurdogan, E., Sorace-Agaskar, C.M., Sun, J., Shah Hosseini, E., Biberman, A., Watts, M.R.: An ultralow power athermal silicon modulator. Nat. Commun. **5**, 1–11 (2014)
10. Davis, T.J., Gómez, D.E., Roberts, A.: Plasmonic circuits for manipulating optical information. Nanophotonics **6**, 543–559 (2017)
11. Lee, H., et al.: Nanoscale conducting oxide PlasMOStor. Nano Lett. **14**, 6463–6468 (2022)
12. Abel, S., et al.: Large Pockels effect in micro-and nanostructured barium titanate integrated on silicon. Nat. Mater. **18**, 42–47 (2019)
13. Boyd, R. W.: Nonlinear Optics. Academic Press (2020)
14. Heni, W., et al.: Nonlinearities of organic electro-optic materials in nanoscale slots and implications for the optimum modulator design. Opt. Express **25**, 2627–2653 (2017)
15. Li, M., Tang, H.X.: Strong pockels materials. Nat. Mater. **18**, 9–11 (2019)
16. Padurariu, L., Curecheriu, L., Buscaglia, V., Mitoseriu, L.: Field-dependent permittivity in nanostructured BaTiO3 ceramics: modeling and experimental verification. Phys. Rev. B **85**, 224111 (2012)
17. Gungor, A.C., Celuch, M., Smajic, J., Olszewska-Placha, M., Leuthold, J.: Electromagnetic and semiconductor modeling of scanning microwave microscopy setups. IEEE J. Multiscale Multiphys. Comput. Tech. **5**, 209–216 (2020)
18. Jin, J.M.: The Finite Element Method in Electromagnetics. Wiley, Hoboken (2015)
19. Gungor, A.C., Ehrengruber, T., Smajic, J., Leuthold, J.: Coupled electromagnetic and hydrodynamic modeling for semiconductors Using DGTD. IEEE Trans. Magn. **57**, 1–5 (2021)
20. Gungor, A.C., Doderer, M., Ibili, H., Smajic, J., Leuthold, J.: Coupled electromagnetic and hydrodynamic semiconductor modeling for Terahertz generation. IEEE Trans. Magn. (2022)

Iterative Charge-Update Schemes for Electro-quasistatic Problems

Fotios Kasolis(✉), Marvin-Lucas Henkel, and Markus Clemens

Chair of Electromagnetic Theory, University of Wuppertal, 42119 Wuppertal, Germany
{kasolis,mhenkel,clemens}@uni-wuppertal.de

Abstract. The electric scalar potential electro-quasistatic field formulation is commonly employed for simulating nonlinear high-voltage problems. To this end, standard iterative nonlinear solvers, such as fixed-point iterations and Newton's method, are used. Here, iterative charge update schemes that possess a constant coefficient matrix throughout the nonlinear iterations are developed, abstract convergence conditions are proved and numerically verified for the fundamental frequency, while their suitability and performance is assessed, in terms of accuracy and computational complexity.

1 Introduction

The electro-quasistatic (EQS) approximation [1] of Maxwell's equations provides a suitable model for various applications in electrical engineering, whenever radiation effects can be neglected and the electric field energy density exceeds the magnetic energy density. A benefit of the EQS approximation is that it enables a scalar potential formulation for the charge continuity equation. Further, since the conductivity is inherent in EQS formulations, it is possible to consider media whose conductivity is a nonlinear function of the field-strength, such as field grading material that are often used in high-voltage engineering devices as means to reduce the field-strength in the vicinity of critical points [2].

2 Problem Description

In the electro-quasistatic (EQS) limit $\partial_t \mathbf{B} \to \mathbf{0}$, Faraday's law implies that the electric field intensity $\mathbf{E}$ is irrotational, and hence, Maxwell's equations reduce to a scalar potential formulation for the continuity equation. More precisely, in the absence of imposed current density, EQS fields are governed by the continuity equation

$$\nabla \cdot (\sigma \nabla \varphi + \varepsilon \nabla \partial_t \varphi) = 0, \tag{1}$$

where $\sigma \geq 0$ is the electric conductivity, $\varepsilon > 0$ is the permittivity, φ is the sought scalar EQS potential, which depends on the time t and on the position vector $\mathbf{r}$, and $\mathbf{E} = -\nabla \varphi$. Provided a time-harmonic excitation, the electric potential φ can be written as

M. van Beurden et al. (Eds.): SCEE 2022, MI 43, pp. 94–101, 2024.
https://doi.org/10.1007/978-3-031-54517-7_11

$\varphi(\mathbf{r},t) = \mathrm{Re}(\phi(\mathbf{r})\mathrm{e}^{\mathrm{i}\omega t})$, where Re(...) denotes the real part of the enclosed expression, $\omega = 2\pi f$ is the angular frequency of the excitation, $\mathrm{i} = \sqrt{-1}$ is the imaginary unit, and ϕ is a complex-valued function that satisfies the Laplace equation

$$\nabla \cdot (\sigma + \mathrm{i}\omega\varepsilon)\nabla\phi = 0. \tag{2}$$

Provided that Eq. (2) holds in an open, bounded, and simply connected domain $\Omega \subset \mathbb{R}^3$ whose boundary $\partial\Omega$ is Lipschitz, such as the one depicted in Fig. 1 (left), a typical boundary value problem for Eq. (2) is formulated by assuming Dirichlet and Neumann boundary conditions of the form $\phi|\Gamma_\mathrm{S} = s$, $\phi|\Gamma_\mathrm{G} = 0$, and $\mathbf{n} \cdot \nabla\phi|\Gamma_\mathrm{I} = 0$. Here, Γ_S is the boundary that supplies a constant potential $s > 0$, Γ_G is grounded, and $\mathbf{n}$ is the outward pointing unit normal on the insulating boundary Γ_I. Further, the computational domain Ω constitutes of $M \geq 1$ sufficiently regular non-overlapping subdomains $\Omega^{(m)}$, where each $\Omega^{(m)}$ with $m \in \{0, 1, \ldots, M-1\}$ is occupied by material with constant material properties. To introduce the variational setting, consider the function spaces $U_\alpha = \{\psi \in H^1(\Omega) : \psi|\Gamma_\mathrm{S} = \alpha, \psi|\Gamma_\mathrm{G} = 0\}$, where $\alpha \in \{0, s\}$, $H^1(\Omega) = H(\mathrm{grad}, \Omega)$ is the space of square-integrable functions whose weak partial derivatives are square-integrable, and the restriction operators $|\Gamma_\mathrm{S}$, $|\Gamma_\mathrm{G}$ are understood in the sense of traces. Then, the variational form of the boundary value problem that constitutes of Eq. (2) and the aforementioned boundary conditions reads as follows.

$$\text{Find } \phi \in U_s \text{ such that } \int_\Omega (\sigma + \mathrm{i}\omega\varepsilon)\nabla\phi \cdot \nabla\psi = 0 \quad \forall\psi \in U_0. \tag{3}$$

Provided the Dirichlet boundary data, and the fact that the permittivity is nonvanishing throughout Ω, problem (3) is well-posed for all $\omega > 0$, as follows from the Lax-Millgram theorem [3].

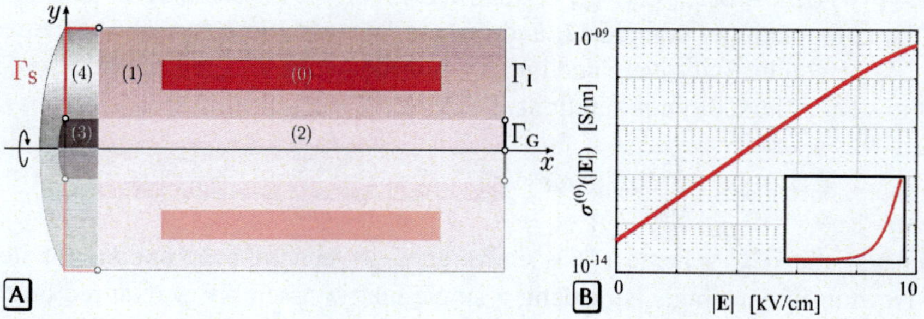

Fig. 1. (A) Conceptual setup of a benchmark device [4] in the EQS limit, with the numbers (m) corresponding to the domains $\Omega^{(m)}$ and (B) a plot of the conductivity function $\sigma^{(0)}$ as a function of the electric field strength $|\mathbf{E}|$, in semilogarithmic scale. The inset shows the same curve using linear scales for both axes.

3 The Iterative Charge-Update Scheme

Equation (2) can be viewed as a combination of the electric divergence law with the time-harmonic continuity equation [5], that is,

$$-\nabla \cdot \varepsilon \nabla \phi = \rho, \quad \mathrm{i}\omega\rho = \nabla \cdot \sigma \nabla \phi, \tag{4}$$

with ρ being the charge density. Based on this observation, an iterative charge-update (ICU) scheme can be introduced by first computing the induced charge density according to the continuity equation, and then using the resulting updated charge density as the right-hand side term for a boundary value problem for the electric divergence law, in order to update the potential. More precisely, consider the variational form

$$\int_\Omega \varepsilon \nabla \phi \cdot \nabla \psi = \int_\Omega \rho \psi \quad \forall \psi \in U_0, \tag{5}$$

which is associated with the boundary value problem for the electric divergence law, and the one that is associated with the continuity equation, that is,

$$\int_\Omega \sigma \nabla \phi \cdot \nabla \psi = -\mathrm{i}\omega \int_\Omega \rho \psi \quad \forall \psi \in U_0. \tag{6}$$

Given an initial guess $\phi_0 \in U_s$, an approximation ϕ_n of ϕ can be introduced by using problem (6) to evaluate the right-hand side term, while afterwards, the potential can be updated using Eq. (5). In particular, let $n = 0$, $e_n = 1$, $0 < \delta < 1$, and choose $\phi_n \in U_s$. While $e_n \geq \delta$, find $\phi_{n+1} \in U_s$ such that

$$\int_\Omega \varepsilon \nabla \phi_{n+1} \cdot \nabla \psi = \frac{\mathrm{i}}{\omega} \int_\Omega \sigma_n \nabla \phi_n \cdot \nabla \psi \quad \forall \psi \in U_0, \tag{7}$$

set $e_{n+1} = \|\phi_{n+1} - \phi_n\|_{0,\Omega} / \|\phi_{n+1}\|_{0,\Omega}$, where $\|\ldots\|_{0,\Omega}$ is the standard L^2-norm throughout the computational domain Ω, and increase the counter n by one. In the resulting Eq. (7), the conductivity-weighted integral is evaluated at the previous iteration, in contrast to standard fixed-point iterations

$$\int_\Omega (\sigma_n + \mathrm{i}\omega\varepsilon) \nabla \phi_{n+1} \cdot \nabla \psi = 0 \quad \forall \psi \in U_0. \tag{8}$$

Whenever the ICU scheme (7) is convergent, its benefit over fixed-point iterations and Newton-like methods is evident; a single matrix assembly is required by the ICU scheme, while only the right-hand-side vector needs to be updated within each iteration. In the following theorem, the condition that is required for the ICU scheme to converge is given.

Theorem 1. *If* $\gamma = \sigma_{\max}/(\omega\varepsilon_{\min}) < 1$, *then the iterative charge-update scheme* (7) *converges to the solution of problem* (3) *in* $H^1(\Omega)$ *for all* $\phi_0 \in U_s$.

Proof. Let $u_n = \phi_n - \phi \in U_0$ and observe that u_n satisfies the variational equation

$$\int_\Omega \varepsilon \nabla u_{n+1} \cdot \nabla \psi = \frac{\mathrm{i}}{\omega} \int_\Omega \sigma_n \nabla \phi_n \cdot \nabla \psi - \frac{\mathrm{i}}{\omega} \int_\Omega \sigma \nabla \phi \cdot \nabla \psi. \tag{9}$$

For $\psi = \overline{u_{n+1}} \in U_0$, with overlines denoting complex conjugation, the latter equation becomes

$$\int_\Omega \varepsilon |\nabla u_{n+1}|^2 = \frac{\mathrm{i}}{\omega} \int_\Omega \sigma_n \nabla \phi_n \cdot \nabla \overline{u_{n+1}} - \frac{\mathrm{i}}{\omega} \int_\Omega \sigma \nabla \phi \cdot \nabla \overline{u_{n+1}}. \tag{10}$$

The left-hand side integral of Eq. (10) satisfies the inequality

$$\int_\Omega \varepsilon |\nabla u_{n+1}|^2 = \sum_{m=0}^{M-1} \varepsilon^{(m)} \int_{\Omega^{(m)}} |\nabla u_{n+1}|^2 \geq \varepsilon_{\min} \|\nabla u_{n+1}\|_{0,\Omega}^2, \tag{11}$$

where $\varepsilon_{\min}$ is the minimum permittivity value throughout Ω. Furthermore, an application of the triangle inequality followed by M applications of the Cauchy-Schwarz inequality to the first term of right-hand side of Eq. (10) yields

$$\left| \frac{\mathrm{i}}{\omega} \int_\Omega \sigma_n \nabla \phi_n \cdot \nabla \overline{u_{n+1}} \right| \leq \frac{\sigma_{\max}}{\omega} \|\nabla \phi_n\|_{0,\Omega} \cdot \|\nabla u_{n+1}\|_{0,\Omega}, \tag{12}$$

where $\sigma_{\max}$ is the maximum conductivity value throughout Ω. A similar argument for the second term on the right-hand side of Eq. (10) results in

$$\left| \frac{\mathrm{i}}{\omega} \int_\Omega \sigma \nabla \phi \cdot \nabla \overline{u_{n+1}} \right| \leq \frac{\sigma_{\max}}{\omega} \|\nabla \phi\|_{0,\Omega} \cdot \|\nabla u_{n+1}\|_{0,\Omega}. \tag{13}$$

By letting $\gamma = \sigma_{\max}/(\omega \varepsilon_{\min})$ and assuming that $\|\nabla u_{n+1}\|_{0,\Omega} \neq 0$, a combination of (10), (11), (12), and (13) results in $\|\nabla u_{n+1}\|_{0,\Omega} \leq \gamma(\|\nabla \phi_n\|_{0,\Omega} + \|\nabla \phi\|_{0,\Omega})$, and hence,

$$\|\nabla u_{n+1}\|_{0,\Omega} \leq \gamma^{n+1}(\|\nabla \phi_0\|_{0,\Omega} + \|\nabla \phi\|_{0,\Omega}). \tag{14}$$

If $\gamma < 1$, then $\|\nabla u_{n+1}\|_{0,\Omega} \to 0$ as $n \to \infty$, which means that $\nabla u_{n+1} \to 0$ almost everywhere in Ω. Provided that Γ_{G} is of nonvanishing Hausdorff measure, an application of the Poincaré inequality followed by an application of (14) yields

$$\|u_{n+1}\|_{0,\Omega} \leq C \|\nabla u_{n+1}\|_{0,\Omega} \leq C\gamma^{n+1}(\|\nabla \phi_0\|_{0,\Omega} + \|\nabla \phi\|_{0,\Omega}) \to 0 \tag{15}$$

as $n \to \infty$, with $C \in (0, +\infty)$ depending only on Ω and Γ_{G}. To summarize, given $\gamma < 1$, it follows that $\|\phi_n - \phi\|_{1,\Omega}^2 = \|\phi_n - \phi\|_{0,\Omega}^2 + \|\nabla \phi_n - \nabla \phi\|_{0,\Omega}^2 \to 0$ as $n \to \infty$, and the proof is complete.

Although the condition on γ, or equivalently the condition $\sigma_{\max} < \omega \varepsilon_{\min}$, appears to be rather hard to satisfy under the presence of perfect electric conductors, there are various ways to overcome most of the difficulties that arise in practical applications, as is demonstrated in the following section.

4 Numerical Explorations

4.1 Reference Simulation

The simulations are performed using the upper half of the cylindrically symmetric device that is depicted in Fig. 1, with homogeneous Neumann boundary data on the boundary that is formed after the truncation. The height of the cylinder is $\ell = 120$ mm, while its radius is $r = 32.5$ mm. The values of the material functions are given in Table 1, while the conductivity of the material in $\Omega^{(0)}$ is modeled by the logistic function

$$\sigma^{(0)}(|\mathbf{E}|) = \frac{\sigma_0}{1 + \mathrm{e}^{-\alpha \cdot (|\mathbf{E}| - E_0)}}, \tag{16}$$

where $\sigma_0 = 10^{-9}$ S/m, $\alpha = 10^{-5}$ m/V, and $E_0 = 10^6$ V/m. The potential that is supplied through Γ_S is $s = 10^5$ V, while the frequency of the excitation is $f = 0.1$ Hz. Mark that the value of the parameter E_0 has been chosen to be approximately 170 V/m higher than s/ℓ, which corresponds to the electric field strength of a plate capacitor whose ℓ-separated plates are at voltage s.

Theorem 1 holds both for linear problems and nonlinear problems whose conductivity is a nonlinear function that is bounded from above, such as the logistic function in Eq. (16). In case of nonlinear problems, a harmonic balance expansion can be used for the transition to the frequency domain, with Theorem 1 being required to be valid for each frequency $\omega_k = k\omega$, with integer $k \geq 1$ and ω being the fundamental frequency. Since the goal here is to verify Theorem 1 and assess the performance of the ICU scheme, it is evident that if $\sigma_{\max}$ throughout the computational domain is smaller than $\omega\varepsilon_{\min}$, then it is also smaller than any k-multiple of $\omega\varepsilon_{\min}$, and hence, only the fundamental frequency needs to be examined in the numerical experiments.

To obtain reference solutions, fixed-point iterations are employed with stopping criterion $e_n < \delta = 10^{-8}$. The simulation is performed using a nearly uniform triangulation whose average edge-length is approximately $1.5 \cdot 10^{-4}$ m. The resulting mesh constitutes of 72 000 vertices, and since the finite element method with Lagrangian elements of unity order is used for the sought scalar potential, the number of vertices coincides with the number of degrees of freedom. The initial guess $\phi_0 \in U_s$ is obtained by solving the associated electrostatic problem, that is, Problem (3) for $\sigma = 0$ everywhere in Ω. The solver requires 13 s and 8 nonlinear iterations to satisfy the given tolerance, with $e_8 \approx 4.5 \cdot 10^{-9}$, while a direct solver is used within each iteration. The results are shown in Fig. 2, where the real and imaginary part of the resulting potential ϕ_8 are shown in (A) and (B), respectively. In the same figure, (C) and (D) depict the electric field strength of the real and imaginary part upon numerical convergence.

Table 1. Material Properties

	$\Omega^{(0)}$	$\Omega^{(1)}$	$\Omega^{(2)}$	$\Omega^{(3)}$	$\Omega^{(4)}$
$\varepsilon_\mathrm{r}^{(m)}$	4.0	4.0	1.0	1.0	4.0
$\sigma^{(m)}$ (S/m)	Eq. (16)	10^{-12}	$4.0 \cdot 10^{-12}$	0.0	10^7

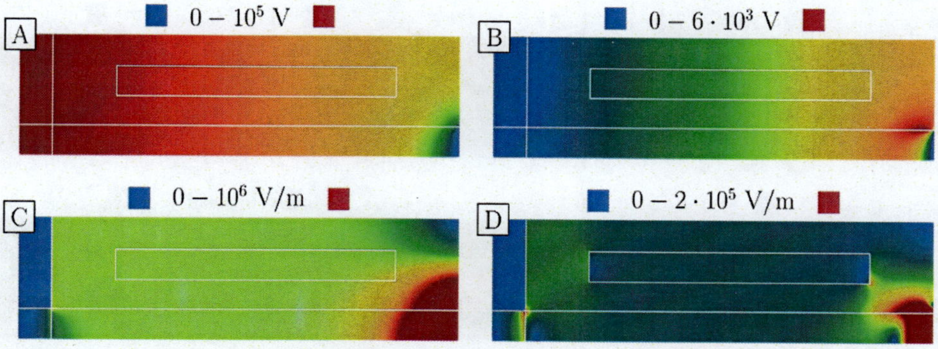

Fig. 2. Reference simulation results that have been obtained with fixed-point iterations. In (A) and (B), the real and imaginary part of the scalar EQS potential is shown, while (C) and (D) depict the electric field strength of the real and imaginary parts, respectively. Mark that the visualization range has been truncated, and hence, the mono-colored regions in the vicinity of the lower right corners of the electric field strength visualizations do not correspond to equal values but to truncated ones.

4.2 The Case of Perfect Electric Conductors

Since $\sigma_{\max} = \sigma^{(4)} = 10^7$ S/m, $\gamma = \sigma_{\max}/(\omega\varepsilon_{\min}) \gg 1$, a direct application of the ICU scheme is not possible. On the other hand, Fig. 2 suggests that the problem in the conducting domain $\Omega^{(4)}$ is nearly static, with $\mathrm{Re}(\phi|\Omega^{(4)}) \approx s$ and $\mathrm{Im}(\phi|\Omega^{(4)}) \approx 0$. Hence, to recover the convergence of the ICU scheme, the nearly perfectly conducting domain $\Omega^{(4)}$ can be removed from the computational domain, while the supplying boundary condition $\phi|\Gamma_{\mathrm{S}} = s$ can be imposed on the newly formed boundaries. The results from this strategy are shown in Fig. 3(A) and (B), with (A) depicting the relative difference between the real part of the potentials that have been obtained with fixed point iterations and the ICU scheme and (B) showing the associated relative difference for the imaginary part. The ICU solver achieves numerical convergence for the same tolerance that has been used for the fixed point iterations, $\delta = 10^{-8}$, after 18 s and 26 iterations. An alternative approach that does not require mesh modifications is show in Fig. 3(C) and (D), where the domain $\Omega^{(4)}$ is kept as part of the computational domain but with modified material parameters so that a nearly static problem is approximated within $\Omega^{(4)}$, as often done for modelling floating potentials [6]. Here, the values $\sigma^{(4)} = 0$ S/m and $\varepsilon^{(4)} = 10^3$ F/m are used. Mark that this approach results in significantly improved accuracy, as depicted in Fig. 3(C) and (D). In the latter case, the ICU solver needs 19 s and 26 iterations for satisfying the tolerance criterion $e_n < \delta = 10^{-8}$.

The latter approach is adopted for the rest of the experiments in this work. Mark that, since the studied device constitutes of weakly conducting material and the conductivity of the perfect electric conductor is set to zero, convergence depends now on the maximum conductivity σ_0 of the nonlinear material, that is, $\gamma = \sigma_0/(\omega\varepsilon_{\min}) < 1$. In the following subsections, numerical verification of the condition $\sigma_0 < \omega\varepsilon_{\min}$ is presented.

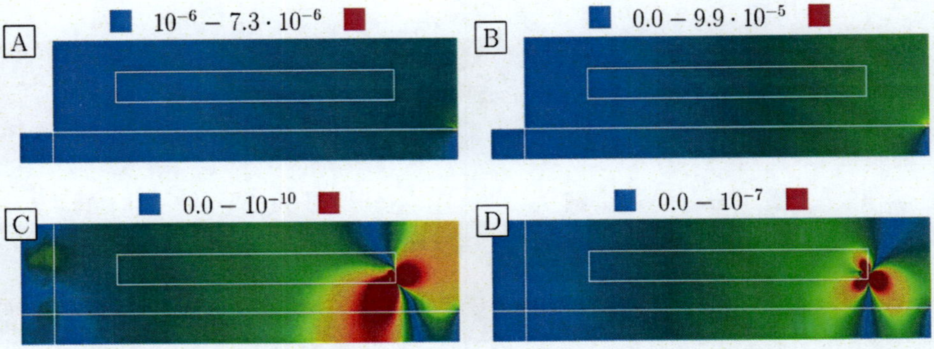

Fig. 3. Relative differences between the real, (A) and (C), and imaginary parts, (B) and (D), of the EQS potentials that have been obtained with fixed point iterations and the ICU scheme. In (A) and (B), the computational domain has been modified, while in (C) and (D) the static problem in $\Omega^{(4)}$ is approximated by setting $\sigma^{(4)} = 0\,\mathrm{S/m}$ and $\varepsilon^{(4)} = 10^3\,\mathrm{F/m}$.

4.3 Numerical Convergence Study

As mentioned below Eq. (16), let $\sigma_0 = 10^{-7}$ S/m, $\alpha = 10^{-5}$ m/V, and $E_0 = 10^6$ V/m; hence, the ICU scheme is expected to be convergent for all frequencies $f > \sigma_0/(2\pi\varepsilon_{\min}) \approx 1800$ Hz, although this estimate is tight only when the maximum value σ_0 is attained. The maximum attained conductivity value is approximately 10^{-9} S/m, which actually results in the lowest frequency being in the order of 10 Hz. In Table 2, the number of iterations that are required by the ICU scheme for $\delta = 10^{-8}$ is depicted for frequencies $f \in \{20, 40, \ldots, 100\}$ Hz together with the total time until convergence.

Table 2. Convergence History and Timing for $\sigma_0 = 10^{-7}$ S/m

	$f = 20$ Hz	$f = 40$ Hz	$f = 60$ Hz	$f = 80$ Hz	$f = 100$ Hz
#1	$3.4 \cdot 10^{-3}$	$1.7 \cdot 10^{-3}$	$1.1 \cdot 10^{-3}$	$8.5 \cdot 10^{-4}$	$6.8 \cdot 10^{-4}$
#2	$6.8 \cdot 10^{-5}$	$1.7 \cdot 10^{-5}$	$7.6 \cdot 10^{-6}$	$4.3 \cdot 10^{-6}$	$2.7 \cdot 10^{-6}$
#3	$7.6 \cdot 10^{-6}$	$9.6 \cdot 10^{-7}$	$2.8 \cdot 10^{-7}$	$1.2 \cdot 10^{-7}$	$6.1 \cdot 10^{-8}$
#4	$2.6 \cdot 10^{-7}$	$1.9 \cdot 10^{-8}$	$3.6 \cdot 10^{-9}$	$1.2 \cdot 10^{-9}$	$4.7 \cdot 10^{-10}$
#5	$1.6 \cdot 10^{-7}$	$4.7 \cdot 10^{-9}$	–	–	–
#6	$1.6 \cdot 10^{-8}$	–	–	–	–
#7	$9.6 \cdot 10^{-9}$	–	–	–	–
Timing	5.3 s	3.8 s	3.0 s	3.0 s	3.0 s

5 Conclusions

An iterative charge-update (ICU) scheme has been introduced as an alternative to the electro-quasistatic (EQS) scalar potential formulation for high-voltage engineering problems and, in particular, for nonlinear problems. The convergence criterion $\sigma_{\max} < \omega\varepsilon_{\min}$ has been derived and strategies have been presented to satisfy this criterion when perfect electric conductors are part of the domain. The proposed ICU scheme avoids reassembling the system matrix for each nonlinear iteration. Instead, only the right-hand side has to be reassembled, and hence, the ICU scheme is expected to be beneficial when the assembly time possesses a significant part of the total computational time, that is, for large-scale problems.

Acknowledgements. This work was supported by the Deutsche Forschungsgemeinschaft (DFG) under grant no. CL143/11-2 and CL143/18-1.

References

1. Egiziano, L., Tucci, V., Petrarca, C., Vitelli, M.: A Galerkin model to study the field distribution in electrical components employing nonlinear stress grading materials. IEEE Trans. Dielectr. Electr. Insul. **6**, 765–73 (1999)
2. Christen, T., Donzel, L., Greuter, F.: Nonlinear resistive electric field grading part 1: theory and simulation. IEEE Electr. Insul. Mag. **26**, 47–59 (2010)
3. Braess, D.: Finite Elements: Theory, Fast Solvers, and Applications in Solid Mechanics. Cambridge University Press, Cambridge (2007)
4. Badics, Z.: Charge density-scalar potential formulation for adaptive time-integration of nonlinear electro-quasistatic problems. IEEE Trans. Mag. **47**, 1338–1341 (2011)
5. Jörgens, C., Clemens, M.: Fast calculation of steady-state charge distribution in high voltage power cables. Int. J. Numer. Model. El. **33**, e2713 (2020)
6. Konrad, A., Graovac, M.: The finite element modeling of conductors and floating potentials. IEEE Trans. Mag. **32**, 4329–4331 (1996)

Check for updates

Electrostatic Forces on Conductors with Boundary Element Methods in 3D

Piyush Panchal(✉) and Ralf Hiptmair

Seminar of Applied Mathematics, ETH Zürich, Zürich, Switzerland
{piyush.panchal,ralf.hiptmair}@sam.math.ethz.ch

Abstract. In this work we are concerned with computing local/global electrostatic forces and torques on perfect electrical conductors using the boundary element method (BEM). Classical boundary based force functionals are not continuous on energy trace spaces and therefore offer low accuracy and convergence rates. Following the work [P. PANCHAL AND R. HIPTMAIR, *Electrostatic Force Computation with Boundary Element Methods*, the SMAI journal of computational mathematics, 8 (2022), pp. 49–74], we derive a similar force expression starting from a floating potential problem for conducting bodies. The computations are done by employing the Virtual Work Principle using shape calculus and the adjoint method. The final expression is structurally simple and can be evaluated without explicitly computing the adjoint solution. It enjoys superior accuracy and convergence rates compared to standard formulas which is demonstrated by means of numerical experiments.

1 Introduction

Numerical computation of Electrostatic forces and torques is of interest in the design of electro-mechanical devices. Classical approaches for computing these quantities rely on either boundary-based, or volume-based formulas (also called egg-shell approach), both of which derive from the Maxwell Stress Tensor [1, Sec. 6.9], [2, Sec. 8.2]. The volume-based approaches tend to be numerically superior to their boundary-based counterpart as seen in [3, Section 1]. Our new boundary integral equation (BIE) constrained shape derivative approach presented in [3] yields a superior boundary-based formula which outperforms even the volume-based approach, making it an attractive option for use with a BEM discretization. In this work we follow a similar approach, starting with the assumption that all bodies are conducting and then employing a floating potential problem [4] posed on an unbounded domain.

1.1 Floating Potential Model Problem

We have two solid perfectly conducting objects occupying bounded Lipschitz domains $D, B \subset \mathbb{R}^3$ as shown in Fig. 1. The complement $M := \mathbb{R}^3 \setminus (\overline{B} \cup \overline{D})$ is filled with a homogeneous isotropic dielectric medium with $\varepsilon \equiv 1$ after rescaling. The solid object B is grounded (at electrostatic potential 0), whereas the solid object D has a known net electric charge $Q \in \mathbb{R}$. Writing ∂B and ∂D for the boundaries of these conducting objects,

M. van Beurden et al. (Eds.): SCEE 2022, MI 43, pp. 102–110, 2024.
https://doi.org/10.1007/978-3-031-54517-7_12

the electrostatic potential u can be obtained as the weak solution in $H^1(M)$[1] of the Laplace boundary value problem (BVP) on the unbounded domain M:

$$\begin{aligned} \Delta u &= 0 && \text{in} \quad M, \\ u &= 0 && \text{on} \quad \partial B, \\ u &= c && \text{on} \quad \partial D, \\ \int_{\partial D} \nabla u \cdot \boldsymbol{n} \, \mathrm{d}S &= -Q && \text{on} \quad \partial D, \end{aligned} \tag{1}$$

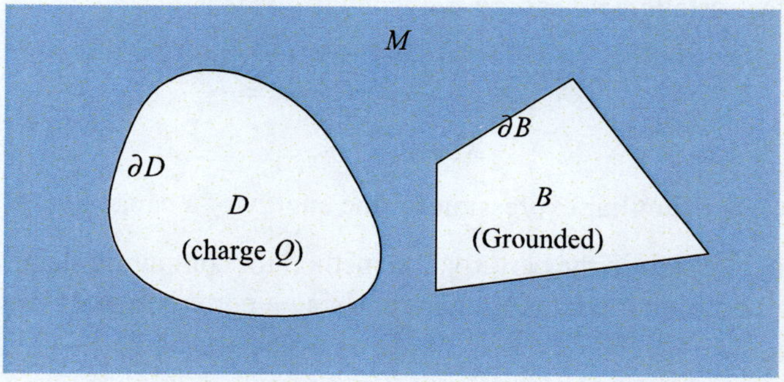

Fig. 1. Geometric setting

where $c \in \mathbb{R}$ is the unknown constant potential of the conducting body D and $\boldsymbol{n}$ is the exterior unit normal vector field on $\Gamma := \partial D \cup \partial B$. The potential u satisfies the decay conditions $u(\boldsymbol{x}) = \mathcal{O}(\|\boldsymbol{x}\|^{-1})$ for $\|\boldsymbol{x}\| \to \infty$.

The floating potential problem (1) can be equivalently formulated on the boundary in terms of traces of the potential u, namely the Dirichlet trace $u|_\Gamma$ and the Neumann trace $\nabla u|_\Gamma \cdot \boldsymbol{n}$. Writing $c\mathbb{1}_{\partial D}$ for the Dirichlet trace and ψ for the Neumann trace, the variational formulation reads: seek $\psi \in H^{-\frac{1}{2}}(\Gamma)$, $c \in \mathbb{R}$ such that

$$\begin{aligned} \int_\Gamma \int_\Gamma \mathrm{G}(\boldsymbol{x},\boldsymbol{y})\psi(\boldsymbol{x})\phi(\boldsymbol{y}) \, \mathrm{d}S(\boldsymbol{y})\mathrm{d}S(\boldsymbol{x}) + c \int_\Gamma \mathbb{1}_{\partial D}(\boldsymbol{x})\phi(\boldsymbol{x}) \, \mathrm{d}S(\boldsymbol{x}) &= 0 \qquad \forall \phi \in H^{-\frac{1}{2}}(\Gamma), \\ d \int_\Gamma \mathbb{1}_{\partial D}(\boldsymbol{x})\psi(\boldsymbol{x}) \, \mathrm{d}S(\boldsymbol{x}) &= -d\,Q \quad \forall d \in \mathbb{R}, \end{aligned} \tag{2}$$

where $\mathrm{G}(\boldsymbol{x},\boldsymbol{y})$ is the fundamental solution for the Laplace operator.

Remark 1. We note that (2) has a saddle-point structure. The double integral in the first equation is precisely the bilinear form associated with the single layer boundary integral operator (BIO) which is bounded and elliptic on $H^{-\frac{1}{2}}(\Gamma)$, and we refer to [5,

[1] We adopt the convention of [5, Sec. 2.3 & Sec. 2.4] for function spaces and Sobolev spaces: $W^{k,p}(\Omega), H^1(\Omega), H^{\frac{1}{2}}(\Omega), L^2(\Omega), C^k(\Omega)$ etc., where Ω denotes a generic domain.

Thm. 3.5.3], [6] for the proof. The bilinear form arising from the fixed charge constraint is bounded on $H^{-\frac{1}{2}}(\Gamma) \times \mathbb{R}$, which is trivial considering the $L^2(\Gamma)$ duality pairing in the expression. It satisfies the stability condition [6, Thm. 3.11] which is trivial to prove for the scalar constraint. Thus from [6, Thm. 3.11] we have the unique solvability of (2).

2 Forces via Shape Differentiation

The Virtual Work Principle [7–11] tells us that the force can be recovered via the *shape derivative* of the total energy. For the electrostatic system (1), the total energy consists of only the electrostatic field energy $\mathscr{E}_F$ which is given as

$$\mathscr{E}_F = \frac{1}{2} \int_{\mathbb{R}^3} \|\nabla u(\boldsymbol{x})\|^2 \, d\boldsymbol{x} = \frac{c\,Q}{2}. \tag{3}$$

Notice that it is the familiar expression for the energy of a capacitor.

Perturbation Method: In the perturbation method for computing shape derivatives [12, Sec. 2.8], we start with a perturbation map. Using a perturbation field $\boldsymbol{v} \in C_0^\infty(\mathbb{R}^3;\mathbb{R}^3)$, it is defined as

$$\mathrm{T}^s_{\boldsymbol{v}} : \mathbb{R}^3 \to \mathbb{R}^3, \quad \mathrm{T}^s_{\boldsymbol{v}}(\boldsymbol{x}) := \boldsymbol{x} + s\,\boldsymbol{v}(\boldsymbol{x}), \quad s \in \mathbb{R}. \tag{4}$$

The implicit function theorem guarantees the existence of a $\delta(\boldsymbol{v})$ such that $\mathrm{T}^s_{\boldsymbol{v}}$ is a C^∞ diffeomorphism for $|s| < \delta(\boldsymbol{v})$. Using the perturbation map, we define a set of admissible domains $\mathscr{A}_{\boldsymbol{v}}(\Omega_0)$ using perturbations of a reference domain $\Omega_0 \subset \mathbb{R}^3$ as $\mathscr{A}_{\boldsymbol{v}}(\Omega_0) := \{\Omega_s := \mathrm{T}^s_{\boldsymbol{v}}(\Omega_0),\ s \in (-\delta(\boldsymbol{v}), \delta(\boldsymbol{v}))\}$. A shape functional J is then a map $J : \mathscr{A}_{\boldsymbol{v}}(\Omega_0) \to \mathbb{R},\ \Omega \mapsto J(\Omega)$. The shape derivative of J at Ω_0 in the direction $\boldsymbol{v}$ is defined as the directional derivative

$$\frac{dJ}{d\Omega}(\Omega_0;\boldsymbol{v}) := \lim_{s\to 0} \frac{J(\mathrm{T}^s_{\boldsymbol{v}}(\Omega_0)) - J(\Omega_0)}{s},$$

if it exists. If, in addition, $\boldsymbol{v} \to \frac{dJ}{d\Omega}(\Omega_0;\boldsymbol{v}) \in \mathbb{R}$ is a distribution on the space of test velocity fields $C_0^\infty(\mathbb{R}^3;\mathbb{R}^3)$, a 1-current as called by de Rham [13, Ch. 3, § 8], then $\Omega \mapsto J(\Omega)$ is called shape-differentiable at Ω_0 and that distribution is the shape derivative. For smooth boundaries, the shape derivatives carry more structure which is captured in the Hadamard Structure Theorem [12, Ch. 9, Thm. 3.6].

BIE on Perturbed Boundary: Let $\Omega_0 := D_0 \cup B_0$ denote the reference domain for our model problem. To make sense of the energy shape functional on the set $\mathscr{A}_{\boldsymbol{v}}(\Omega_0)$, we consider our model problem on the elements of this set. For all values of the shape parameter s, the conducting object $B_s := \mathrm{T}^s_{\boldsymbol{v}}(B_0)$ is grounded and $D_s := \mathrm{T}^s_{\boldsymbol{v}}(D_0)$ carries the net charge Q. The variational form of the floating potential problem on $\Omega_s := D_s \cup B_s$ reads: seek $(\psi_s, c_s) \in V_s := H^{-\frac{1}{2}}(\Gamma_s) \times \mathbb{R}$ such that

$$\mathrm{A}(s;(\psi_s,c_s),(\phi,d)) = -d\,Q \quad \forall (\phi,d) \in V_s, \tag{5}$$

where $\Gamma_s := \partial\Omega_s$ and $\mathrm{A} : (-\delta(\boldsymbol{v}), \delta(\boldsymbol{v})) \times V_s \times V_s \to \mathbb{R}$ is defined as

$$\mathrm{A}(s;(\psi,c),(\phi,d)) := \int\limits_{\Gamma_s}\int\limits_{\Gamma_s} \mathrm{G}(\boldsymbol{x},\boldsymbol{y})\psi(\boldsymbol{x})\phi(\boldsymbol{y})\,\mathrm{d}S(\boldsymbol{y})\mathrm{d}S(\boldsymbol{x})$$
$$+d\int\limits_{\Gamma_s} \mathbb{1}_{\partial D}(\boldsymbol{x})\psi(\boldsymbol{x})\,\mathrm{d}S(\boldsymbol{x}) + c\int\limits_{\Gamma_s} \mathbb{1}_{\partial D}(\boldsymbol{x})\phi(\boldsymbol{x})\,\mathrm{d}S(\boldsymbol{x}).$$

The electrostatic field energy for the s configuration is denoted by $\mathscr{E}_F(s)$ and taking cue from (3) it is defined as

$$\mathscr{E}_F(s) := \frac{c_s\,Q}{2}. \tag{6}$$

Equivalent BIE on Reference Boundary: To compute shape derivative, we need an equivalent formulation on the reference boundary Γ_0. This is achieved using a transformation and a pullback. In the first step we transform all integrals on Γ_s to Γ_0 using the perturbation map (4) and the identity [12, Ch. 9, Sec. 4.2, eq. 4.9]:

$$\int\limits_{\Gamma_s} f\,\mathrm{d}S = \int\limits_{\Gamma_0} f \circ \mathrm{T}^s_{\boldsymbol{v}}\ \omega_s\,\mathrm{d}S, \quad \boldsymbol{y} = \mathrm{T}^s_{\boldsymbol{v}}(\boldsymbol{x}),$$

where the Jacobian of transformation is given as $\omega_s := \|\mathbf{C}(\mathrm{D}\mathrm{T}^s_{\boldsymbol{v}})\,\boldsymbol{n}_0\|$, $\boldsymbol{n}_0$ is the exterior unit normal vector field on Γ_0 and $\mathbf{C}(\mathbf{A})$ denotes the cofactor matrix of $\mathbf{A}$. The matrix $\mathrm{D}\mathrm{T}^s_{\boldsymbol{v}}$ is the Jacobian matrix and is defined as $[\mathrm{D}\mathrm{T}^s_{\boldsymbol{v}}]_{i,j} = \partial_j[\mathrm{T}^s_{\boldsymbol{v}}]_i$, $i,j = 1,2,3$, where $[\mathrm{T}^s_{\boldsymbol{v}}]_i$ denotes the ith component of $\mathrm{T}^s_{\boldsymbol{v}}$. The next step is to get rid of function spaces on Γ_s which we accomplish using a pullback for surface charge densities:

$$\hat{\psi} \in H^{-\frac{1}{2}}(\Gamma_0): \quad \hat{\psi} := (\psi \circ \mathrm{T}^s_{\boldsymbol{v}})\,\omega_s, \quad \psi \in H^{-\frac{1}{2}}(\Gamma_s). \tag{7}$$

Since $\mathrm{T}^s_{\boldsymbol{v}}$ is a smooth mapping and $\omega_s \in L^\infty(\Gamma_0)$, the trace spaces are preserved under pullback. We skip the details since the procedure is identical to [3, Sec. 4] and write the equivalent formulation to (5): we seek $\hat{\psi}_s, c_s \in V_0 = H^{-\frac{1}{2}}(\partial\Gamma_0) \times \mathbb{R}$ such that

$$\hat{\mathrm{A}}(s;(\hat{\psi}_s,c_s),(\hat{\phi},d)) = -d\,Q \quad \forall(\hat{\phi},d) \in V_0, \tag{8}$$

where $\hat{\mathrm{A}} : (-\delta(\boldsymbol{v}),\delta(\boldsymbol{v})) \times V_0 \times V_0 \to \mathbb{R}$ is defined as

$$\hat{\mathrm{A}}(s;(\hat{\psi},c),(\hat{\phi},d)) := \int\limits_{\Gamma_0}\int\limits_{\Gamma_0} \mathrm{G}(\mathrm{T}^s_{\boldsymbol{v}}(\boldsymbol{x}),\mathrm{T}^s_{\boldsymbol{v}}(\boldsymbol{y}))\ \hat{\psi}(\boldsymbol{x})\ \hat{\phi}(\boldsymbol{y})\,\mathrm{d}S(\boldsymbol{x})\,\mathrm{d}S(\boldsymbol{y})$$
$$+c\int\limits_{\partial D_0} \hat{\psi}(\boldsymbol{x})\,\mathrm{d}S(\boldsymbol{x}) + d\int\limits_{\partial D_0} \hat{\phi}(\boldsymbol{x})\,\mathrm{d}S(\boldsymbol{x}).$$

Since the field energy $\mathscr{E}_F(s)$ from (6) depends on the solution c_s to (8) it will be useful to define the following linear functional

$$\hat{J} : V_0 \to \mathbb{R}, \quad \hat{J}((\hat{\psi},c)) := \frac{cQ}{2}. \tag{9}$$

Constrained Shape Derivative via Adjoint Method: For computing the energy shape derivative with the variational constraint (8) we use the well known adjoint approach from literature [14, Sect. 1.6.4]. We start by defining the Lagrangian $\mathscr{L}: (-\delta(\boldsymbol{v}),\delta(\boldsymbol{v}))\times V_0\times V_0\to\mathbb{R}$,

$$\mathscr{L}(s;(\hat{\psi},c),(\hat{\phi},d)) := \hat{\mathrm{A}}(s;(\hat{\psi},c),(\hat{\phi},d)) + dQ + \hat{J}((\hat{\psi},c)). \tag{10}$$

We observe that by plugging in the state solution $(\hat{\psi},c)=(\hat{\psi}_s,c_s)$ we get

$$\mathscr{L}(s;(\hat{\psi}_s,c_s),(\hat{\phi},d)) = \hat{J}((\hat{\psi}_s,c_s)) = \mathscr{E}_F(s) \quad \forall(\hat{\phi},d)\in H^{-\frac{1}{2}}(\Gamma_0)\times\mathbb{R}. \tag{11}$$

From the above expression, the energy shape derivative can be calculated as

$$\frac{d\mathscr{E}_F}{ds}(0) = \frac{\partial\mathscr{L}}{\partial s}(0;(\hat{\psi}_0,c_0),(\hat{\zeta},e)) = \frac{\partial\hat{\mathrm{A}}}{\partial s}(0;(\hat{\psi}_0,c_0),(\hat{\zeta},e)), \tag{12}$$

where $(\hat{\zeta},e)\in H^{-\frac{1}{2}}(\Gamma_0)\times\mathbb{R}$ solves the adjoint equation

$$< \frac{\partial\mathscr{L}}{\partial(\hat{\psi},c)}(0;(\hat{\psi}_0,c_0),(\hat{\zeta},e));(\hat{\phi},d) >= 0 \quad \forall(\hat{\phi},d)\in H^{-\frac{1}{2}}(\Gamma_0)\times\mathbb{R}. \tag{13}$$

Using the symmetry of $\hat{\mathrm{A}}$ the above expression simplifies to

$$\hat{\mathrm{A}}(0;(\hat{\zeta},e),(\hat{\phi},d)) = -\frac{dQ}{2} \quad \forall(\hat{\phi},d)\in H^{-\frac{1}{2}}(\Gamma_0)\times\mathbb{R}. \tag{14}$$

Comparing with (8) we immediately see that the adjoint solution is the scaled state solution and is given as $(\hat{\zeta},e)=\frac{1}{2}(\hat{\psi}_0,c_0)$ which yields the final form of the shape derivative. It represents the force computed in the direction of $\boldsymbol{v}$ and is denoted as $F(\boldsymbol{v})$. Thus we have

$$F(\boldsymbol{v}) := \frac{1}{2}\int_{\Gamma_0}\int_{\Gamma_0}(\nabla_x \mathrm{G}(\boldsymbol{x},\boldsymbol{y})\cdot\boldsymbol{v}(\boldsymbol{x}) + \nabla_y \mathrm{G}(\boldsymbol{x},\boldsymbol{y})\cdot\boldsymbol{v}(\boldsymbol{y}))\ \hat{\psi}_0(\boldsymbol{x})\ \hat{\psi}_0(\boldsymbol{y})\ \mathrm{d}S(\boldsymbol{x})\mathrm{d}S(\boldsymbol{y}). \tag{15}$$

A detailed analysis of the expression (15) has already been done in [3, Sect. 4.4] where it appears as $\mathbf{T_2}$ in [3, Eq. 4.17]. Since it is a continuous, bilinear mapping on $H^{-\frac{1}{2}}(\Gamma_0)\times H^{-\frac{1}{2}}(\Gamma_0)\to\mathbb{R}$, its Galerkin approximation enjoys superconvergence [3, Prop. 1.2].

For computing electrostatic force on conductors, (15) has two major advantages over the shape derivative formula derived in the companion work [3, Eq. 4.17]. The first advantage is that there is no need to explicitly compute an adjoint solution since it turns out to be a scaled state solution. Second advantage is its considerable simplicity while retaining the superior numerical performance of [3, Eq. 4.17] which will be demonstrated next.

3 Numerical Experiments

3.1 Implementation

The numerical implementation is done in the MATLAB-based *Gypsilab* framework and is available online[2] Given a triangular mesh $\mathcal{M}$ of the boundary Γ (we drop the subscript notation), the trace space $H^{-\frac{1}{2}}(\Gamma)$ is approximated using the space $S_0^{-1}(\mathcal{M})$ which consists of piece-wise constant functions supported on the elements of the mesh. For solving the state problem (8) at $s = 0$, the single-layer boundary integral operator is assembled using semi-analytic integration technique available in Gypsilab. Given an orthonormal basis $\{\beta_1, \beta_2, ..., \beta_N\}$ of $S_0^{-1}(\mathcal{M})$, let $\boldsymbol{\psi} \in \mathbb{R}^N$ represent the coefficients for the basis expansion of the Galerkin solution to (8). Then for a given direction $\boldsymbol{v}$, the approximate force in that direction $F^h(\boldsymbol{v})$ can be computed as

$$F^h(\boldsymbol{v}) = \boldsymbol{\psi}^T \, \mathbf{M} \, \boldsymbol{\psi}, \tag{16}$$

where the entries of $\mathbf{M} \in \mathbb{R}^{N,N}$ are given as

$$[\mathbf{M}]_{i,j} := \frac{1}{2} \int_\Gamma \int_\Gamma (\nabla_x \mathrm{G}(\boldsymbol{x},\boldsymbol{y}) \cdot \boldsymbol{v}(\boldsymbol{x}) + \nabla_y \mathrm{G}(\boldsymbol{x},\boldsymbol{y}) \cdot \boldsymbol{v}(\boldsymbol{y})) \, \beta_i(\boldsymbol{x}) \, \beta_j(\boldsymbol{y}) \, \mathrm{d}S(\boldsymbol{x}) \mathrm{d}S(\boldsymbol{y}). \tag{17}$$

The above singular integral is evaluated using the Sauter and Schwab quadrature technique for triangular panels [5, Ch. 5] using a tensor product quadrature rule with 3^4 quadrature points.

Forces and torques can be computed by plugging-in appropriate velocity fields into the shape derivative formula (15). Cartesian components of the net electrostatic force $\mathbf{F} = (F_1, F_2, F_3)$ acting on the object D can be computed as

$$F_k = F(\boldsymbol{x} \mapsto \mathbf{e}_k \, \chi(\boldsymbol{x})), \tag{18}$$

where $\xi \in C^\infty(\Gamma)$, $\xi|_{\partial B} \equiv 0$, $\xi|_{\partial D} \equiv 1$ and $\mathbf{e}_k$ are the unit vectors pointing along the axes of the cartesian coordinate system. The net torque T about a point $\mathbf{x}_0 \in \mathbb{R}^3$ and axis $\mathbf{a} \in \mathbb{R}^3$ can be computed as

$$T = F(\boldsymbol{x} \mapsto \mathbf{a} \times (\boldsymbol{x} - \boldsymbol{x}_0) \, \chi(\boldsymbol{x})). \tag{19}$$

3.2 Force and Torque Convergence Results

To study convergence a quasi-uniform sequence of mesh partitions $(\mathcal{M})_h$, consisting of triangular panels of Γ with decreasing meshwidth h was employed. Net forces $((F_1^2 + F_2^2 + F_3^2)^{\frac{1}{2}})$ and torques were computed using the Maxwell Stress Tensor based boundary formula [3, Eq. 1.3] for comparison. These computations are done for two different geometries shown in Fig. 2:

---

[2] Code available at https://gitlab.ethz.ch/ppanchal/gypsilab.

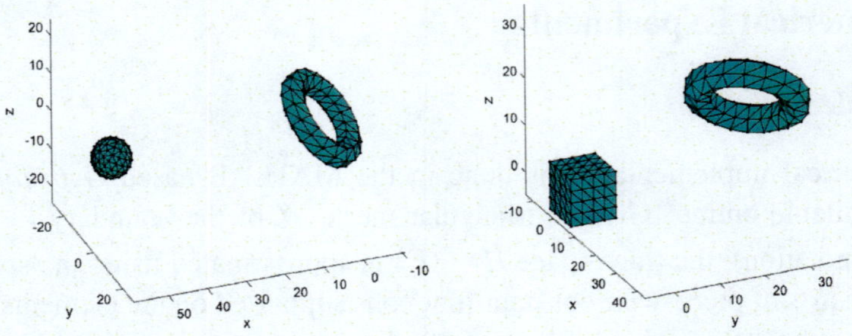

(a) Sphere and torus (b) Cube and torus

Fig. 2. Coarsest mesh of the geometries in the numerical experiments

1. Cube and Torus: The cube shaped conductor D is $(-5,5)^3$ while the grounded torus shaped conductor B has $R = 10$ and $r = 3$, where r is the radius of the tube and R is the distance between center of tube and center of torus. The cube is centered at the origin while the torus is centered at $[25,25,25]$. The torus is rotated about its center by a random orthogonal matrix provided in the code. Reference values are computed using (15) at a refinement level of $h = 0.42$. The results are computed using $Q = 10^2$ and plotted in Fig. 3. We observe the superiority of the shape derivative formula compared to the boundary formula. The difference is stark for the domain with sharp corners which agrees with the observations in [3].

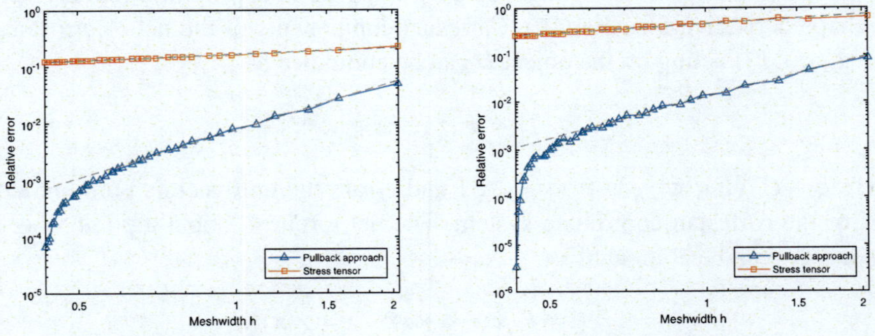

(a) Force computation (b) Torque computation

Fig. 3. Error plots for Cube Torus geometry. Dashed lines represent the linear regression fit.

2. Sphere and Torus: The conductor D is a torus shaped domain with $R = 10$ and $r = 3$, and the grounded conductor B is a sphere of radius 5. The sphere is centered at $[54,0,0]$ whereas the torus is centered at the origin. The torus is rotated about its center by a random orthogonal matrix provided in the code. Reference values are computed using (15) at a refinement level of $h = 0.36$ and using $Q = 10^2$. The results are plotted in Fig. 4. We see that the shape derivative formula is slightly superior to the Maxwell

Stress Tensor formula in absolute accuracy and asymptotic rate of convergence for both force and torque computation for this smooth case (Table 1).

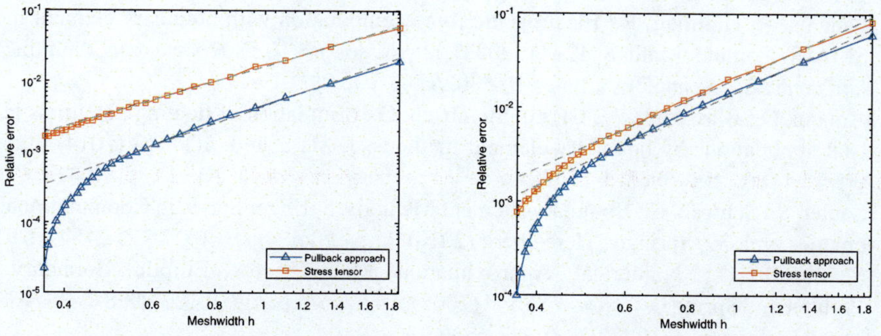

(a) Force computation (b) Torque computation

Fig. 4. Error plots for Sphere Torus geometry. Dashed lines represent the linear regression fit.

Table 1. Fitted asymptotic rates of algebraic convergence

Method	Torus D, sphere B	Cubic D, Torus B	Torus D, sphere B	Cubic D, Torus B
Pullback approach	2.55	2.77	2.51	2.85
Stress tensor	2.14	0.35	2.43	0.65

(a) Forces (b) Torques

4 Conclusion

In this work we derived a simple and stable boundary based expression for computing net force and torque on conducting objects which is suitable for use with BEM. This was achieved by interpreting the Virtual Work Principle via shape calculus and defining the force field as the shape derivative of the electrostatic field energy. The obtained expression (15) turned out to be a continuous force functional on the energy trace spaces, allowing for the superconvergence of its Galerkin approximation, which we observed in the numerical experiments. The derived formula carries a few advantages compared to its companion formula derived for a Dirichlet BVP [3, Eq. 4.17]. It contains only one weakly singular integral and does not require the explicit computation of the adjoint solution.

References

1. Jackson, J.: Classical Electrodynamics, 3rd edn. Wiley, Hoboken (1998)
2. Griffiths, D.J.: Introduction to Electrodynamics. Pearson, London (2013)
3. Panchal, P., Hiptmair, R.: Electrostatic force computation with boundary element methods. SMAI J. Comput. Math. **8**, 49–74 (2022). https://doi.org/10.5802/smai-jcm.79, https://smai-jcm.centre-mersenne.org/articles/10.5802/smai-jcm.79/
4. Amann, D., Blaszczyk, A., Of, G., Steinbach, O.: Simulation of floating potentials in industrial applications by boundary element methods. J. Math. Ind. **4**(1), 13 (2014). https://doi.org/10.1186/2190-5983-4-13
5. Sauter, S., Schwab, C.: Boundary Element Methods. Springer Series in Computational Mathematics, vol. 39. Springer, Heidelberg (2010). https://doi.org/10.1007/978-3-540-68093-2
6. Steinbach, O.: Numerical Approximation Methods for Elliptic Boundary Value Problems. Springer, New York (2008). https://doi.org/10.1007/978-0-387-68805-3, http://dx.doi.org/10.1007/978-0-387-68805-3
7. Bossavit, A.: Forces in magnetostatics and their computation. J. Appl. Phys. **67**(9), 5812–5814 (1990). https://doi.org/10.1063/1.345972
8. Carpentier, A., Galopin, N., Chadebec, O., Meunier, G., Guérin, C.: Application of the virtual work principle to compute magnetic forces with a volume integral method. Int. J. Numer. Modell.: Electron. Netw. Devices Fields **27**(3), 418–432 (2014). https://doi.org/10.1002/jnm.1957, https://onlinelibrary.wiley.com/doi/abs/10.1002/jnm.1957
9. Coulomb, J.: A methodology for the determination of global electromechanical quantities from a finite element analysis and its application to the evaluation of magnetic forces, torques and stiffness. IEEE Trans. Magn. **19**(6), 2514–2519 (1983)
10. Henrotte, F., Hameyer, K.: A theory for electromagnetic force formulas in continuous media. IEEE Trans. Magn. **43**(4), 1445–1448 (2007). https://doi.org/10.1109/TMAG.2007.892457
11. Henrotte, F., Hameyer, K.: Computation of electromagnetic force densities: Maxwell stress tensor vs. virtual work principle. J. Comput. Appl. Math. **168**(1–2), 235–243 (2004). https://doi.org/10.1016/j.cam.2003.06.012, http://dx.doi.org/10.1016/j.cam.2003.06.012
12. Delfour, M., Zolésio, J.P.: Shapes and Geometries. Advances in Design and Control, vol. 22, 2nd edn. SIAM, Philadelphia (2010)
13. de Rham, G.: Differentiable Manifolds. Grundlehren der Mathematischen Wissenschaften [Fundamental Principles of Mathematical Sciences], vol. 266. Springer, Berlin (1984). https://doi.org/10.1007/978-3-642-61752-2. Forms, currents, harmonic forms, Translated from the French by F. R. Smith, With an introduction by S. S. Chern
14. Hinze, M., Pinnau, R., Ulbrich, M., Ulbrich, S.: Optimization with PDE Constraints. Mathematical Modelling: Theory and Applications, vol. 23. Springer, New York (2009). https://doi.org/10.1007/978-1-4020-8839-1

25 Years Computational Electromagnetics @ SCEE

Ursula van Rienen[1,2,3(✉)]

[1] Institute of General Electrical Engineering, Faculty of Computer Science and Electrical Engineering, University of Rostock, Rostock, Germany
ursula.van-rienen@uni-rostock.de

[2] Department Life, Light and Matter, Interdisciplinary Faculty, University of Rostock, Rostock, Germany

[3] Department of Ageing of Individuals and Society, Interdisciplinary Faculty, University of Rostock, Rostock, Germany

Abstract. From the beginning, Computational Electromagnetics (CEM) was among the core themes of the conference series on Scientific Computing in Electrical Engineering (SCEE). This invited contribution to the 25th anniversary of the SCEE sheds light on some selected highlights in Computational Electromagnetics as presented during the past 25 years. In this context, CEM comprises different challenges to applied mathematics, such as the treatment of partial differential equations or the numerical solution of linear systems.

After some overview of the number of CEM contributions over the years and the type of discretisation methods employed, one example contribution is described for each edition of the conference series. Typically, these were invited papers.

1 General Introduction

Computational Electromagnetics comprises different challenges to applied mathematics, such as solving partial differential equations derived from Maxwell's equations or the numerical solution of linear systems. Thus, a close interaction between mathematicians and electrical engineers is of fundamental importance for the progress of CEM. One aim of the SCEE conferences is to inform engineers about current methods from applied mathematics and, on the other hand, to make mathematicians aware of difficulties posed to their methods by problems arising in CEM. The score is as comprehensive as both sciences reach; therefore, only some highlights can be set, while the contributions over the last 25 years covered many more topics. Many of them can be found in the previous post-conference books.

First, let us regard an abundance distribution. Contributions could be attributed to CEM based on the programmes and abstract books of the previous SCEE conferences[1] Even though in most years, no classification was made, especially for the posters, contributions were selected according to our best knowledge and sorted out generously if something referred to Circuit Design, Coupled Problems or Mathematical Methods, for

[1] see https://scee-conferences.org/pages/foundation-publications.

M. van Beurden et al. (Eds.): SCEE 2022, MI 43, pp. 111–122, 2024.
https://doi.org/10.1007/978-3-031-54517-7_13

example. Since there is a great deal of overlap in the topical areas of the SCEE and some topics only evolved from a few single CEM contributions in the beginning to an explicit topical area, this classification is not unique. The showcase position of CEM always occupying at least around 30 % of all contributions is valid over the course of the years, as visible in Fig. 1.

A further classification of the 302 contributions, previously filtered out as CEM, was achieved by screening the titles and abstracts. It was pretty straightforward for 40 % of the contributions. Figure 1 shows this share's distribution of discretisation methods. It can be observed that most of the contributions concern the Finite Element Method (FEM), including Discontinuous Galerkin (DG). The second-highest share regards the Finite Integration Technique (FIT), followed by Boundary Element Methods (BEM), partly coupled with FEM (BEM+FEM). The remaining contributions regard the Finite Difference Time Domain (FDTD), Finite Volume Method (FVM), and Particle-in-Cell method (PIC and FVPIC). The prominence of FEM, FIT, and BEM has been visible over the years.

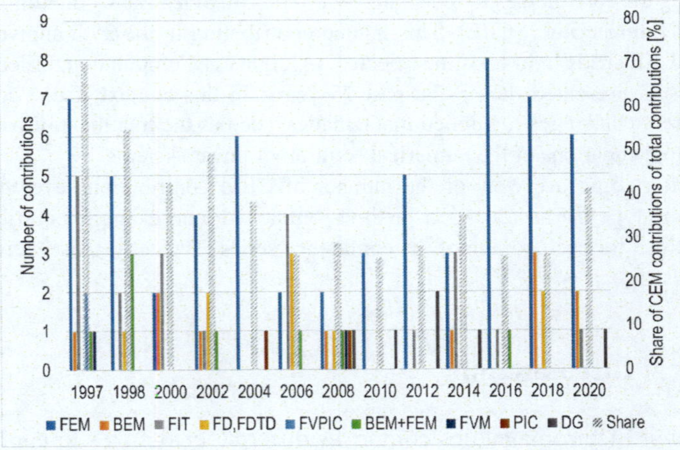

Fig. 1. Left Axis: Distribution of discretisation methods in the share of 302 CEM contributions from 1997 till 2020: The Finite Element Method (FEM), Boundary Element Methods (BEM), Finite Integration Technique (FIT), Finite Difference and Finite Difference Time Domain (FD, FDTD), Finite Volume Particle-in-Cell (FVPIC), BEM coupled with FEM (BEM+FEM), Finite Volume Method (FVM), Particle-in-Cell (PIC), and Discontinuous Galerkin (DG). Right axis: Share of CEM contributions in the total number of SCEE contributions per year, displayed as grey streaked bars. The support by Hendrikje Raben in this evaluation is highly acknowledged.

Common to all these methods is the starting point in Maxwell's equations. In differential and integral form, they read as

$$
\begin{aligned}
\nabla\times\vec{E} &= -\frac{\partial\vec{B}}{\partial t} \qquad & \oint_{\partial A}\vec{E}\cdot d\vec{s} &= -\iint_{\partial V}\frac{\partial\vec{B}}{\partial t}\cdot d\vec{A} \\
\nabla\times\vec{H} &= \frac{\partial\vec{D}}{\partial t}+\vec{J} \qquad & \oint_{\partial A}\vec{H}\cdot d\vec{s} &= \iint_{\partial V}\left(\frac{\partial\vec{D}}{\partial t}+\vec{J}\right)\cdot d\vec{A} \\
\nabla\cdot\vec{D} &= \rho \qquad & \oint_{\partial V}\vec{D}\cdot d\vec{A} &= \iiint_{V}\rho\, dV \\
\nabla\cdot\vec{B} &= 0 \qquad & \oint_{\partial V}\vec{B}\cdot d\vec{A} &= 0
\end{aligned}
\tag{1}
$$

Well-known, Maxwell's equations (1) comprise Faraday's law linking the electric field $\vec{E}$ with the time derivative of the magnetic flux density $\vec{B}$ (top row), the Ampère-Maxwell law relating the magnetic field $\vec{H}$ with the current density $\vec{J}$ and the time derivative of the electric flux density $\vec{D}$ (second row). In addition, Gauss' laws for electric and magnetic fields (third and fourth row) provide electric charges ρ as electric sources and state the source-freeness of magnetic fields. Several of the discretisation methods for Maxwell's equations mentioned above are well-understandably described in [1].

1.1 Finite Element Method

Before showing examples of the wide variety of invited talks and some selected other contributions over the 25 years in chronological but otherwise arbitrary order, we will briefly shed light on the FEM and the FIT as primary discretisation methods for Maxwell's equations.

In the numerical solution of differential equations by the FEM, the solution domain is decomposed into smaller subdomains called finite elements, for example, triangles in 2D or tetrahedra in 3D.

Regard the equation $L[f]=s$ with an operator L, the source s, and the unknown function f in the region Ω. The main idea behind the FEM is approximating the solution f by a linear combination of basis functions $\varphi_i, i=1,\dots,n$ (usually low-order polynomials) associated with, for example, local element nodes, edges or faces:

$$
f(\vec{r}) \approx \sum_{i=1}^{n} f_i\,\varphi_i(\vec{r}) \tag{2}
$$

Their coefficients $f_i, i=1,\dots,n$ are obtained from a variational problem. The weak form is derived from the strong form of the partial differential equation by applying suitable test functions to the entire domain. Then, the basic steps are as follows:

Minimise the residual $r=L[f]-s$. In the so-called weak sense, it shall be zero; a weighted average is set to zero.

Choose weighting functions $w_i, i=1,\dots,n$ for weighting the residual r. In Galerkin's method, the weighting functions equal the basis functions, $w_i=\varphi_i$.

Set the weighted residuals to zero and solve for the unknowns f_i; that is, solve the set of equations

$$\langle w_i, r\rangle = \int_{\Omega} w_i r d\Omega = 0, i = 1, ..., n \tag{3}$$

With its topologically irregular meshes, the FEM is very flexible in its discretisation; curved elements are possible. The dual coupling of electromagnetic fields demands specific function rooms; the so-called Nédélec edge elements allow the construction of such finite element spaces very elegantly.

Nowadays, automated FEM modelling is enabled by open-source tools such as FEniCS[2] or NGSolve[3]. Let us regard an example. The strong form (left) and the corresponding weak form (right) of the homogeneous electro-quasistatic equation read as

$$\nabla \cdot (\hat{\sigma} \nabla \phi) = 0, \qquad \int \hat{\sigma} \nabla \phi \nabla v d\Omega = \int 0 v d\Omega, \tag{4}$$

with the complex permittivity $\hat{\sigma}$, the (complex) electric potential ϕ and the test function v. The resulting code implemented in NGSolve becomes easily readable:

```
a = BilinearForm(fes)
a += sigma*grad(phi)*grad(v)*dx
f = LinearForm(fes)
f += 0 * v * dx
```

"fes" is a suitable function space. This formulation is generally valid and enables, for example, to process patient-specific brain models as described in Fig. 2.

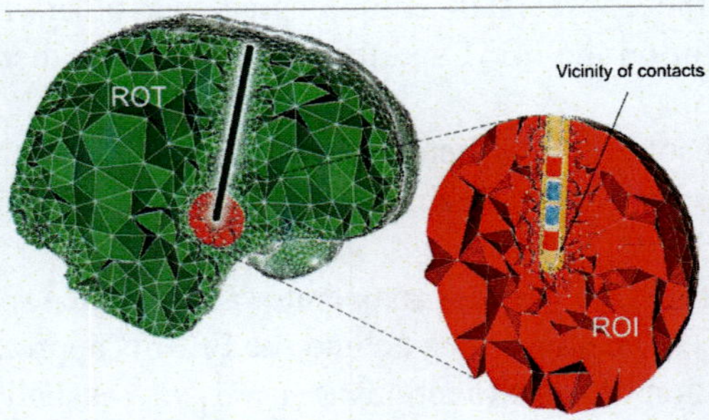

Fig. 2. Initial human brain discretisation in open-source platform OSS-DBS [2] with electrodes for Deep Brain Stimulation (DBS), a standard therapy in later stages of Parkinson's disease. The region of interest (ROI), the vicinity of the electrode contacts and the rest of the tissue (ROT), are three different domains with sub-meshes of different element size requirements. Additionally, sub-meshes are defined for the electrode contacts and the encapsulation layer around them. Controlled surface refinement of the contacts – active contacts in red, floating conductors in blue - leads to a highly dense sub-mesh around the electrode.

[2] fenicsproject.org.
[3] ngsolve.org.

1.2 Finite Integration Technique

In a very 'natural way', the FIT represents Maxwell's Equations in their integral form on some suitable grid, respectably some appropriate pair of grids. Though explaining it for a Cartesian grid, the FIT has also been realised for several other grids, such as triangular and further non-orthogonal grids. The FIT was developed in 1977 by Thomas Weiland [3] as a method with the same scope as Maxwell's Equations. It was developed independently from Yee's FDTD method [4] for time-domain problems dating to 1966. The FIT development aimed to achieve a consistent numerical solution for Maxwell's Equations. Here, we see the derivation of the discrete induction law using the integral state variables, that is, the electric grid voltages along the edges of an elementary grid area and the magnetic grid fluxes normal to the facets. The integrals are thus transferred directly to the grid cells, and the corresponding system of linear equations follows immediately (Fig. 3).

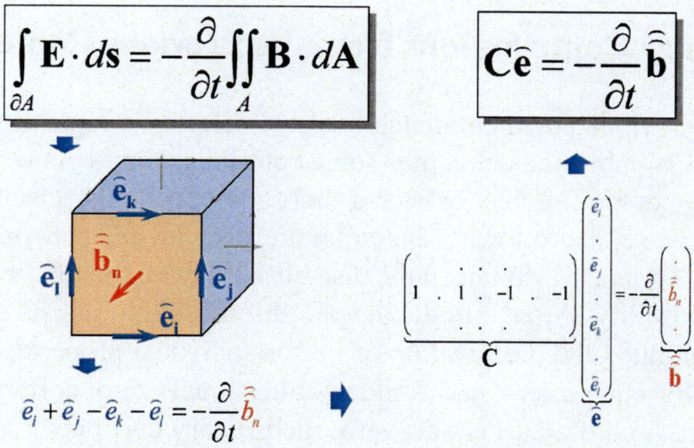

Fig. 3. Derivation of the discrete induction law using the electric grid voltages along the edges of an elementary grid area, the magnetic grid fluxes normal to the facets, and the corresponding system of linear equations.

A dual grid is introduced to reflect the interrelations between the electric and magnetic fields. It then holds the magnetic grid 'voltages', that is, line integrals over the magnetic field along the edges and the electric grid flux. In addition, the material relations are transferred to the grid pair.

Without going into more detail, this procedure finally leads to the so-called Maxwell-Grid-Equations, a discrete analogue for Maxwell's equations in the form of a set of matrix equations. The linear operators $\mathbf{C}$ and $\tilde{\mathbf{C}}$ and $\mathbf{S}$ and $\tilde{\mathbf{S}}$ correspond to a discrete curl operator and a discrete divergence operator, respectively. The operators $\mathbf{M}$ refer to the material operators which link the quantities on the primary grid to those on the dual grid. More can be found, e.g., in [5]. Figure 4 displays the grid for a PIC simulation.

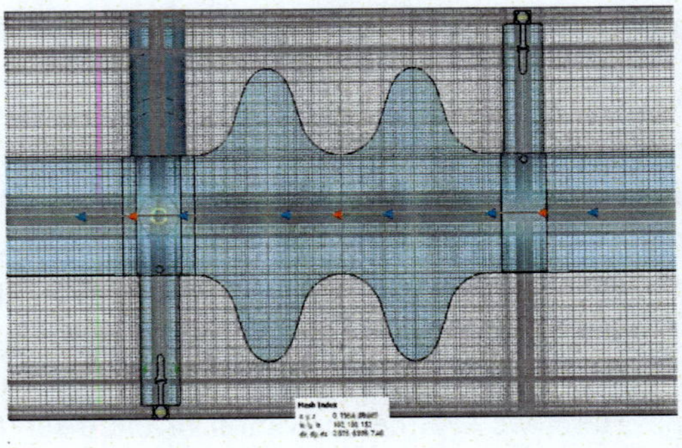

Fig. 4. 2D slice of a hexahedral mesh of a two-cell cavity for the FCC-ee [6] assembled with fundamental power coupler and two Higher Order Mode couplers. Model for a Particle-in-Cell (PIC) simulation. The particles pass the cavity on its axis (see blue and red arrows indicating this path). The figure was created by Sosoho-Abasi Udongwo using CST Studio Suite® 2020.

2 Exemplary Contributions from the Previous Conferences

Based on the previous post-conference books, I selected a representative diversity of the invited papers published there, plus some contributed ones. As very theoretical contributions were not too suitable for such a short glimpse, I only selected two of them. If not stated otherwise, the following examples are from invited papers.

1997 Peter Thoma [5] explained the discretisation with the FIT and showed its wide range of applicability. After introducing the discretisation and its operators in more detail, he highlighted the preservation of vector-analytical properties by the FIT operators, such as SC equals zero, and S-tilde C-tilde equals zero, corresponding with curl grad equals zero and div curl equals zero; the primary curl operator equals the transpose of the dual curl operator. This relation builds the basis for the consistency of FIT. Starting from the Maxwell Grid Equations (MGE), linear-algebraic and differential-algebraic problems result in the different problem classes known from electrodynamics. The resulting large-scale problems from numerical linear algebra demand modern solution methods (as is the case for the resulting equations of the other discretisation methods). The practical application examples covered all problem classes resulting from Maxwell's equations.

1998 Thomas Weiland introduced the FIT and its MGE, including a dozen examples of provable properties of MGE, such as charge conservation, energy conservation, and stability of the time discretisation. He highlighted advanced discretisation techniques such as non-orthogonal grids in 2D and 3D, consistent and recursive sub-grids and the Perfect Boundary Approximation technique (PBA). As topical challenges in electromagnetic computations, he discussed user-friendly codes, CAD input and output, and high-speed- and parallel computing. A wide range of practice-relevant examples, for example, electromagnetic compatibility (EMC) applications, was provided. As the use of mobile phones rapidly grew, the EMC issue rapidly gained importance given the users and the use scenarios up to a scenario with a driver in the car and a human head model employing the first stable, consistent and recursive local sub-gridding algorithm.

2000 Leszek Demkowicz [7] reviewed the main ideas behind constructing variable-order edge elements: adaptive hp FE modelling for the time-harmonic Maxwell's equations. Moreover, he discussed the possibility of extending the construction to Nédélec's elements of the first kind. He elaborated on the power of varying the order of approximation p and the mesh size h, such as an exponential rate of convergence even for very irregular solutions: Using elements with higher-order p strongly reduces the dispersion error. Small elements, that is, small h, capture small geometrical details. Last, he derived the De Rham diagram comprising the exact sequences on the continuous and discrete levels together with the hp-interpolation operators. It corresponds to the classical diagram for a uniform approximation order, entailing the Nédélec and Raviart-Thomas elements. He further elaborated on various numerical aspects and open problems at that time.

FEM can handle complex geometries due to the underlying variational principles, but it is cost-intensive, for example, in mesh generation. 2002 Igor Tsukerman [8] introduced a new method with desired characteristics such as Finite Difference data structures and schemes based on FE variational principles, the ability to treat curved interfaces on relatively coarse regular rectangular or hexahedral grids with high accuracy without having to resolve small geometric details, the feasibility of h- and p-refinement, means in the discretisation and the polynomial order, and an optimal convergence rate. Generalised FEM by partition-of-unity, Discontinuous Galerkin methods, and FIT are some of the then-existing techniques most closely related to this wish list. The newly introduced "Finite Element Difference" method, FED, is equally applicable for general FE meshes. Still, due to the choice of approximation functions, it is applied here to regular hexahedral or rectangular meshes. For problems with small particles, for example, in nanotechnological applications, the FED can avoid the usually necessary high mesh refinement around the particles since the special approximation functions with jumps represent the behaviour of the potential well.

Electrical machines often require the solution of coupled or multi-physics problems, for example, sliding interfaces, capturing thermal limitations, or coupling to electric circuits, partly with nonlinear components. Such problems result in weakly or strongly coupled problems. Complex geometries pose challenges to FEM, often related to mesh generation demanding a search for more efficient alternatives. In computational mechanics, meshless methods demonstrated their strengths. 2004 Dave Rodger [9] employed the meshless local Petrov Galerkin method applying small local domains around points (nodes) and satisfying the weak form locally with different possible functions such as radial basis functions or the Heaviside step function. Two examples were shown: rigid body motion in a linear actuator and heating in an induction machine. For the simple linear actuator, the usual FEM Galerkin procedure results in one set of equations for the armature and another for the core with its surrounding air. They are coupled using a Lagrange sliding interface. The concept of master elements taking precedence over slave elements ensures that only armature elements are used to model the overlapping area. The displacement is estimated in each time step.

2006 Francois Henrotte [10] introduced an energy-based formulation of electromagnetism, which, compared to the classical theory, results in a stronger connection with the universal principles of thermodynamics. This representation of electromagnetism results in a flow chart for the energy flow. The diagram consists of energy reservoirs containing, amongst others, electric and magnetic energy. A co-moving time derivative captures possible motion or deformation of the solution domain. The advantages of this formulation have been examined from a numerical point of view. The governing equations are preserved in a form directly usable by the FEM and convex analysis. Furthermore, all terms retain a clear physical significance supporting the definition of coupling terms in multi-physics modelling and providing meaningful criteria for parameter identification. Model reduction approaches were discussed as well. The energy-based theory provides operative concepts clarifying issues such as hysteresis modelling. A suitable decomposition of the force acting on the induction naturally yields a vectorial hysteresis model based on an accurate physical description, which can be reasonably employed in a 3D model, even if the parameter identification is based on uniaxial quasi-static measurements.

Again 2006, Irina Munteanu [11] started with a brief review after almost 30 years of FIT to solve electromagnetic field problems. For example, she highlighted that the FIT was the first eigenmode algorithm reliably eliminating spurious modes, underlined the various gridding options and pointed out the then-recent model order reduction in conjunction with the FIT. She emphasised that due to its versatility, FIT was probably the 3D numerical method at the time that covered the most comprehensive possible range of simulation requirements in the enormously diverse field of RF and microwave simulation. The challenge stems from the many different types of components, the great variety of geometric complexity, the level of detail, and the different characteristics of materials for the problems in both narrow-band and wide-band applications, from the MHz to the multi-GHz range. The time-domain solvers are the preferred solvers to obtain broadband results with a single run. A typical time-domain application is the full 3D simulation of a 30-metre aircraft illuminated by a plane wave at 500 MHz. Despite the relatively large number of 9 million cells back then, the simulation with the efficient FIT/PBA time-domain algorithm took less than two hours on a regular PC. EMC concerns and novel medical imaging techniques have increasingly required simulations with human body models. These are highly inhomogeneous, contain many different dispersive tissues and usually require a fine mesh in the time-domain simulations.

Against the background of ever-shorter development cycles in the modern electronics industry, the importance of simulations as an alternative to traditional prototypes and measurements started growing. Wireless high-frequency consumer devices, such as mobile phones with their combinations of other devices such as cameras and sensors, were becoming more and more complex with simultaneous miniaturisation making compliance with EMC regulations very challenging. However, numerical analysis at an early design stage can predict potential EMC problems well before building physical prototypes. Challenging are the typically two orders of magnitude between the characteristic dimensions of the module or device and the printed circuit board (PCB) details, thus characterising a multi-scale problem. For comparison with given standards, absolute values of the magnetic field emissions radiated from a PCB must be obtained, posing a second challenge. 2008 Sergey Yuferev [12] solved the multi-scale problem by iteratively combining 2.5D and 3D field calculation. In this manner, real industrial EMC problems of wireless devices could be simulated that could not be analysed exclusively with 3D simulators because of the high computational effort. Secondly, using a simple enough reference PCB to allow accurate numerical modelling, a methodology was used to calibrate electromagnetic sources in numerical models using measured data by solving the inverse problem and thus determining the absolute values of radiated magnetic field emissions without detailed knowledge of the source's characteristics.

Processes in the so-called low-frequency range, where wave propagation does not play a role, are usually treated with electro- or magnetostatics or electro- or magnetoquasistatic models requiring specific expertise. In addition, arrangements with coupled inductive/capacitive effects cannot be simulated with the quasi-static models. 2010 Jörg Ostrowski [13] presented a stabilised full-wave Maxwell formulation in the time domain based on a stabilised full-wave frequency domain formulation by Ralf Hiptmair. Instability occurs in the so-called stationary limit, that is, for ω approaching zero in the frequency domain or huge time steps Δt. The approach splits the computational domain into a conducting and a non-conducting subdomain. For stabilisation, in the Coulomb-gauged approach with vector and scalar potential, the scalar potential is divided into two parts, one of which is constant in the conducting region, whereby a corresponding function space is added to the two usual function spaces. For numerical experiments, a conformal Galerkin FE discretisation was performed. The resulting solution is not unique, so at each time step, a singular system of equations with a consistent right-hand side has to be solved, which was done with the preconditioned BiCGstab method. For a simple plate capacitor, the stabilised method was compared with a commercial solver. While the latter becomes unstable after a particular moment, the stabilised method provides an accurate solution as validated with a frequency domain solution, despite a time step that is three orders of magnitude larger. In addition to this elementary example, a practical lightning impulse test simulation was also described.

With his contributed paper presented in 2012, Maximilian Wiesmüller [14] shall represent the solution of complex problems from the application. In a typical on-load tap-changer (OLTC) in a power transformer, the influence of the transformer and the tap leads on the OLTC insulation was studied with a comprehensive model simulated with a commercial solver. Curved second-order tetrahedral finite elements were used. A multi-stage simulation was carried out that used adaptive mesh refinement in the second stage after identifying critical parts in the tap selector. The importance of always mapping the error estimation and refinement to the original CAD geometry was emphasised and visualised using the simple example of a sphere with a non-zero potential above a grounded plate. Different evaluation methods, also in oil, were considered. The influence of the transformer and the tap leads on the OLTC insulation proved small in areas critical for a dielectric breakdown legitimating the usual design optimisation and test procedures.

Modern computers allow very accurate models and simulation results so that inaccuracies caused by uncertainties receive increasing attention. Stochastic approaches such as the Monte Carlo or perturbation methods via approximation methods to truncated polynomial chaos expansions are common. 2014 Stéphane Clénet [15] considered a magnetostatic problem as a deterministic problem in a vector potential formulation with the Galerkin FEM. For uncertain input parameters, this uncertainty propagates into the output parameters. The input parameters are modelled as random variables, and the random output parameters are characterised. Non-intrusive methods encapsulate the deterministic model in an environment of stochastic procedures. Approximation methods were presented, particularly a non-intrusive projection method. Therein, the number of multivariate polynomials grows exponentially (curse of dimensionality), which can, however, be prevented with sparse grids such as the Smolyak cubature. Intrusive methods require access to the deterministic model. Their solution is sought in the tensor product of the standard FE space with a space approximating the random variables. A short review of applications of uncertainty analysis from CEM was given.

Vacuum interrupters are protective devices for switching nominal currents and interrupting fault currents, relying on the vacuum insulation in the gap between two electrodes to withstand the voltage. When interrupting current, a vacuum arc is formed. Numerical simulations of the motion of vacuum arcs are critical for improving vacuum interrupters' performance but pose a challenge to standard computational fluid dynamics methods based on the Eulerian approach. In a full arc simulation, the movement of the plasma arc must be tracked in a strongly coupled multi-physics problem. In an initial study in 2016, Massimiliano Cremonesi [16] solved the conservation equations using a Lagrangian FE approach. A forward Euler scheme was used for the temporal integration of the equations for mass, momentum and energy, leading to an explicit solution scheme, while the conservation of current is to be computed implicitly. The nodal coordinates are updated in each time step following the fluid velocity. A simplified 2D arc model was studied, and the results were validated with commercial software.

Efficient electric machines are of great importance for e-mobility. The desired properties depend strongly on the application areas and are also influenced by the design of the machines. 2018 Peter Gangl [17] employed multipatch isogeometric analysis (IgA) for the simulation and shape optimisation of the electrical machines. IgA represents an alternative to the FEM. Here continuous Galerkin IgA was used. The same basis functions represent the geometry of the computational domain and the solution of the partial differential equations, making it particularly interesting for design optimisation procedures. Non-overlapping domain decomposition (DD) methods were used for a fast solution of the large algebraic equation systems. The DD matches well with the multipatch representation of the computational domain and parallelisation of the DD solvers. Numerical experiments showed excellent scaling behaviour. The studied device was an interior, permanent magnet electric motor. The shape of the motor should be optimised with an interior point line-search optimiser to maximise the smoothness of the motor's rotation, that is, the so-called runout performance.

2020 Herbert Eggers [18] investigated two discretisation strategies for Maxwell's equations for wave propagation through linear dispersive media: one based on the traditional leapfrog time integration scheme FDTD, the other on convolution quadrature. The polarisation is expressed in the time domain using the convolution theorem for the Laplace-transform, and the susceptibility kernel is written as a superposition of simple Debye functions. In the integral for the polarisation, an appropriate convolution quadrature is used. Both are equivalent for certain classes of problems and preserve the underlying energy-dissipation structure of the problem. An advantage of the convolution quadrature is its independence of the number of internal states of the memory part of the polarisation and its applicability to fairly general dispersive materials. An efficient implementation is possible by employing the so-called fast-and-oblivious convolution quadrature. Their test problem considered the propagation of an electromagnetic pulse across the interface between air and human tissue. A five-pole Debye model characterised the dielectric response of the tissue. They used a plane wave setting, leading to a 1D wave propagation problem, achieving excellent agreement between the two schemes.

3 Conclusion

These insights into the CEM contributions at SCEE over the last 25 years hopefully motivate picking up the earlier post-conference books, mostly online, to read more about the presented examples and many other exciting contributions. The SCEE has proven itself as a conference that allows the presentation and discussion of new cross-disciplinary results linking electrical engineering and mathematics in a relatively wide variety, with representatives from industry and in an almost familiar atmosphere. This characteristic profile makes the SCEE stand out and complement many other, less inter-disciplinary conference series in either of the fields.

References

1. Rylander, T., Bondeson, A., Ingelström, P.: Computational Electromagnetics. Texts in Applied Mathematics, vol. 51, 2nd edn. Springer, Cham (2013)
2. Butenko, K., Bahls, C., Schröder, M., Köhling, R., van Rienen, U.: OSS-DBS: open-source simulation platform for deep brain stimulation with a comprehensive automated modeling. PLoS Comput. Biol. **16**(7), e1008023 (2020). https://doi.org/10.1371/journal.pcbi.1008023
3. Weiland, T.: Eine Methode zur Lösung der Maxwellschen Gleichungen für sechskomponentige Felder auf diskreter Basis. AEU-ARCH ELEKTRON UB **31**, 116–120 (1977)
4. Yee, K.S.: Numerical solution of initial boundary value problems involving Maxwell's equations in isotropic media. IEEE Trans. Antenn. Propag. **14**, 302–307 (1966)
5. Clemens, M., Thoma, P., Weiland, T., van Rienen, U.: Computational electro-magnetic field calculation with the finite-integration method. Surv. Math. Ind. **8**, 213–232 (1998)
6. Abada, A., Abbrescia, M., AbdusSalam, S.S., et al.: FCC-ee: the lepton collider. Eur. Phys. J. Spec. Top. **228**, 261–623 (2019). https://doi.org/10.1140/epjst/e2019-900045-4
7. Demkowicz, L.: Edge finite elements of variable order for Maxwell's equations. In: van Rienen, U., Günther, M., Hecht, D. (eds.) Scientific Computing in Electrical Engineering. LNCSE, vol. 18, pp. 15–34. Springer, Heidelberg (2001). https://doi.org/10.1007/978-3-642-56470-3_2
8. Tsukerman, I.: Toward generalized finite element difference methods for electro-and magnetostatics. In: Schilders, W.H.A., ter Maten, E.J.W., Houben, S.H.M.J. (eds.) Scientific Computing in Electrical Engineering. MATHINDUSTRY, vol. 4, pp. 58–77. Springer, Berlin Heidelberg (2004). https://doi.org/10.1007/978-3-642-55872-6_5
9. Rodger, D., Lai, H.C., Coles, P.C., Hill-Cottingham, R.J., Vong, P.K., Viana, S.: Finite element modelling of electrical machines and actuators. In: Anile, A.M., Ali, G., Mascali, G. (eds.) Scientific Computing in Electrical Engineering. TECMI, vol. 9, pp. 159–168. Springer, Heidelberg (2006). https://doi.org/10.1007/978-3-540-32862-9_23

Mathematical and Computational Methods

Machine Learning Techniques to Model Highly Nonlinear Multi-field Dynamics

Ruxandra Barbulescu[1(✉)], Gabriela Ciuprina[2], Anton Duca[2], and L. Miguel Silveira[3]

[1] INESC-ID, Rua Alves Redol 9, 1000-029 Lisbon, Portugal
ruxi@inesc-id.pt

[2] Politehnica University of Bucharest, Spl. Independentei 313, 060042 Bucharest, Romania
gabriela@lmn.pub.ro, anton.duca@upb.ro

[3] INESC-ID and IST Técnico Lisboa, Universidade de Lisboa, Av. Rovisco Pais 1, 1049-001 Lisbon, Portugal
lms@inesc-id.pt

Abstract. Modelling the dynamics of the membrane displacement in a micromachined beam fixed at both ends for different applied voltages is important for real applications. The strong nonlinearities involved and the interaction between multiple physical fields make this task challenging for classical modelling and model reduction approaches. In this work we search for a simplified, yet accurate, data-driven models, based on different recurrent neural network architectures, using only peripheral input-output information of the original system. The main goal is to find the most suitable neural network architecture having the smallest number of hidden units that provides low error of the minimum gap dynamics for different applied voltages. We show that these black-box models, with only 4 hidden units, are able to accurately reproduce the original system's response to a variety of different stimuli, and a strategy to make them parameter aware is proposed.

1 Introduction

Since their creation in research labs in the 1950's, Micro-Electro-Mechanical Systems (MEMS) and their Radio-Frequency (RF) variety have seen a wide range of applications, from sensors to switches, vehicle controls, pacemakers and even games. The basic structure of many RF-MEMS switches is based on a beam suspended like a bridge across a substrate, which is pulled down by a force (such as an electrostatic force) and eventually contacts a dielectric on the substrate thus blocking the signal (Fig. 1). The pull-in voltage and the response time are some of the most important parameters of electrostatically-actuated MEMS switches. Both the response time and the force needed to pull-in the membrane depend nonlinearly on its displacement and this dependence is the result of coupled electro-mechanical-fluid systems interaction.

One of the devices synthesising this mechanism is a micromachined beam fixed at both ends, often used as a benchmark for model reduction algorithms [1] and even studied as a pressure sensor [2]. A physics-aware model reduction approach is proposed in [3], where the low-order model is built by progressively adding physical phenomena. The dynamics of the bridge benchmark is described in detail in Sect. 2. The reduced

M. van Beurden et al. (Eds.): SCEE 2022, MI 43, pp. 125–132, 2024.
https://doi.org/10.1007/978-3-031-54517-7_14

model in [3] reproduces with high-fidelity the dependence between the pull-in voltage and the membrane displacement as well as the dynamic behaviour but, in the latter case, only a few basic input stimuli are considered in the modelling and reduction processes.

In all these approaches, the difficulty of modelling and producing simplified representations comes from the nonlinearity of the system and the interaction of more than one physical field. In particular, adding the air damping phenomena makes the system highly nonlinear. Physics-awareness can be both a plus and a minus, while gaining specificity and physical interpretation, one sacrifices generalization.

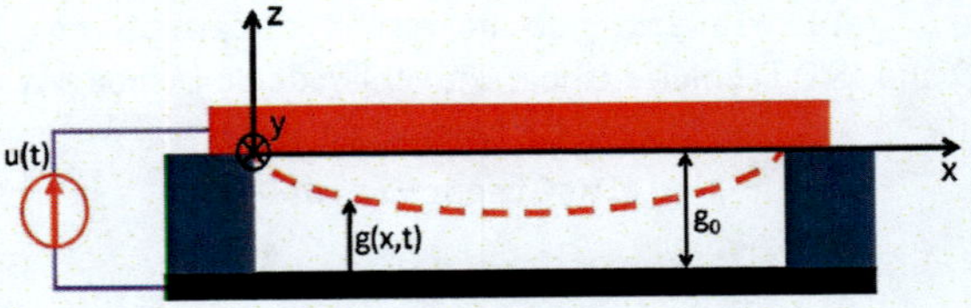

Fig. 1. The bridge benchmark (extracted from [3]).

In recent years, machine learning type models have been successfully used in various fields to tackle strongly nonlinear problems with considerable success. In this work, we use machine learning techniques to model the dynamics of the membrane displacement in the bridge benchmark, using only input-output information of the original full-order system. We generate data-driven, black-box models assuming no prior knowledge of the original system's structure and constitutive equations, which can further be explained using specific interpretation techniques for neural networks [4]. We train recurrent neural networks on datasets representing the system's response (membrane's minimum gap) to different input voltage signals (different shapes and magnitudes). We compare these architectures in terms of their properties and accuracy in reproducing the output of the original system. We show that with a recurrent layer of only 4 hidden units, it is possible to accurately reproduce the original system's response to a variety of different stimuli. We further show that we can generate parameter-aware models, which are able to predict with fidelity the system's behaviour for different values of specific parameters.

2 The Bridge Benchmark Dynamics

The bridge benchmark is a polysilicon beam of length $l = 610\,\mu\text{m}$, width $w = 40\,\mu\text{m}$ and height $h = 2.2\,\mu\text{m}$ suspended like a bridge over a silicon substrate. The initial gap is $g_0 = 2.3\,\mu\text{m}$. The mechanism is described by the strongly coupled 1D Euler's beam equation (1) and 2D Reynolds' squeeze film damping equation (2):

$$EI\frac{\partial^4 g}{\partial x^4} - S\frac{\partial^2 g}{\partial x^2} = F_{\text{elec}} - \rho\frac{\partial^2 g}{\partial t^2} + F_{\text{air}}, \tag{1}$$

$$\text{div}\left(\left(1+6\frac{\lambda}{g}\right)g^3 p\,(\text{grad}(p))\right) = 12\mu\frac{\partial(pg)}{\partial t}, \tag{2}$$

where $g(x,t)$ is the unknown gap (the displacement is $g_0 - g(p_m,t)$, where p_m is the middle point of the membrane $p_m = l/2$), $E = 149\,\text{GPa}$ is the Young modulus, $I = wh^3/12$ is the inertial moment, $S/(hw) = -3.7\,\text{MPa}$ is the initial stress, ρ is the mass per unit of length $\left(\rho/(hw) = 2330\,\text{kg/m}^3\right)$, $F_{\text{elec}}(x,t) = -\varepsilon_0 w v(t)/\left(2g^2(x,t)\right)$ is the electric force per unit of length (ε_0 is the air permittivity), $F_{\text{air}} = \int_0^w (p - p_a)\text{d}y$ is the damping force per unit of length, $p(x,y,t)$ is the unknown pressure, $p_a = 1.013 \cdot 10^5\,\text{Pa}$ is the environment pressure, $\lambda = 0.064\,\mu\text{m}$ is the mean free path of air and $\mu = 1.82 \cdot 10^{-5}\,\text{kg/(m} \cdot \text{s)}$ is the air viscosity.

Since $l \gg w$, the deflection is assumed uniform across the width. Moreover, the deformation is symmetrical along the length of the membrane, therefore we can consider as quantity of interest the middle point $p_m = l/2$, where the gap is minimum. This point would also be the first touching the dielectric in case the membrane is pulled down completely. Figure 2 shows the membrane's displacement at different moments in time and the minimum gap for a periodic impulse.

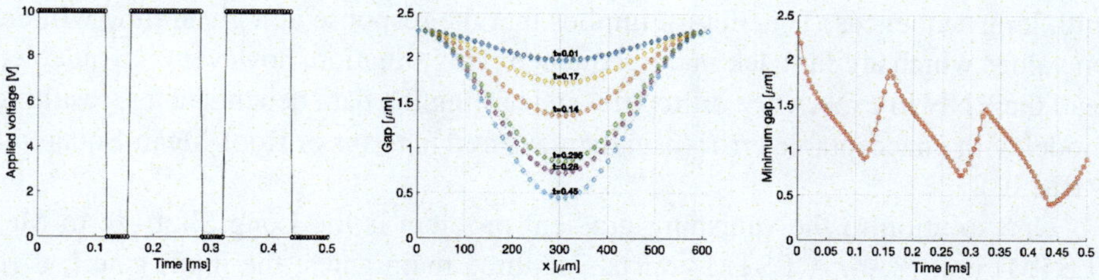

Fig. 2. Example of membrane displacement, generated with the original code from [3]). Left: Input – applied voltage $v(t)$. Middle: Displacement $g(x,t)$ at different moments in time (t in [ms]). Right: Output – minimum gap $g(p_m,t)$ in time.

3 Neural Networks Models

Recurrent Neural Networks (RNNs) [5–7] are a family of neural networks used for processing sequential data. Compared to their predecessors – the Feedforward Neural Networks (FFNNs) – the RNNs allow neurons in a given layer to form connections among themselves, thus being particularly adept to processing sequences of values $\mathbf{x}_1, \ldots, \mathbf{x}_t$ of equal or variable length. In this work we create models based on three architectures: the simple RNN [6], the Long Short-Term Memory (LSTM) unit [8,9] and the Gated Recurrent Unit (GRU) [10]. The structures of their cells are presented comparatively in Fig. 3.

In the simplest form of RNNs (Fig. 3-left), the prediction at a certain time point $\hat{\mathbf{y}}_t$ depends on the hidden state of the cell at the current time point $\mathbf{h}_t$, which in turn depends of the hidden state at the previous time point $\mathbf{h}_{t-1}$. In the following equations $\mathbf{V}$, $\mathbf{W}$ and $\mathbf{U}$ are matrices of weights, $\mathbf{b}$ and $\mathbf{c}$ arrays of biases and ϕ is an activation function, usually the hyperbolic tangent:

$$\mathbf{h}_t = \phi\left(\mathbf{V}\mathbf{x}_t + \mathbf{U}\mathbf{h}_{t-1} + \mathbf{c}\right), \qquad \hat{\mathbf{y}}_t = \mathbf{W}\mathbf{h}_t + \mathbf{b}.$$

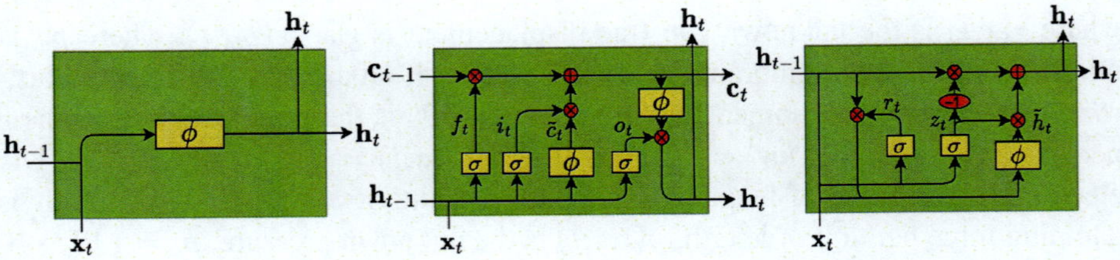

Fig. 3. Structure of a cell in the three architectures: simple RNN, LSTM, GRU (adapted from [11]).

The simple RNNs however are known to suffer from various issues, a delicate one being the vanishing gradient [12], which happens when long term components go exponentially fast to norm 0, making it impossible for the model to learn the correlation between temporally distant events. In our case, for a faithful reproduction of the dynamics, the simulations of the original model require the use of fine time steps, leading to datasets with long sequences. This in turn implies that the response at a given time will depend on values which are far back in the sequence. This situation, however unavoidable, may lead the RNN to experience difficulties in learning the data dependencies, resulting in a model with unacceptable error (usually measured in terms of Root-Mean-Squared Error – RMSE).

One solution to the vanishing gradient problem is the Long Short-Term Memory (LSTM) unit [8,9]. A LSTM consists of three main gates: the input gate $\mathbf{i}_t \in (0,1)^h$ that controls whether the cell state is updated or not, where h is the number of hidden units, the forget gate $\mathbf{f}_t \in (0,1)^h$ defining how the previous memory cell affects the current one and the output gate $\mathbf{o}_t \in (0,1)^h$, which controls how the hidden state is updated. The usage of gates is a major difference from the simple RNNs, since besides the hidden state $\mathbf{h}_t \in (-1,1)^h$, the LSTM also outputs a cell state $\mathbf{c}_t \in \mathbb{R}^h$ to the next LSTM unit, as shown in Fig. 3-center. The computation of the cell state is based on the candidate cell state $\tilde{\mathbf{c}}_t \in (-1,1)^h$. The vanishing gradient problem is partially solved by the LSTM units by allowing gradients to also flow *unchanged*. The LSTM mechanism is described by the following equations, where the learned parameters are the weights $\mathbf{W}_* \in \mathbb{R}^{h\times d}$ and $\mathbf{U}_* \in \mathbb{R}^{h\times h}$, and the biases $\mathbf{b}_* \in \mathbb{R}^h$, where d is the number of input features:

$$\mathbf{i}_t = \sigma\left(\mathbf{W}_i\mathbf{x}_t + \mathbf{U}_i\mathbf{h}_{t-1} + \mathbf{b}_i\right), \qquad \tilde{\mathbf{c}}_t = \phi\left(\mathbf{W}_c\mathbf{x}_t + \mathbf{U}_c\mathbf{h}_{t-1} + \mathbf{b}_c\right),$$

$$\mathbf{f}_t = \sigma\left(\mathbf{W}_f\mathbf{x}_t + \mathbf{U}_f\mathbf{h}_{t-1} + \mathbf{b}_f\right), \qquad \mathbf{c}_t = \mathbf{f}_t \circ \mathbf{c}_{t-1} + \mathbf{i}_t \circ \tilde{\mathbf{c}}_t,$$

$$\mathbf{o}_t = \sigma\left(\mathbf{W}_o\mathbf{x}_t + \mathbf{U}_o\mathbf{h}_{t-1} + \mathbf{b}_o\right), \qquad \mathbf{h}_t = \mathbf{o}_t \circ \phi\left(\mathbf{c}_t\right).$$

σ and ϕ are the logistic sigmoid and the hyperbolic tangent activation functions, respectively. The operator $\circ$ denotes the Hadamard product (element-wise product).

A simpler unit composed of only two gates is the Gated Recurrent Unit (GRU) [10], proposed in 2014. The GRU is described by:

$$\mathbf{z}_t = \sigma\left(\mathbf{W}_z\mathbf{x}_t + \mathbf{U}_z\mathbf{h}_{t-1} + \mathbf{b}_z\right), \qquad \hat{\mathbf{h}}_t = \phi\left(\mathbf{W}_h\mathbf{x}_t + \mathbf{U}_h\left(\mathbf{r}_t \circ \mathbf{h}_{t-1}\right) + \mathbf{b}_h\right),$$

$$\mathbf{r}_t = \sigma(\mathbf{W}_r\mathbf{x}_t + \mathbf{U}_r\mathbf{h}_{t-1} + \mathbf{b}_r), \qquad \mathbf{h}_t = \left(\mathbf{1} - \mathbf{z}_t\right) \circ \mathbf{h}_{t-1} + \mathbf{z}_t \circ \hat{\mathbf{h}}_t,$$

where the weights $\mathbf{W}_* \in \mathbb{R}^{h\times d}$ and the biases $\mathbf{b}_* \in \mathbb{R}^h$ are learned parameters. The GRU (Fig. 3-right) is only composed of two gates, the update gate $\mathbf{z}_t \in (0,1)^h$ and the reset gate $\mathbf{r}_t \in (0,1)^h$. The update gate controls how much of the past information needs to be passed along to the future, while the reset gate is used to decide how much information the model should forget. The GRU only outputs the hidden state $\mathbf{h}_t \in \mathbb{R}^h$ computed based on the candidate hidden state $\hat{\mathbf{h}}_t \in (-1,1)^h$.

4 Results

Our first objective is, from a model reduction perspective, to find the most suitable architecture and the smallest neural network (in terms of hidden units) that provides good predictions. We therefore search for a nonlinear approximation $\mathscr{F}$ of the minimum gap $\hat{g}(p_m,t)$, for different applied voltages $v(t)$: $\hat{g}(p_m,t) = \mathscr{F}(v(t))$.

Data. We simulate the high-fidelity model from [3] described in Sect. 2 for 0.5 ms, with a time step of 5 µs and different input signals to generate snapshots of the system's dynamical behaviour. Each snapshot contains pairs of input/output (I/O) data for 100 points, namely the input voltage's variation in time $v(t)$ and the membrane's minimum gap $g(p_m,t)$ (where $p_m = l/2$ is the middle point of the membrane along the length), which takes values in the range $[0, 2.3]$ µm. In the cases when the membrane is totally pulled down, the minimum gap is set to 0 for the remaining simulation time. We use scaled values (the gap is in µm, time is in ms), since the supervised learning uses an absolute error metric (the RMSE – Root-Mean-Squared Error), whose very low values might misdirect the training process.

We generate 40 snapshots, each with I/O values for 100 time moments, therefore amounting to 4000 pairs of I/O values, and divide them into three sets of data: the training set 50%, the validation set 25% and the test set 25%. The first two sets are used to compute the learned parameters, then the model is evaluated against the test set, containing data unseen before. The number of snapshots as well as their nature were chosen as to sufficiently sample the training and test distributions. Numerical tests showed that this choice was suitable for this benchmark.

Table 1. The average RMSE out of ten simulations, for RNN with 16 and 64 hidden units and for LSTM and GRU with 8 and 32 hidden units, for the iteration with the smallest validation loss.

	RNN-16	LSTM-8	GRU-8	RNN-64	LSTM-32	GRU-32
Training	0.0856	0.0376	0.0304	0.1728	0.0243	0. 0505
Validation	0.1335	0.0759	0.0896	0.1928	0.1781	0.1378

Implementation. The implementation is done in Python's libraries Keras and Tensorflow. We use a Normalization layer that shifts and scales the inputs into a distribution centered around 0 with standard deviation of 1, with the mean and variance adapted to the data. For a consistent comparison, we fixed the *hyperparameters* for all three

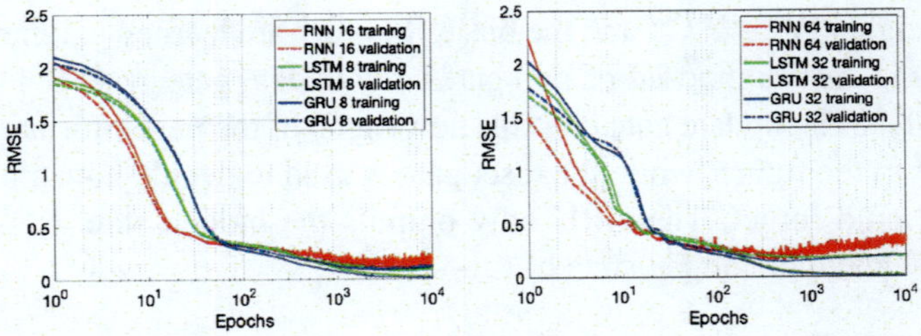

Fig. 4. Training and validation RMSE averaged over 10 runs, for different architectures and sizes.

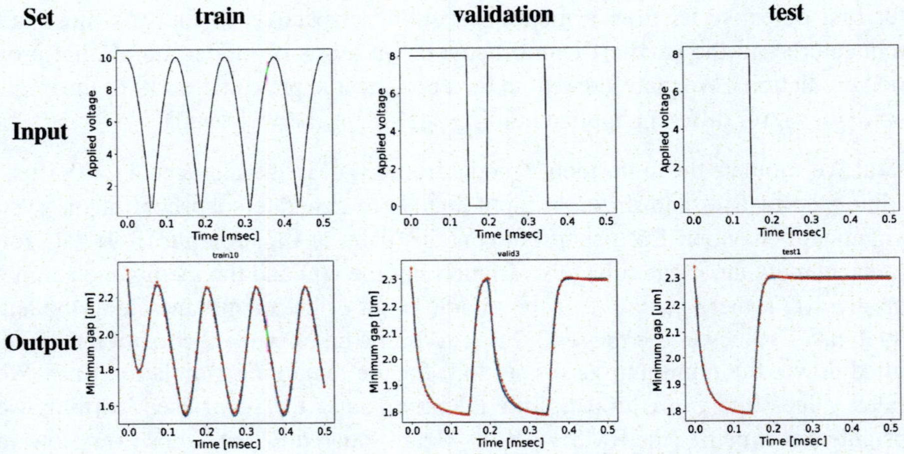

Fig. 5. Real (blue) and predicted (red) minimum gap for the GRU model with 8 hidden units extracted the training, validation and test sets.

architectures to the same previously optimized values, as follows: 1) We ordered the hyperparameters decreasingly based on their expected influence on the model's performance; 2) We set the most important one and so on; to set one, we kept fixed all the other hyperparameters and trained with different values for it. For the hyperparameters that are interdependent – for example batch size with sequence length, we did a grid search for different combinations of these. Our search in the hyperparameters space led to the following: the optimizer was set to Adam, the loss function is the RMSE, and the learning rate is 0.005. We trained the models for 10000 epochs (the number of passes of the training dataset through the algorithm), choosing the number of hidden units (neurons in the recurrent layer) so that the total number of parameters is comparable between models, e.g. a RNN with 16 hidden units (305 parameters) corresponds to a LSTM with 8 (328 parameters) and a GRU with also 8 hidden units (249 parameters).

Figure 4 shows the variation of the RMSE over the epochs, comparatively. Despite the lower complexity, the GRU performed best overall (see Table 1). In fact, with a recurrent layer as small as 4 hidden units, the RMSE is 0.0345 μm for the training and 0.0597 μm for the validation set. Figure 5 shows examples from the three sets, the input and the corresponding real and predicted outputs for the GRU with 8 units.

Parameter-Aware Models. A second objective is parameter-awareness, i.e. the ability of the neural network model to take into account geometrical characteristics and other parameters of the system that impact the output. We identified three important parameters: the membrane length l and width w, and the air viscosity μ. A series of potential values for each are listed in Table 2. We are now looking for an approximation that takes into account these parameters, of the form $\hat{g}(p_m, t) = \mathscr{F}(v(t), l, w, \mu)$.

Table 2. Different parameter values. The air viscosity is dependent on the temperature.

Parameter	Air viscosity μ $\left[\frac{kg}{m \cdot s}\right]$	Membrane length l [μm]	Membrane width w [μm]
Reference value	$1.82 \cdot 10^{-5}$ (15°)	610	40
Other values considered	$1.73 \cdot 10^{-5}$ (0°)	590	36
	$1.78 \cdot 10^{-5}$ (10°)	600	38
	$1.85 \cdot 10^{-5}$ (25°)	620	42
	$1.90 \cdot 10^{-5}$ (35°)	630	44

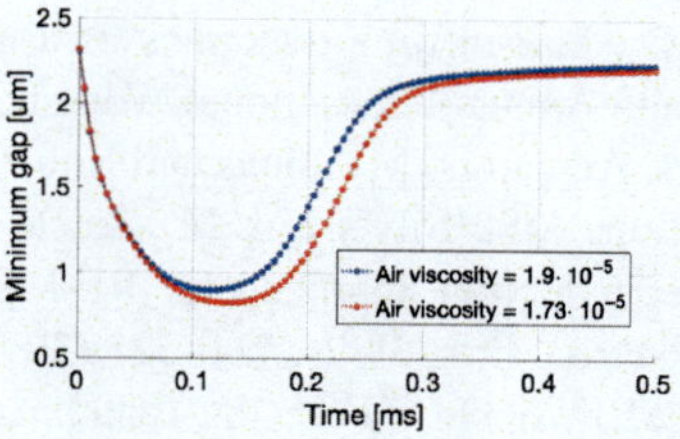

Fig. 6. Predicted minimum gap for two cases with the same input and fixed membrane length and width, but different air viscosity.

We generated new datasets containing the original 40 input-output data and random combinations of these values in a total of 500 examples (out of 5000 possible combinations) divided in 300 for training, 100 for validation and 100 for test). The 500 examples took more than 30h to generate with the original code. Using this data, we trained a GRU and re-optimized the hyperparameters.

The RMSEs obtained are of the same order of magnitude as for the previous case. Figure 6 shows the predicted output in two cases where the input is the same, as well as two of the parameters, the length and the width, but the air viscosity is different. The model successfully captures the difference in the output for the different values of air viscosity.

5 Conclusions

In this work we create data-driven models for the dynamics of the minimum gap in the bridge benchmark, using different recurrent neural network architectures. We show that a GRU layer with only 4 hidden units accurately reproduces the output for various different stimuli, and we further propose a strategy to make the model parameter-aware. The main advantage of this model is the ability to accurately predict the response to various different stimuli and for different parameters. Moreover, once the neural network is trained, the prediction is done instantaneously. The source code, datasets and results are publicly available at https://github.com/ruxandrab/beam. Our next focus is to model the second half of the mechanism – the opening of the switch, as well as look into physics-informed neural networks, by embedding physical constraints that would allow both feature preservation and subsequent interpretation of the low-order models.

Acknowledgements. This work was supported by European Funds "Recovery and Resilience Plan - Comp. 5" included in the NextGenerationEU program, under project n° 62 - "Responsible AI" and Portuguese national funds, under projects UIDB/50021/2020, PTDC/EEI-EEE/31140/2017.

References

1. Rewienski, M., White, J.: A trajectory piecewise-linear approach to model order reduction and fast simulation of nonlinear circuits and micromachined devices. IEEE Trans. Comput.-Aided Des. Integrated Circ. Syst. **22**(2), 155–170 (2003)
2. Gupta, R.J., Senturia, S.D.: Pull-in time dynamics as a measure of absolute pressure. In: Proceedings IEEE the Tenth Annual International Workshop on MEMS. An Investigation of Micro Structures, Sensors, Actuators, Machines and Robots, pp. 290–294. IEEE (1997)
3. Ciuprina, G., Ioan, D., Lup, A.S., Silveira, L.M., Duca, A., Kraft, M.: Simplification by pruning as a model order reduction approach for RF-MEMS switches. COMPEL- Int. J. Comput. Math. Electr. Electron. Eng. **39**(2), 511–523 (2019)
4. Ismail, A.A., Gunady, M., Bravo, H.C., Feizi, S.: Benchmarking deep learning interpretability in time series predictions. arXiv preprint arXiv:2010.13924 (2020)
5. Elman, J.L.: Finding structure in time. Cogn. Sci. **14**(2), 179–211 (1990)
6. Rumelhart, D.E., Hinton, G.E., Williams, R.J.: Learning representations by back-propagating errors. Nature **323**, 533–536 (1986)
7. Werbos, P.J.: Generalization of backpropagation with application to a recurrent gas market model. Neural Netw. **1**(4), 339–356 (1988)
8. Gers, F.A., Schmidhuber, J., Cummins, F.: Learning to forget: continual prediction with LSTM. Neural Comput. **12**(10), 2451–2471 (2000)
9. Hochreiter, S., Schmidhuber, J.: Long short-term memory. Neural Comput. **9**(8), 1735–1780 (1997)
10. Cho, K., van Merrienboer, B., Gülçehre, Ç., Bougares, F., Schwenk, H., Bengio, Y.: Learning phrase representations using RNN encoder-decoder for statistical machine translation. CoRR, abs/1406.1078 (2014)
11. Barbulescu, R., Mestre, G., Oliveira, A., Silveira, L.M.: Learning the dynamics of realistic models of *C. elegans* nervous system with RNNs. Sci. Rep. **13**(467) (2023)
12. Bengio, Y., Simard, P., Frasconi, P.: Learning long-term dependencies with gradient descent is difficult. IEEE Trans. Neural Netw. **5**(2), 157–166 (1994)

Port-Hamiltonian Systems' Modelling in Electrical Engineering

Andreas Bartel[1], Markus Clemens[2], Michael Günther[1(✉)], Birgit Jacob[3], and Timo Reis[4]

[1] IMACM, Chair of Applied Mathematics, Bergische Universität Wuppertal, Gaußstraße 20, 42119 Wuppertal, Germany
{bartel,guenther}@uni-wuppertal.de

[2] IMACM, Chair of Electromagnetic Theory, Bergische Universität Wuppertal, Rainer-Gruenter-Straße 21, 42119 Wuppertal, Germany
clemens@uni-wuppertal.de

[3] IMACM, Chair of Functional Analysis, Bergische Universität Wuppertal, Gaußstraße 20, 42119 Wuppertal, Germany
bjacob@uni-wuppertal.de

[4] Fakultät für Mathematik und Naturwissenschaften, Chair of System Theory and PDEs, TU Ilmenau, PF 10 05 65, 98684 Ilmenau, Germany
timo.reis@tu-ilmenau.de

Abstract. The port-Hamiltonian (pH) modelling framework allows for models that preserve essential physical properties such as energy conservation or dissipative inequalities. If all subsystems are modelled as pH systems and the inputs are related to the output in a linear manner, the overall system can be modelled as a pH system, too, which preserves the properties of the underlying subsystems. If the coupling is given by a skew-symmetric matrix, as usual in many applications, the overall system can be easily derived from the subsystems without the need of introducing dummy variables and therefore artificially increasing the complexity of the system. Hence the framework of pH systems is especially suitable for modelling multiphysical systems.

In this paper, we show that pH systems are a natural generalization of Hamiltonian systems, define coupled pH systems as ordinary and differential-algebraic equations. To highlight the suitability for electrical engineering applications, we derive pH models for MNA network equations, electromagnetic devices and coupled systems thereof.

1 Port-Hamiltonian Systems Modelling in a Nutshell

Port-Hamiltonian (pH) systems are a generalization of Hamiltonian systems

$$\dot{x} = J \cdot \nabla H(x), \quad x(0) = x_0 \tag{1}$$

with $x = (p, q)$ consisting of generalized position $q(t) \in \mathbb{R}^n$ and momentum $p(t) \in \mathbb{R}^n$ (where $t \in [0, T]$), the skew-symmetric matrix J given by

$$J = \begin{bmatrix} 0 & -I \\ I & 0 \end{bmatrix}$$

M. van Beurden et al. (Eds.): SCEE 2022, MI 43, pp. 133–143, 2024.
https://doi.org/10.1007/978-3-031-54517-7_15

and the Hamiltonian $H(x) = H(p, q) = U(p) + V(q)$ given as the sum of potential and kinetic energy, which maps $\mathbb{R}^n \times \mathbb{R}^n \to \mathbb{R}$ and is twice continuously differentiable. The Hamiltonian flow $\varphi(t; x_0))$, i.e., the solution of (1) at time point t, starting at the initial value $x(0) = x_0$, is characterized by four geometric properties:

1. *Preservation of the Hamiltonian*:
$$\frac{\mathrm{d}}{\mathrm{d}t} H(\varphi(t; x_0)) = (\nabla H(\varphi(t; x_0)))^\top J(\nabla H(\varphi(t; x_0))) = 0.$$
2. *Time-reversibility*:
$$\rho \circ \varphi(t; x_0) \circ \rho \circ \varphi(t; x_0) = x_0,$$
with $\rho(p, q) = (-p, q)$, which is a direct consequence of the ρ-reversibility of the Hamiltonian flow: $\rho \circ J\nabla H(\varphi(t; x_0))) = -J\nabla H(\rho \circ \varphi(t; x_0)))$.
3. *Symplectic structure of the Hamiltonian flow*:
$$\Psi(t)^\top J^{-1} \Psi(t) = J^{-1}, \qquad \Psi(t) := \tfrac{\partial \varphi(t; x_0)}{\partial x_0},$$
which is a direct consequence of the skew-symmetry of J.
4. *Volume-preservation*:
$$(\det \Psi(t))^2 = 1,$$
which follows immediately from the symplectic structure in 3.

First generalization step: arbitrary skew-symmetric matrices J

If we replace in (1) J by an arbitrary skew-symmetric matrix, the Hamiltonian is still preserved. As x will loose its characterization as generalized positions and momenta of classical mechanics, time-reversibility will generally not hold anymore. However, the symplectic structure of the flow still holds in the case of a regular J, and volume preservation is still a consequence of the Hamiltonian flow.

Second generalization step: adding dissipation to the system

Allowing the flow to become dissipative, we may generalize (1) to the dissipative Hamiltonian system

$$\dot{x} = (J - R) \cdot \nabla H(x), \quad x(0) = x_0 \tag{2}$$

with $R \geq 0$ being symmetric and positive semi-definite. In this case, the flow will neither be symplectic nor volume preserving, and the preservation of the Hamiltonian is replaced by the dissipativity condition

$$\frac{\mathrm{d}}{\mathrm{d}t} H(x(t)) = (\nabla H(x))^\top \dot{x} = -(\nabla H(x))^\top R \nabla H(x) \leq 0$$
$$\Rightarrow H(x(t)) = H(x_0) - \int_0^t (\nabla H(x(\tau)))^\top R \nabla H(x(\tau))\, \mathrm{d}\tau \leq H(x_0).$$

Third generalization step: coupling to the environment via inputs and outputs

Allowing for inputs and outputs to couple the system to the environment, we end up with linear pH system characterized by

$$\begin{aligned}\dot{x} &= (J - R) \cdot \nabla H(x) + Bu(t), \quad x(0) = x_0,\\ y &= B^\top \nabla H(x)\end{aligned}$$

with inputs $u(t) \in \mathbb{R}^p$, outputs $y(t) \in \mathbb{R}^p$ and port-matrices $B \in \mathbb{R}^{n \times p}$. The dissipativity inequality now reads

$$\begin{aligned}\frac{\mathrm{d}}{\mathrm{d}t} H(x(t)) &= (\nabla H(x))^\top \dot{x} = -(\nabla H(x))^\top R \nabla H(x) + (\nabla H(x))^\top Bu(t))\\ &= -(\nabla H(x))^\top R \nabla H(x) + y(t)^\top u(t) \leq y(t)^\top u(t)\\ \Rightarrow H(x(t)) &= H(x_0) - \int_0^t (\nabla H(x(\tau)))^\top R \nabla H(x(\tau))\, \mathrm{d}\tau + \int_0^t y(\tau)^\top u(\tau) \mathrm{d}\tau\\ &\leq H(x_0) + \int_0^t y(\tau)^\top u(\tau) \mathrm{d}\tau.\end{aligned}$$

Fourth generalization step: pH-DAE systems

Linear pH systems can be easily generalized to port-Hamiltonian differential-algebraic equations (pH-DAEs) given by

$$\frac{\mathrm{d}}{\mathrm{d}t}(Ex) = (J - R) \cdot z(x) + Bu(t), \quad x(0) = x_0, \tag{3a}$$

$$y = B^\top z(x) \tag{3b}$$

with a possibly singular matrix $E \in \mathbb{R}^{n \times n}$ and the nonlinear mapping $z : \mathbb{R}^n \to \mathbb{R}^n$ fulfilling the compatibility condition $E^\top z = \nabla H$. Now the dissipativity condition reads

$$\begin{aligned}H(x(t)) &= H(x_0) - \int_0^t z(x(\tau))^\top R \nabla z(x(\tau))\, \mathrm{d}\tau + \int_0^t y(\tau)^\top u(\tau)\, \mathrm{d}\tau\\ &\leq H(x_0) + \int_0^t y(\tau)^\top u(\tau)\, \mathrm{d}\tau.\end{aligned}$$

The key point in pH modelling is the following: there is an easy way to couple arbitrary many pH-DAE system such that the overall system is still a pH-DAE system, which preserves a dissipativity inequality.

Let us consider r autonomous pH-DAE systems

$$\frac{\mathrm{d}}{\mathrm{d}t}(E_i x_i) = (J_i - R_i) z_i(x_i) + B_i u_i, \tag{4a}$$

$$y_i = B_i^\top z_i(x_i) \tag{4b}$$

with r Hamiltonians $H_1, H_2, \ldots, H_r$ and compatibility conditions $E_i^\top z_i = \nabla H_i$. If the inputs and outputs fulfill a linear interconnection relation $Mu + Ny = 0$ for the aggregated input $u = (u_1, u_2, \ldots, u_r)$ and output $y = (y_1, y_2, \ldots, y_r)$, it has been shown in [13] that one can write the aggregated system as a joint pH-DAE system as

$$\frac{\mathrm{d}}{\mathrm{d}t}\left(\begin{bmatrix} E\,0\,0 \\ 0\,0\,0 \\ 0\,0\,0 \\ 0\,0\,0 \end{bmatrix}\begin{bmatrix} x \\ \hat{u} \\ \hat{y} \end{bmatrix}\right) = \begin{bmatrix} J-R & B & 0 & 0 \\ -B^\top & 0 & I_m & -M^\top \\ 0 & -I_m & 0 & -N^\top \\ 0 & M & N & 0 \end{bmatrix}\begin{bmatrix} z(x) \\ \hat{u} \\ \hat{y} \\ 0 \end{bmatrix} + \begin{bmatrix} 0 \\ 0 \\ I_m \\ 0 \end{bmatrix} u,$$
$$y = \hat{y},$$

with $z(x)^\top = (z_1(x_1)^\top, z_2(x_2)^\top, \ldots, z_r(x_r)^\top)$, new dummy variables $\hat{u}, \hat{y}$ and setting $X = \mathrm{diag}(X_1, X_2, \ldots, X_r)$ for $X \in \{E, J, R, B\}$. This coupling property of pH-DAE systems makes the pH modelling framework well suited for multi-physical applications.

Now, we consider external, time dependent inputs. To this end, we split the inputs and outputs into external (bar-notation) and internal ones (hat-notation), i.e., $B_i u_i$ is split into $\bar{B}_i\bar{u}_i + \hat{B}_i\hat{u}_i$. Then, the subsystem (4) reads

$$\frac{\mathrm{d}}{\mathrm{d}t}(E_i x_i) = (J_i - R_i) z_i(x_i) + \hat{B}_i\hat{u}_i + \bar{B}_i\bar{u}_i, \tag{5a}$$
$$\hat{y}_i = \hat{B}_i^\top z_i(x_i), \tag{5b}$$
$$\bar{y}_i = \bar{B}_i^\top z_i(x_i). \tag{5c}$$

For the coupling relation (of the internal quantities) $\hat{u} + C\hat{y} = 0$ with a skew-symmetric matrix $C = -C^\top$ (which often arises in application), these systems can be written as a joint pH-DAE system in condensed form [8]:

$$\frac{\mathrm{d}}{\mathrm{d}t}(Ex) = (\tilde{J} - R) z(x) + \bar{B}\bar{u}, \tag{6a}$$
$$\bar{y} = \bar{B}^\top z(x) \tag{6b}$$

with the condensed skew-symmetric matrix $\tilde{J} = J - \hat{B}C\hat{B}^\top$. Note that in this case all internal coupling modelled via the port-matrices $\hat{B}_i$ has now been transferred into the off-block diagonal elements of the skew-symmetric matrix $\tilde{J}$, i.e., $-\hat{B}C\hat{B}^\top$.

A systems theoretic treatment of pH systems goes back to Bernhard Maschke and Arjan van der Schaft (see [12, 14] for an overview), where nonlinear systems governed by ordinary differential equations are treated. For simplicity of presentation, we will (a) not follow the differential geometric path via Dirac structures, (b) neglect a feed-through from input to output and (c) only consider finite dimensional systems, i.e., ordinary (ODEs) and differential-algebraic equations (DAEs), but no partial differential equations (PDEs). For simulation, the latter are usually transformed into ODEs and DAEs by spatial semi-discretization. For a differential geometric setting of pH systems see [15] and an introduction into pH-PDEs see [11].

The paper is organized as follows: In the next, section we introduce pH-DAEs, which allow for a general nonlinear dissipative part. A pH-DAE formulation of the modified nodal analysis (MNA) network equations is derived in Sect. 3, and for electromagnetic devices in Sect. 4. Section 5 discusses formulations of pH systems of coupled EM/circuit systems, which allow for monolithic as well as weak coupling simulation approaches. Section 6 finishes with conclusions.

2 pH-DAE Systems

When dealing with applications in electrical engineering, the concept of pH modelling has to be generalized to coupled differential-algebraic equations, which (a) allow for a general nonlinear resistive part $r(z)$ instead of a quasilinear setting Rz as in the approach of [13] and (b) has only to be accretive on a subspace $\mathcal{V} \subset \mathbb{R}$ according to the constraints of the system.

A differential-algebraic equation of the form

$$\begin{aligned} \frac{\mathrm{d}}{\mathrm{d}t} Ex(t) &= Jz(x(t)) - r(z(x(t))) + Bu(t), \\ y(t) &= B^\top z(x(t)) \end{aligned} \tag{7}$$

is called a *port-Hamiltonian differential-algebraic equation* (pH-DAE) [8] if the following holds:

- $E \in \mathbb{R}^{n\times n}$, $J \in \mathbb{R}^{n\times n}$ and $B \in \mathbb{R}^{n\times m}$, $z, r : \mathbb{R}^n \to \mathbb{R}^n$.
- There exists a subspace $\mathcal{V} \subset \mathbb{R}^n$ with the following properties:
  - (i) for all intervals $\mathcal{I} \subset \mathbb{R}$ and functions $u : \mathcal{I} \to \mathbb{R}^m$ such that (7) has a solution $x : \mathcal{I} \to \mathbb{R}^n$, it holds $z(x(t)) \in \mathcal{V}$ for all $t \in \mathcal{I}$.
  - (ii) J is skew-symmetric on $\mathcal{V}$. That is, $v^\top Jw = -w^\top Jv$ for all $v, w \in \mathcal{V}$.
  - (iii) r is accretive on $\mathcal{V}$. That is, $v^\top r(v) \geq 0$ for all $v \in \mathcal{V}$.
- There exists some function $H \in C^1(\mathbb{R}^n, \mathbb{R})$ such that $\nabla H(x) = E^\top z(x)$ for all $x \in z^{-1}(\mathcal{V})$.

Remark 1. a) The pH-DAE (7) system provides the usual energy balance

$$\frac{\mathrm{d}}{\mathrm{d}t} H(x(t)) = -z(x(t))^\top r(z(x(t)))) + y(t)^\top u(t) \leq y(t)^\top u(t).$$

b) pH-DAE subsystems now read

$$\frac{\mathrm{d}}{\mathrm{d}t} E_i x_i(t) = J_i z_i(x_i(t)) - r_i\big(z_i(x_i(t))\big) + B_i u_i(t), \tag{8a}$$

$$y_i(t) = B_i^\top z_i\big(x_i(t)\big) \tag{8b}$$

instead of (4), and if they are coupled by a skew-symmetric coupling relation $\hat{u} + C\hat{y} = 0$ with a skew-symmetric matrix $C = -C^\top$ as before, they can be condensed into an overall pH-DAE system

$$\frac{\mathrm{d}}{\mathrm{d}t} Ex = \hat{J}z - r + \bar{B}\bar{u}, \tag{9a}$$

$$\bar{y} = \bar{B}^\top z \tag{9b}$$

with the skew-symmetric matrix $\hat{J}$ again given by $\hat{J} = J - \hat{B}\hat{C}\hat{B}^\top$.

3 Electrical Networks

We consider the classical charge-/flux oriented MNA network equations [8,9]

$$\frac{\mathrm{d}}{\mathrm{d}t}\begin{bmatrix}0\,0\,0\,A_C\,0\\0\,0\,0\;\;0\;\;I\\0\,0\,0\;\;0\;\;0\\0\,0\,0\;\;0\;\;0\\0\,0\,0\;\;0\;\;0\end{bmatrix}\begin{bmatrix}e\\\jmath_L\\\jmath_V\\q_C\\\phi_L\end{bmatrix}=\begin{bmatrix}0&-A_L&-A_V&0\,0\\A_L^\top&0&0&0\,0\\A_V^\top&0&0&0\,0\\0&0&0&0\,0\\0&0&0&0\,0\end{bmatrix}\begin{bmatrix}e\\\jmath_L\\\jmath_V\\q_C\\\phi_L\end{bmatrix}$$

$$-\begin{bmatrix}A_R g(A_R^\top e)\\0\\0\\q_C-q(A_C^\top e)\\\phi_L-\phi(\jmath_L)\end{bmatrix}+\begin{bmatrix}-A_I&0\\0&0\\0&-I\\0&0\\0&0\end{bmatrix}\begin{bmatrix}\imath(t)\\v(t)\end{bmatrix}$$

with $e, \jmath_L$ and $\jmath_V$ denoting node potentials and currents through flux storing elements and voltages sources, q_C and Φ_l charge and flux-storing elements, $i(t)$ and $v(t)$ independent current and voltage sources, the resistive currents g and the incidence matrices A_C, A_L, A_R, A_V, A_I for charge- and flux storing elements, resistive elements, voltage and current sources, and seek a formulation as a pH-DAE system. For this, we need the following assumptions, which naturally occur in circuit simulation, see [8]:

(a) **Soundness.** The circuit graph has at least one branch and is connected. Furthermore, it contains neither V-loops nor I-cutsets. Equivalently, A_V and $(A_C\, A_R\, A_L\, A_V)^\top$ have full column rank.

(b) **Passivity.** The functions q, ϕ and g fulfill

  (i) $q : \mathbb{R}^{n_C} \to \mathbb{R}^{n_C}$ and $\phi : \mathbb{R}^{n_L} \to \mathbb{R}^{n_L}$ are bijective, continuously differentiable, and their Jacobians

$$\widetilde{C}(u_C) := \frac{\mathrm{d}q}{\mathrm{d}u_C}(u_C), \qquad \widetilde{L}(\jmath_L) := \frac{\mathrm{d}\phi}{\mathrm{d}\jmath_L}(\jmath_L)$$

  are symmetric and positive definite for all $u_C \in \mathbb{R}^{n_C}$, $\jmath_L \in \mathbb{R}^{n_L}$.

  (ii) $g : \mathbb{R}^{n_R} \to \mathbb{R}^{n_R}$ is continuously differentiable, and its Jacobian has the property that $\frac{\mathrm{d}g}{\mathrm{d}u_R}(u_R) + \frac{\mathrm{d}g}{\mathrm{d}u_R}(u_R)^\top$ is positive definite for all $u_R \in \mathbb{R}^{n_R}$.

If $q : \mathbb{R}^{n_C} \to \mathbb{R}^{n_C}$ and $\phi : \mathbb{R}^{n_L} \to \mathbb{R}^{n_L}$ fulfill these assumptions, then there exist twice continuously differentiable and non-negative functions $V_C : \mathbb{R}^{n_C} \to \mathbb{R}$, $V_L : \mathbb{R}^{n_L} \to \mathbb{R}$ with the following property: the gradients of V_C and V_L are, respectively, the inverse functions of q and ϕ. That is,

$$\forall q_C \in \mathbb{R}^{n_C} : \nabla V_C(q_C) = q^{-1}(q_C), \quad \forall \phi_L \in \mathbb{R}^{n_L} : \nabla V_L(\phi_L) = \phi^{-1}(\phi_L).$$

With this setting, the pH-DAE MNA network equations can now be derived as follows: we first eliminate the equation $\phi_L - \phi(\jmath_L) = 0$: $\jmath_L = \phi^{-1}(\phi_L)$; secondly, we replace the equation $q_C - q(A_C^\top e) = 0$ by $A_C^\top e - q^{-1}(q_C) = 0$. We end up with

$$\underbrace{\frac{\mathrm{d}}{\mathrm{d}t}\begin{bmatrix} A_C & 0 & 0 & 0 \\ 0 & I & 0 & 0 \\ 0 & 0 & 0 & 0 \\ 0 & 0 & 0 & 0 \end{bmatrix}}_{E:=}\underbrace{\begin{bmatrix} q_C \\ \phi_L \\ e \\ \jmath_V \end{bmatrix}}_{x:=} = \underbrace{\begin{bmatrix} 0 & -A_L & 0 & -A_V \\ A_L^\top & 0 & 0 & 0 \\ 0 & 0 & 0 & 0 \\ A_V^\top & 0 & 0 & 0 \end{bmatrix}}_{J:=}\underbrace{\begin{bmatrix} e \\ \phi^{-1}(\phi_L) \\ q^{-1}(q_C) \\ \jmath_V \end{bmatrix}}_{z(x)}$$
$$- \underbrace{\begin{bmatrix} A_R g(A_R^\top e) \\ 0 \\ A_C^\top e - q^{-1}(q_C) \\ 0 \end{bmatrix}}_{r(z(x)):=} + \underbrace{\begin{bmatrix} -A_I & 0 \\ 0 & 0 \\ 0 & 0 \\ 0 & -I \end{bmatrix}}_{B:=}\underbrace{\begin{bmatrix} \imath(t) \\ v(t) \end{bmatrix}}_{u(t):=}, \tag{10}$$

which is a pH-DAE of type (7) with subspace $\mathcal{V}$ and Hamiltonian $H(x)$ given by $H(x) = V_C(q_C) + V_L(\phi_L), \mathcal{V} = \left\{ \left(e,\, \jmath_L,\, u_C,\, \jmath_V\right)^\top \in \mathbb{R}^n \middle| \, A_C^\top e = u_C \right\}$.

Remark 2. a) The pH-DAE formulation shares the index properties of charge/flux-oriented MNA network equations, if the assumption on soundness and passivity hold: the index is one if, and only if, it neither contains LI-cutsets nor CV-loops except for C-loops; otherwise it is two.

b) If r subcircuits given as pH-DAE MNA network equations are coupled via voltage/current sources, the overall system can be written as a pH-DAE MNA of type (10).

4 Electromagnetic Devices

In [5], the Maxwell grid equations for an electromagnetic device have been developed as a linear pH-DAE system provided that (a) the three-dimensional domain of the device is connected, bounded and surrounded by perfectly conducting material, (b) the permittivity ϵ, the permeability μ are symmetric positive definite, and the conductivity σ is symmetric positive semi-definite, and (c) finite integration technique [6] has been used for the spatial discretization with orthogonal staggered cells:

$$\begin{bmatrix} M_\mu & 0 \\ 0 & M_\epsilon \end{bmatrix} \frac{\mathrm{d}}{\mathrm{d}t} \begin{bmatrix} \hat{h} \\ \hat{e} \end{bmatrix} = \left(\begin{bmatrix} 0 & -C \\ C^\top & 0 \end{bmatrix} - \begin{bmatrix} 0 & 0 \\ 0 & M_\sigma \end{bmatrix} \right) \begin{bmatrix} \hat{h} \\ \hat{e} \end{bmatrix} + \begin{bmatrix} 0 \\ X_S \end{bmatrix} \hat{u}_2, \tag{11a}$$

$$\hat{y}_2 = \begin{bmatrix} 0 \\ X_S \end{bmatrix}^\top \begin{bmatrix} \hat{h} \\ \hat{e} \end{bmatrix} = X_S^\top \hat{e}. \tag{11b}$$

Here C denotes the discrete curl operator, the material matrices M_ϵ, M_μ and M_σ represent the discretized permittivity, permeability and conductivity distributions, $\hat{e}$ is vector of the electric mesh voltages e, $\hat{h}$ the vector of the magnetic mesh voltages h, and the (dual grid facet) source current $\hat{u}_2$ as input. This input is allocated at positions X_S. In fact, X_S maps the interior mesh links onto the exterior mesh nodes. Furthermore, the respective electric mesh voltage $\hat{y}_2$ forms the output. The Hamiltonian of the electromagnetic device is given by $H_1 = \frac{1}{2}(\tilde{e}^\top M_\epsilon \tilde{e} + \tilde{h}^\top M_\mu \tilde{h})$.

5 Coupled EM/circuit System

When coupling an electromagnetic device with an electric circuit, it remains only to define the inputs, outputs and the coupling equation. For the circuit, the electromagnetic device produces the current $\jmath_E$ flowing into the network, which is assembled at the respective nodes of the circuit via an incidence matrix A_E. Hence the circuit part reads (where we split inputs again in external inputs $\imath$, v, and internal ones):

$$\frac{\mathrm{d}}{\mathrm{d}t}\begin{bmatrix} A_C & 0 & 0 & 0 \\ 0 & I & 0 & 0 \\ 0 & 0 & 0 & 0 \\ 0 & 0 & 0 & 0 \\ 0 & 0 & 0 & 0 \end{bmatrix}\begin{bmatrix} q_C \\ \phi_L \\ e \\ \jmath_V \\ \jmath_E \end{bmatrix} = \begin{bmatrix} 0 & -A_L & 0 & -A_V & -A_E \\ A_L^\top & 0 & 0 & 0 & 0 \\ 0 & 0 & 0 & 0 & 0 \\ A_V^\top & 0 & 0 & 0 & 0 \\ A_E^\top & 0 & 0 & 0 & 0 \end{bmatrix}\begin{bmatrix} e \\ \jmath_L \\ u_C \\ \jmath_V \\ \jmath_E \end{bmatrix} \tag{12a}$$
$$- \begin{bmatrix} A_R g(A_R^\top e) \\ 0 \\ A_C^\top e - u_C \\ 0 \\ 0 \end{bmatrix} + \begin{bmatrix} -A_I & 0 \\ 0 & 0 \\ 0 & 0 \\ 0 & -I \\ 0 & 0 \end{bmatrix}\begin{bmatrix} \imath(t) \\ v(t) \end{bmatrix} + \begin{bmatrix} 0 \\ 0 \\ 0 \\ 0 \\ 1 \end{bmatrix}\hat{u}_1,$$

$$\begin{bmatrix} \bar{y}_{1,1} \\ \bar{y}_{1,2} \\ \hat{y}_1 \end{bmatrix} = \begin{bmatrix} -A_I & 0 & 0 \\ 0 & 0 & 0 \\ 0 & 0 & 0 \\ 0 & -I & 0 \\ 0 & 0 & 1 \end{bmatrix}^\top \cdot \begin{bmatrix} e \\ \jmath_L \\ u_C \\ \jmath_V \\ \jmath_E \end{bmatrix} = \begin{bmatrix} -A_I^\top e \\ -\jmath_V \\ \jmath_E \end{bmatrix} \tag{12b}$$

with the Hamiltonian: $H_2 = V_C(q_C) + V_L(\phi_L)$.

The coupling is as follows [5]: the input $\hat{u}_1$ (of the electric circuit) is given by the voltage drop at the electromagnetic device, which reads $\hat{u}_1 = -X_S^\top \tilde{e} = -\hat{y}_2$; on the other hand, the input $\hat{u}_2$ (of the magnetic device) is given by the current $\hat{u}_2 = \jmath_E = \hat{y}_1$. Overall, we get the following skew-symmetric relation between inputs and outputs:

$$0 = \begin{bmatrix} \hat{u}_1 \\ \hat{u}_2 \end{bmatrix} + \begin{bmatrix} 0 & I \\ -I & 0 \end{bmatrix}\begin{bmatrix} \hat{y}_1 \\ \hat{y}_2 \end{bmatrix}. \tag{13}$$

As we have a system consisting of two pH-DAE systems (11) and (12) with a skew-symmetric linear coupling condition (13), the overall system can be written as a condensed pH-DAE system (9) with Hamiltonian $H = H_1 + H_2$ and enlarged matrices as above.

6 Simulation Strategies

Generally, for simulating the coupled EM/circuit system numerically, two approaches are feasible:

a) *Monolithic approach.* The condensed system (9) can be solved by any integration scheme suitable for index-1 and index-2 systems, depending on the index. To preserve the dissipation inequality also on a discrete level, collocation schemes [13] and discrete gradient schemes tracing back to [7] are the methods-of choice. This strategy is also referred to as strong coupling.

b) *Monolithic multirate approach.* In fact, we are facing models, where the subsystems can have widely separated time scales. This can create so-called multirate potential, where it is beneficial to employ schemes, which use inherent step sizes for each subsystem. In this way, each subsystem can be sampled on its time scale. See e.g. [2, 10].

c) *Weak coupling.* Since the coupling equations is merely the one-to-one identification of output and input, we can insert this. Furthermore, omitting outputs due to external sources, we have

$$
\frac{\mathrm{d}}{\mathrm{d}t}\begin{bmatrix} A_C & 0 & 0 & 0 \\ 0 & I & 0 & 0 \\ 0 & 0 & 0 & 0 \\ 0 & 0 & 0 & 0 \\ 0 & 0 & 0 & 0 \end{bmatrix}\begin{bmatrix} q_C \\ \phi_L \\ e \\ \jmath_V \\ \jmath_E \end{bmatrix} = \begin{bmatrix} 0 & -A_L & 0 & -A_V & -A_E \\ A_L^\top & 0 & 0 & 0 & 0 \\ 0 & 0 & 0 & 0 & 0 \\ A_V^\top & 0 & 0 & 0 & 0 \\ A_E^\top & 0 & 0 & 0 & 0 \end{bmatrix}\begin{bmatrix} e \\ \jmath_L \\ u_C \\ \jmath_V \\ \jmath_E \end{bmatrix}
- \begin{bmatrix} A_R g(A_R^\top e) \\ 0 \\ A_C^\top e - u_C \\ 0 \\ 0 \end{bmatrix} + \begin{bmatrix} -A_I & 0 \\ 0 & 0 \\ 0 & 0 \\ 0 & -I \\ 0 & 0 \end{bmatrix}\begin{bmatrix} \imath(t) \\ v(t) \end{bmatrix} - \begin{bmatrix} 0 \\ 0 \\ 0 \\ 0 \\ 1 \end{bmatrix}\hat{y}_2, \tag{14a}
$$

$$
\hat{y}_1 = \jmath_E \tag{14b}
$$

and

$$
\begin{bmatrix} M_\mu & 0 \\ 0 & M_\epsilon \end{bmatrix}\frac{\mathrm{d}}{\mathrm{d}t}\begin{bmatrix} \tilde{h} \\ \tilde{e} \end{bmatrix} = \left(\begin{bmatrix} 0 & -C \\ C^\top & 0 \end{bmatrix} - \begin{bmatrix} 0 & 0 \\ 0 & M_\sigma \end{bmatrix}\right)\begin{bmatrix} \tilde{h} \\ \tilde{e} \end{bmatrix} + \begin{bmatrix} 0 \\ X_S \end{bmatrix}\hat{y}_1 \tag{15a}
$$

$$
\hat{y}_2 = X_S \tilde{e}. \tag{15b}
$$

Here dynamic iteration schemes [1] are the methods-of choice, as due to the ODE-DAE coupling no stability constraints occur [4]. In addition, each step of a Jacobi or Gauß-Seidel iteration scheme defines a pH-DAE system by its own [8].

Operator splitting approaches are not generally feasible for differential-algebraic equations, which can easily be seen for the linear pH-DAE (3a) with $z(x) = x$ and $B = 0$. A Lie-Trotter splitting approach may read

$$
\frac{\mathrm{d}}{\mathrm{d}t}(Ex) = Jx, \quad x(0) = x_0,
$$

$$
\frac{\mathrm{d}}{\mathrm{d}t}(Ew) = -Rw. \quad w(0) = x(T),
$$

allowing for using a symplectic integrator for the first step, and a dissipative scheme for the second one. However, the matrix pencil $\{E, J\}$ or $\{E, R\}$ may be singular and thus not define a unique solution for the respective subproblem, even if the matrix pencil $\{E, J - R\}$ of the overall system is regular. Even if this does not happen, the first problem, for example, may not allow for a unique solution for arbitrary choices of consistent initial values. For

$$E = \mathrm{diag}(1,0,1), \quad J = \begin{bmatrix} 0 & -1 & 0 \\ 1 & 0 & 0 \\ 0 & 0 & 0 \end{bmatrix}, \quad R = \mathrm{diag}(0,1,1), \quad x_0 = \begin{bmatrix} 1 \\ -1 \\ 0 \end{bmatrix},$$

all matrix pencils $\{E, J - R\}$, $\{E, J\}$ and $\{E, R\}$ are regular, but the first step yields $x_1 \equiv 0 \neq 1$.

One may overcome the problem by rewriting the DAE in terms of an underlying ODE and subsequent algebraic variables given by explicit evaluations. For network equations a branch oriented loop-cutset approach is an option for defining such a pH-DAE system, see [5]. Another way to avoid the problems above is to follow an operator splitting based approach for dynamic iteration. In the latter case, no stability problems occur and a monotone convergence can be obtained [3].

7 Conclusions

Port-Hamiltionian (pH) systems provide a modelling framework which preserves essential physical properties. It is especially suited for multiphysical applications, as the proper coupling of pH subsystems yields an overall pH system. In electrical engineering, we have shown that electrical networks and electromagnetic devices can be written as pH systems, and coupled EM/circuit system yield coupled pH systems with a skew-symmetric coupling, which can be rewritten as an overall pH system. For simulation, a monolithic approach is suitable for the former, and weak coupling methods for the latter. There are still many unresolved questions, such as how to adequately integrate distributed ports into the pH system's modeling.

Acknowledgements. Michael Günther is indebted to the funding given by the European Union's Horizon 2020 research and innovation programme under the Marie Sklodowska-Curie Grant Agreement No. 765374, ROMSOC.

References

1. Arnold, M., Günther, M.: Preconditioned dynamic iteration for coupled differential-algebraic equations. BIT **41**, 1–25 (2001)
2. Bartel, A., Günther, M.: Multirate schemes – an answer of numerical analysis to a demand from applications. In: Günther, M., Schilders, W. (eds.) Novel Mathematics Inspired by Industrial Challenges, pp. 5–27. Springer (2022). https://doi.org/10.1007/978-3-030-96173-2_1

3. Bartel, A., Günther, M., Jacob, B., et al.: Operator splitting based dynamic iteration for linear differential-algebraic port-Hamiltonian systems. Numer. Math. **155**, 1–34 (2023). https://doi.org/10.1007/s00211-023-01369-5
4. Bartel, A., Brunk, M., Günther, M., Schöps, S.: Dynamic iteration for coupled problems of electric circuits and distributed devices. SIAM J. Sci. Comput. **35**(2), B315–B335 (2013). https://doi.org/10.1137/120867111
5. Diab, M.: Splitting methods for partial differential-algebraic systems with application on coupled field-circuit DAEs. Ph.D. thesis, Humboldt Universität zu Berlin (2022)
6. Weiland, T.: Time domain electromagnetic field computation with finite difference methods. Int. J. Numer. Model. Electr. Netw. Devices Fields **9**, 295–319 (1996)
7. Gonzalez, O.: Time integration and discrete Hamiltonian systems. J. Nonlinear Sci. **6**, 1432–1467 (1996)
8. Günther, M., Bartel, A., Jacob, B., Reis, T.: Dynamic iteration schemes and port-Hamiltonian formulation in coupled DAE circuit simulation. Int. J. Circuit Theory Appl. **49**, 430–452 (2021)
9. Günther, M., Feldmann, U.: CAD based electric circuit modeling I: mathematical structure and index of network equations. Surv. Math. Ind. **8**, 97–129 (1999)
10. Günther, M., Sandu, A.: Multirate generalized additive Runge Kutta methods. Numer. Math. **133**, 497–524 (2016). https://doi.org/10.1007/s00211-015-0756-z
11. Jacob, B., Zwart, H.J.: Linear Port-Hamiltonian Systems on Infinite-Dimensional Spaces. Birkhäuser Verlag, Basel (2012)
12. Jeltsema, D., van der Schaft, A.J.: Port-Hamiltonian systems theory: an introductory overview. Found. Trends Syst. Control **1**(2–3), 173–387 (2014)
13. Mehrmann, V., Morandin, R.: Structure-preserving discretization for port-Hamiltonian descriptor systems. In: 2019 IEEE 58th Conference on Decision and Control (CDC), pp. 6863–6868 (2019)
14. van der Schaft, A.J.: Port-Hamiltonian systems: network modeling and control of nonlinear physical systems. In: Schlacher, K., Irschnik, H. (eds.) Advanced Dynamics and Control of Structures and Machines. CISM, vol. 444, pp. 127–167. Springer, Vienna (2004). https://doi.org/10.1007/978-3-7091-2774-2_9
15. van der Schaft, A.: Port-Hamiltonian systems: an introductory survey. In: Sanz-Sole, M., Soria, J., Varona, J.L., Verdera, J. (eds.) Proceedings of the International Congress of Mathematicians, vol. III, pp. 1339–1365. European Mathematical Society Publishing House (2006)

Large-Scale $\mathscr{H}_2$ Optimization for Thermo-Mechanical Reliability of Electronics

Pascal den Boef(✉), Jos Maubach, Wil Schilders, and Nathan van de Wouw

Eindhoven University of Technology, Eindhoven, The Netherlands
p.d.boef@tue.nl

Abstract. Optimization of transient models is required in several domains related to thermo-mechanical reliability of electronics, such as Prognostic Health Monitoring (PHM) and design optimization. A novel framework for efficient (local) parameter optimization of transient models in the $\mathscr{H}_2$ norm is proposed. The optimization is feasible for large-scale transient models because it approximates the gradient using physics-based model order reduction (MOR), in contrast to existing approaches that typically use data-driven surrogate models such as neural networks. To demonstrate the framework an optimal fixed-order virtual sensor for PHM of a Ball Grid Array (BGA) is numerically determined.

1 Introduction

Electronics have become an indispensable part of modern appliances. Their safety is critical in certain applications such as automotive, where electronics are expected to operate reliably under both thermal and mechanical loads. To achieve this, transient simulation models are used for various stages of the lifecycle of an electronic product. Often, optimization of these models plays a role. Two examples are 1) minimization of mechanical stress during the design of a product and 2) minimization of the prediction error of a virtual sensor that predicts mechanical stress during operation for PHM.

These examples demonstrate the importance of optimization of transient models. However, these models are typically generated using Finite Element Method (FEM) and, hence, are large-scale. As a result, much attention has been paid in literature to replace these high-fidelity models by inexpensive surrogate models, of which an overview is now provided.

In the design phase, a Response Surface Model (RSM) is used in [1] as a surrogate model to capture the transient behavior of chip stress during thermal cycling as a function of several geometric design parameters. Subsequently, an optimization is carried out over the RSM. The authors of [2] apply a recurrent neural network to model the solder joint reliability of a glass wafer level chip-scale package as a function of design parameters.

For PHM, in [3], different neural network architectures are employed to predict the stress distribution of electronic packages during thermal cycling based on past observations. In [4], a neural network trained from FEM snapshots is used as virtual sensor for internal mechanical stresses with the goal of detecting delamination.

M. van Beurden et al. (Eds.): SCEE 2022, MI 43, pp. 144–151, 2024.
https://doi.org/10.1007/978-3-031-54517-7_16

Based on the above examples, it is clear that transient thermo-mechanical models play a key role in electronics reliability assessment. However, the surrogate models employed in the literature for design optimization and predictor models for PHM, such as neural networks and RSM, are not physics-based. Therefore, large amounts of high-quality data is required (which is often not available in the context of microelectronics) and generalization to situations not present in the data is poor.

Alternatively, this paper provides a physics-based approach for the optimization of transient models. Namely, projection-based model reduction is employed to accelerate the optimization. The proposed technique is general in the sense that it can readily be applied to a wide class of problems, including design optimization and PHM as highlighted here.

The structure of the article is as follows. Section 2 provides background on $\mathcal{H}_2$ model reduction, an important ingredient in the proposed optimization framework which is treated in Sect. 3. Numerical results for a virtual sensor optimization for a BGA model are presented in Sect. 4. Concluding remarks are given in Sect. 5.

2 Reduction in the $\mathcal{H}_2$ Norm

The $\mathcal{H}_2$ norm is a norm for transient systems that are linear, time-invariant, strictly proper and asymptotically stable. Let this set of systems be denoted by $\mathcal{R}$. The $\mathcal{H}_2$ norm is induced by the $\mathcal{H}_2$ inner product, which is defined as follows for $G, H \in \mathcal{R}$:

$$\langle G,H\rangle_{\mathcal{H}_2} = \frac{1}{2\pi}\int_{-\infty}^{\infty} \operatorname{tr}\left(G(\mathrm{i}\omega)H(\mathrm{i}\omega)^*\right)\mathrm{d}\omega, \tag{1}$$

where $G(s)$ and $H(s)$ denote, by slight abuse of notation, the transfer functions of G and H evaluated at complex frequency s. Then, $\|G\|_{\mathcal{H}_2} = \sqrt{\langle G,G\rangle_{\mathcal{H}_2}}$.

The $\mathcal{H}_2$ norm is attractive as a measure for transient systems for several reasons. First, it can be seen as the average magnitude of the frequency response of G over all frequencies. Weighting filters may be used to emphasize frequency ranges of interest for a particular problem. The physical interpretation of the $\mathcal{H}_2$ makes it an appropriate choice for many engineering problems. Second, extensions of the $\mathcal{H}_2$ norm for non-linear systems have been proposed, such as bilinear systems [5]. Third, the $\mathcal{H}_2$ norm has mathematical properties that make it possible to design efficient model reduction schemes that are optimal in this norm, as will be seen next.

The $\mathcal{H}_2$ norm is commonly used to find an optimal approximation to some large-scale system G by a much smaller reduced-order model (ROM):

$$\mu^* = \arg\min_{\mu} \left\|G - \widehat{G}(\mu)\right\|_{\mathcal{H}_2}^2, \tag{2}$$

with μ parameterizing the search space of ROMs.

In literature, many methods for solving (2) for large-scale G have been proposed. Gradient-based methods are proposed in [6–8]. Another class of method is based on discretization of the $\mathcal{H}_2$ norm [9]. A third type of method is based on iteratively constructing reduced-order models that, upon convergence, satisfy a set of necessary optimality conditions [10].

3 General $\mathcal{H}_2$ Optimization

As shown in the previous section, there is a wide selection of techniques for solving the standard $\mathcal{H}_2$ reduction problem (2). These methods can be applied to large-scale problems arising, for instance, from FEM discretization. However, the motivating examples encountered in Sect. 1 indicate the need to solve a more general $\mathcal{H}_2$ problem:

$$\mu^* = \arg\min_{\mu} \|G(\mu)\|^2_{\mathcal{H}_2}, \tag{3}$$

with G a large-scale transient system parameterized in a parameter vector $\mu \in \mathbb{R}^{n_\mu}$. The physical interpretation of μ depends on the problem. In the context of design optimization, μ represents material and geometric parameters, such as the shape of solder balls or spacing between components on a Printed Circuit Board (PCB). When considering virtual sensor design, μ represents, e.g., state-space coefficients of the virtual sensor model.

Solving (3) is challenging because, similar to (2), the cost function is 1) large-scale, and 2) typically non-convex. In addition, while the optimization variables of the cost function of the reduction problem (2) are associated with the small-scale reduced order model $\widehat{G}$, this is not the case with (3).

To solve (3), a gradient-based technique is proposed. The gradient can be conveniently written using the $\mathcal{H}_2$ inner product. Let μ_i be the ith component of μ. Define $f(\mu) := \|G(\mu)\|^2_{\mathcal{H}_2}$. Then, the partial derivative of f w.r.t. μ_i may be compactly written using the properties of the inner product as

$$\frac{\partial f}{\partial \mu_i}(\mu) = 2\mathrm{Re}\left(\left\langle G(\mu), \frac{\partial G}{\partial \mu_i}(\mu)\right\rangle_{\mathcal{H}_2}\right) = 2\left(\left\langle G(\mu), \frac{\partial G}{\partial \mu_i}(\mu)\right\rangle_{\mathcal{H}_2}\right) \tag{4}$$

where $\mathrm{Re}(\cdot)$ denotes the real part of the argument. The second inequality holds since the considered transient systems, $G(\mu)$ and $\frac{\partial G}{\partial \mu_i}(\mu)$ are real. The gradient of f is

$$\nabla f(\mu) = 2\left[\left\langle G(\mu), \tfrac{\partial G}{\partial \mu_1}(\mu)\right\rangle_{\mathcal{H}_2} \cdots \left\langle G(\mu), \tfrac{\partial G}{\partial \mu_{n_\mu}}(\mu)\right\rangle_{\mathcal{H}_2}\right]^{\mathrm{T}}. \tag{5}$$

The $\mathcal{H}_2$ inner product (1) is only defined if both arguments belong to $\mathcal{R}$. Although this assumption was made for $G(\mu)$, it can be shown that its parameter-gradient system inherits, among others, these properties if the state-space coefficients of $G(\mu)$ are differentiable functions of μ, in which case its transfer function may be written as

$$G(\mu; s) = C(\mu)(sE(\mu) - A(\mu))^{-1}B(\mu). \tag{6}$$

The following theorem shows that for systems of the form (6) the parameter-gradient system inherits all required properties for (5) to exist.

Theorem 1. *Let $G(\mu) \in \mathcal{R}$ be a parameterized transient system with transfer function of the form (6). Furthermore, let $\mu \in \mathbb{D} \subset \mathbb{R}^{n_\mu}$ with $\mathbb{D}$ an open subset of $\mathbb{R}^{n_\mu}$. Then, (5) exists for all $\mu \in \mathbb{D}$.*

Proof. Using the identity from matrix calculus $\frac{\mathrm{d}F^{-1}}{\mathrm{d}x} = -F^{-1}\frac{\mathrm{d}F}{\mathrm{d}x}F^{-1}$, the parameter-gradient of $G(\mu)$ w.r.t. μ_i is derived as

$$\begin{aligned}\frac{\partial G}{\partial \mu_i}(\mu;s) = &\frac{\partial C}{\partial \mu_i}(\mu)(sE(\mu)-A(\mu))^{-1}B(\mu)\\ &+C(\mu)(sE(\mu)-A(\mu))^{-1}\frac{\partial B}{\partial \mu_i}(\mu)\\ &-C(\mu)(sE(\mu)-A(\mu))^{-1}\left(s\frac{\partial E}{\partial \mu_i}(\mu)-\frac{\partial A}{\partial \mu_i}(\mu)\right)(sE(\mu)-A(\mu))^{-1}B(\mu).\end{aligned} \tag{7}$$

The first two terms on the right-hand side of (7) are transfer functions of the same form as $G(\mu)$ that have furthermore the same pole locations as G. Thus, they are asymptotically stable. The last term is slightly more complicated, with additional appearances of the complex frequency s. However, it is equivalent to

$$\begin{bmatrix}0 & -C\end{bmatrix}\left(s\begin{bmatrix}E & 0\\ \frac{\partial E}{\partial \mu_i} & E\end{bmatrix}-\begin{bmatrix}A & 0\\ \frac{\partial A}{\partial \mu_i} & A\end{bmatrix}\right)^{-1}\begin{bmatrix}B\\ 0\end{bmatrix}, \tag{8}$$

where dependence on μ has been omitted for brevity. The equivalence can be easily checked using the Schur complement. From this equivalence, it can be seen that the third term of (7) admits a descriptor state-space representation with twice the order of $G(\mu)$ and block-triangular E and A matrices as given in (8). From this block-triangular representation, it can be readily seen that the poles of this third term coincide with the poles of $G(\mu)$ with double multiplicity. Since the above decomposition holds for all components of μ, properness and asymptotic stability of the parameter-gradient systems is guaranteed and, thus, the gradient (5) exists. □

It can be seen from (7) that the state order of the parameter-gradient systems is at most $n+n+2n=4n$.

3.1 Gradient-Based Optimization

In the previous section, it was shown that for a wide-class of systems, the cost function of (3) is differentiable. This motivates the application of a gradient-based technique. A large class of gradient-based methods exist (see, e.g., [11] for an overview). However, in this work only steepest descent is considered, i.e., if the parameter in the k-th iterate is denoted by $\mu^{(k)}$, then

$$\mu^{(k+1)} = \mu^{(k)} - \alpha^{(k)}\nabla f(\mu^{(k)}) \tag{9}$$

with $\alpha^{(k)}$ the step size. Note that, as f is non-convex in realistic scenarios, proper initialization of (9) is required, for example using Bayesian optimization.

3.2 Approximating the Gradient

Computation of $\left\langle G, \frac{\partial G}{\partial \mu_i}\right\rangle_{\mathcal{H}_2}$ $(i=1,\ldots,n_\mu)$ involves the solution of 2 generalized Sylvester equations [12, Lemma 2.1.5] with computational complexity $\mathcal{O}(n^3)$. Thus, computation of (5) is infeasible for large-scale $G(\mu)$. However, the following theorem shows that the inner product can be computed efficiently if one of the 2 arguments is of low order.

Theorem 2. *Let $G, H \in \mathscr{R}$. If G has a sparse descriptor state-space representation of order n and H has a state-space representation of order $r \ll n$, then computing $\langle G, H \rangle_{\mathscr{H}_2}$ can be done with complexity $\mathscr{O}(nr + r^3) = \mathscr{O}(nr)$.*

Proof. See [13].

The significance of Theorem 2 is that if only 1 of the arguments of the inner product is reduced (e.g., by $\mathscr{H}_2$ model reduction), then the calculation is efficient. Such a reduction can be performed element-wise to approximate (5). A choice can be made to reduce $G(\mu)$, the parameter-gradient systems $\frac{\partial G}{\partial \mu_i}(\mu)$ $(i = 1, \ldots, n_\mu)$ or both. Table 1 lists the computational complexity of these options. It shows that all choices have the same complexity. At the same time, introducing fewer errors leads to a more accurate gradient approximation. Thus, it is expected that the optimization converges fastest if only 1 of the arguments is reduced.

4 Numerical Comparison

The proposed optimization is tested for a PHM use case using a transient thermo-mechanical simulation model of a BGA. The geometry can be seen in Fig. 1. Using the MATLAB Partial Differential Equation Toolbox, a FEM semi-discretization of the geometry using linear basis functions was performed, resulting in a sparse descriptor state-space model (G_{BGA}) of order $n = 544$ modelling the transfer from die power dissipation to 2 outputs: 1) the vertical component of the stress in the solder ball labeled "SQ10" (which cannot be measured and should be predicted); and 2) a temperature sensor located directly underneath the solder ball. Measurements of the die power and temperature at the sensor are assumed available, corrupted by additive white noise.

Table 1. Possible choices for element-wise approximation of (5). A reduction scheme of complexity $\mathscr{O}(nr)$ is used with $r \ll n$.

Form	Reduction cost	Evaluation cost	Total cost
$\left\langle G, \frac{\partial G}{\partial \mu_i} \right\rangle$	(no approximation)	$\mathscr{O}(n_\mu n^3)$	$\mathscr{O}(n_\mu n^3)$
$\left\langle \widehat{G}, \frac{\partial G}{\partial \mu_i} \right\rangle$	$\mathscr{O}(nr)$	$\mathscr{O}(n_\mu nr)$	$\mathscr{O}(n_\mu nr)$
$\left\langle G, \widehat{\frac{\partial G}{\partial \mu_i}} \right\rangle$	$\mathscr{O}(n_\mu nr)$	$\mathscr{O}(n_\mu nr)$	$\mathscr{O}(n_\mu nr)$
$\left\langle \widehat{G}, \widehat{\frac{\partial G}{\partial \mu_i}} \right\rangle$	$\mathscr{O}(n_\mu nr)$	$\mathscr{O}(r^3)$	$\mathscr{O}(n_\mu nr)$

A fixed-order ($r = 4$) virtual sensor (with state-space coefficients parametrized in μ) should be designed to predict the solder ball stress in an $\mathcal{H}_2$ optimal way.

The following techniques for obtaining a virtual sensor are compared:

1. K: a reference virtual sensor that is found by synthesizing a Kalman filter for G_{BGA} directly. Although this virtual sensor is of order n, it is $\mathcal{H}_2$ optimal and gives a lower bound on the prediction error achievable with fixed-order virtual sensors. Note that synthesizing a full-order Kalman filter costs $\mathcal{O}(n^3)$ and thus is infeasible for large n.
2. K_r: Iterative Rational Krylov Approximation (IRKA) is applied to obtain an r-th order ROM $\widehat{G}_{\text{BGA}} \approx G_{\text{BGA}}$ and, subsequently, a Kalman filter is synthesized for this approximate model.
3. $K_{r,\text{opt1}}$: the virtual sensor obtained using the proposed optimization framework. The gradient is approximated by projecting, in each iteration, $G(\mu)$ using IRKA with order 20. Note that this order can be selected independent of the order of the virtual sensor. The search is initialized from K_r and the step size in each iteration is determined using an Armijo line search [11]. The iteration is stopped when the step size is smaller than 10^{-10}.
4. $K_{r,\text{opt2}}$: the same procedure for $K_{r,\text{opt1}}$ is followed, but both $G(\mu)$ and the parameter-gradient systems are reduced to approximate the gradient.

The prediction error for all virtual sensors is displayed in Fig. 2. As expected, the lowest error (0.40) is achieved by K. Furthermore, the proposed method finds fixed-order sensors with lower error than K_r (the Kalman filter synthesized on the ROM of G_{BGA}).

For every 100 iterations, the relative 2-norm error of the gradient approximation is calculated for $K_{r,\text{opt1}}$ and $K_{r,\text{opt2}}$. It is on average, respectively, 0.095 and 0.765. Surprisingly, despite the higher gradient error, $K_{r,\text{opt2}}$ achieves the lowest prediction error. One possible explanation for this is that the additional gradient noise pushes the optimizer towards a better local minimum.

Note that in some frequency regions, K_r has lower error than $K_{r,\text{opt1}}$, $K_{r,\text{opt2}}$ or even K: the $\mathcal{H}_2$ characterizes the *average* frequency response. Thus, higher errors in certain regions may be compensated by lower errors elsewhere. In practical applications, including weighting filters allows the engineer to steer the objective and ensure low error is achieved in the important frequency bands.

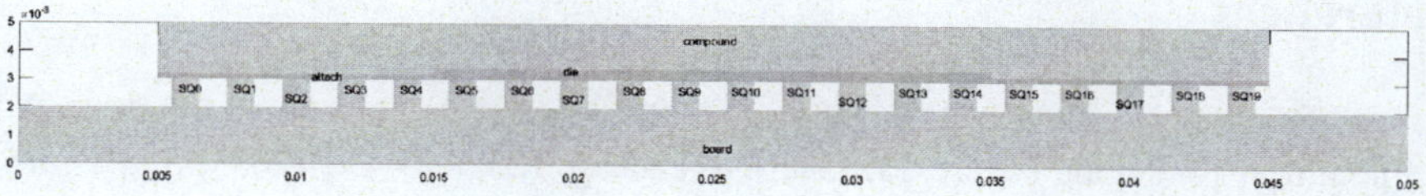

Fig. 1. Geometry of BGA model. Solder balls are denoted by "SQ".

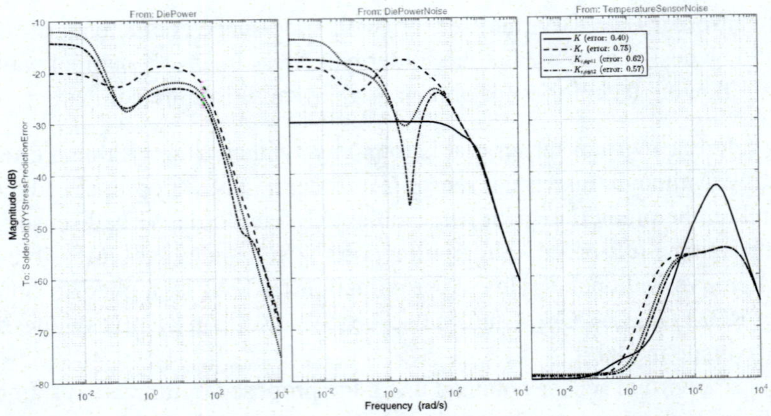

Fig. 2. Bode magnitude diagram comparing the prediction error of the 4 virtual sensors. The transfer from die power to prediction error is not shown for the full-order Kalman filter (K) as it is zero in theory and only limited by machine precision.

5 Conclusions

A novel gradient-based low-complexity technique is proposed for the optimization of large-scale parametric transient models in the $\mathscr{H}_2$ norm. The approach can be applied to a wide class of problems for which the transient cost can be expressed in terms of the $\mathscr{H}_2$ norm. In particular, it was shown how the approach can be applied to relevant problems in the area of thermo-mechanical reliability analysis of electronics.

The method was demonstrated on the synthesis of a reduced-order virtual sensor for PHM of a BGA package.

As next steps, the method should be extended to non-linear models including temperature-dependent material parameters and visco-plasticity. In addition, due to space restrictions, an analysis of the effect the gradient approximation has on convergence will be published in a future article.

Acknowledgements. This work has been funded in part by ITEA under the COMPAS project (ITEA project 19037).

References

1. van Driel, W.D., Zhang, G.Q., Janssen, J.H.J., Ernst, L.J.: Response surface modeling for nonlinear packaging stresses. J. Electron. Packag. **125**, 490–497 (2003). https://doi.org/10.1115/1.1604149
2. Yuan, C.C.A., Lee, C.C.: Solder joint reliability modeling by sequential artificial neural network for glass wafer level chip scale package. IEEE Access **8**, 143,494–143,501 (2020). https://doi.org/10.1109/ACCESS.2020.3014156
3. Majd, M., Meszmer, P., Priscaru, A., Gromala, P.J., Wunderle, B.: Stress prognostics for encapsulated standard packages by neural networks using data from in-situ condition monitoring during thermal shock tests. In: 2020 21st International Conference on Thermal, Mechanical and Multi-Physics Simulation and Experiments in Microelectronics and

Microsystems (EuroSimE), pp. 1–10. IEEE (2020). https://doi.org/10.1109/EuroSimE48426.2020.9152697

4. Prisacaru, A., et al.: Towards virtual twin for electronic packages in automotive applications. Microelectron. Reliab. **122**, 114,134 (2021). https://doi.org/10.1016/j.microrel.2021.114134. https://linkinghub.elsevier.com/retrieve/pii/S0026271421001001
5. Zhang, L., Lam, J.: On h_2 model reduction of bilinear systems. Automatica **38**, 205–216 (2002). https://doi.org/10.1016/S0005-1098(01)00204-7. https://linkinghub.elsevier.com/retrieve/pii/S0005109801002047
6. Beattie, C.A., Gugercin, S.: A trust region method for optimal h2 model reduction. In: Proceedings of the 48h IEEE Conference on Decision and Control (CDC) held jointly with 2009 28th Chinese Control Conference, pp. 5370–5375. IEEE (2009). https://doi.org/10.1109/CDC.2009.5400605
7. Yang, P., Jiang, Y.L.: H2 optimal model reduction of coupled systems on the grassmann manifold. Math. Model. Anal. **22**, 785–808 (2017). https://doi.org/10.3846/13926292.2017.1381863. http://journals.vgtu.lt/index.php/MMA/article/view/927
8. Petersson, D.: A nonlinear optimization approach to h2-optimal modeling and control (2013)
9. Hokanson, J.M., Magruder, C.C.: H2-optimal model reduction using projected nonlinear least squares. SIAM J. Sci. Comput. **42**, A4017–A4045 (2020). https://doi.org/10.1137/19M1247863. https://epubs.siam.org/doi/10.1137/19M1247863
10. Güğercin, S., Antoulas, A.C., Beattie, C.: H2 model reduction for large-scale linear dynamical systems. SIAM J. Matrix Anal. Appl. **30**, 609–638 (2008). https://doi.org/10.1137/060666123. http://epubs.siam.org/doi/10.1137/060666123
11. Nocedal, J., Wright, S.: Numerical Optimization, 2nd edn. Springer, New York (2006). https://doi.org/10.1007/978-0-387-40065-5
12. Antoulas, A.C., Beattie, C.A., Güğercin, S.: Interpolatory Methods for Model Reduction. Society for Industrial and Applied Mathematics (2020). https://doi.org/10.1137/1.9781611976083. https://epubs.siam.org/doi/book/10.1137/1.9781611976083
13. Breiten, T., Beattie, C., Güğercin, S.: Near-optimal frequency-weighted interpolatory model reduction. Syst. Control Lett. **78**, 8–18 (2015). https://doi.org/10.1016/j.sysconle.2015.01.005. https://linkinghub.elsevier.com/retrieve/pii/S0167691115000067

Check for updates

Data-Driven Model Order Reduction of Parameterized Dissipative Linear Time-Invariant Systems

Tommaso Bradde(✉), Alessandro Zanco, and Stefano Grivet-Talocia

Politecnico di Torino, Turin, Italy
{tommaso.bradde,alessandro.zanco,stefano.grivet}@polito.it

Abstract. We introduce a framework for data-driven model order reduction of parameterized LTI systems with guaranteed uniform dissipativity. The strategy casts the problem into a multivariate rational fitting scheme that formally preserves the bounded realness of the model response. The formulation relies on the solution of a semi-definite program arising from a rational parameterization based on Bernstein polynomials. The models can be employed in system-level simulations both in the frequency and time domain.

1 Introduction

Parameterized Reduced Order Models (pROMs) of dissipative systems are valuable tools for enabling fast simulation and optimization of complex electrical components depending on a number of free design parameters. These models reproduce the input-output behavior of the underlying structure and its dependency on the parameters, making use of a minimal set of explanatory variables. Use of pROMs drastically reduces simulation time requirements at the system level, especially for what concerns transient analyses.

In order to be fully exploitable within large system-level simulations, pROMs of physically passive structures must be compliant with the dissipativity property of the reference system for all the admissible parameters values. Even if accurate, pROMs that do not exhibit this property may be the root cause of spurious numerical instabilities, that compromise the validity of the results.

By restricting the focus on Linear Time-Invariant (LTI) systems, this contribution presents a novel data-driven framework for generating pROMs that are dissipative by construction. Differently from recent approaches based on port-Hamiltonian realizations [1,7], the proposed approach represents the model as a rational transfer function with parameterized coefficients. Based on this structure, the model identification stage involves the solution of a sequence of constrained convex rational fitting problems. By representing the parameterized coefficients of the model transfer function as Bernstein polynomials expansions, we show that the involved infinite-dimensional frequency domain conditions for dissipativity can be formulated as Linear Matrix Inequalities (LMI) of finite dimension, which are enforced in polynomial time making use of robust convex optimization solvers.

M. van Beurden et al. (Eds.): SCEE 2022, MI 43, pp. 152–158, 2024.
https://doi.org/10.1007/978-3-031-54517-7_17

2 Background and Notation

In the following, s will denote the Laplace variable, $\mathscr{S}_n$ the set of symmetric matrices of size n. The symbol $\otimes$ will stand for the Kronecker product, while the superscripts $\top$, $\star$ and $*$ will denote transposition, hermitian transposition and complex conjugation, respectively. The functions $b_{\boldsymbol{\ell}}^{\bar{\boldsymbol{\ell}}}(\mathbf{x}), \mathbf{x} \in [0,1]^d \subset \mathbb{R}^d$ are multivariate Bernstein polynomials whose degree in each scalar variable is collected in the d-dimensional multi-index $\bar{\boldsymbol{\ell}} = (\ell_1, \ldots, \ell_d)$. Accordingly, $\boldsymbol{\ell}$ is an identifier for each component of this basis and $\mathscr{I}_{\bar{\boldsymbol{\ell}}}$ is the set of admissible multi-indices spanning the basis.

2.1 Problem Statement

Our goal is to generate a pROM of a P-port dissipative LTI system depending on a set of external (normalized) parameters $\boldsymbol{\vartheta} \in \Theta = [0,1]^d \subset \mathbb{R}^d$. We assume that the equations describing the reference system are not known in closed form, but that samples of its parameterized input-output frequency response $\breve{H}(s, \boldsymbol{\vartheta}) \in \mathbb{C}^{\mathrm{P}\times\mathrm{P}}$ are made available by real or virtual high-fidelity measurements

$$\breve{H}_{k,m} = \breve{H}(j\omega_k, \boldsymbol{\vartheta}_m), \quad 1 \le k \le \bar{k},\ 1 \le m \le \bar{m}, \tag{1}$$

retrieved for fixed frequency-parameter configurations. The problem is thus to synthesize a small order transfer function $\mathsf{H}(s, \boldsymbol{\vartheta})$ fitting the available samples

$$\mathsf{H}(j\omega_k, \boldsymbol{\vartheta}_m) \approx \breve{H}_{k,m}, \quad k = 1, \ldots, \bar{k}, \quad m = 1, \ldots, \bar{m} \tag{2}$$

and at the same time preserving the dissipativity property of the underlying system. For parameterized LTI P-port systems, dissipativity can be characterized in terms of the associated transfer function. Given the following conditions

1. $\mathsf{G}(s, \boldsymbol{\vartheta})$ regular for $\Re\{s\} > 0 \quad \forall \boldsymbol{\vartheta} \in \Theta$
2. $\mathsf{G}^*(s, \boldsymbol{\vartheta}) = \mathsf{G}(s^*, \boldsymbol{\vartheta}) \quad \forall s \in \mathbb{C}, \forall \boldsymbol{\vartheta} \in \Theta$
3. a. $\mathbb{I}_{\mathrm{P}} - \mathsf{G}^\star(s, \boldsymbol{\vartheta})\mathsf{G}(s, \boldsymbol{\vartheta}) \succeq 0$ for $\Re\{s\} > 0, \forall \boldsymbol{\vartheta} \in \Theta$ Scattering
   b. $\mathsf{G}^\star(s, \boldsymbol{\vartheta}) + \mathsf{G}(s, \boldsymbol{\vartheta}) \succeq 0$ for $\Re\{s\} > 0, \forall \boldsymbol{\vartheta} \in \Theta$ Immittance

a parameterized transfer function $\mathsf{G}(s, \boldsymbol{\vartheta})$ in immittance representation is Positive Real (PR) if it satisfies conditions 1, 2, 3b, while a transfer function in scattering representation is Bounded Real (BR) if it satisfies 1, 2, 3a. The poles of PR or BR transfer functions are always stable, as required by condition 1. Immittance transfer functions are also classified as Strictly Positive Real (SPR) if they satisfy condition 1 also for $\Re\{s\} = 0$ and

$$\mathsf{G}^\star(j\omega, \boldsymbol{\vartheta}) + \mathsf{G}(j\omega, \boldsymbol{\vartheta}) \succ 0, \forall \omega \in \{\mathbb{R} \cup \infty\}, \forall \boldsymbol{\vartheta} \in \Theta \tag{3}$$

in place of 3b. A SPR transfer function exhibits no poles nor zeros on the imaginary axis [8].

Models with (S)PR or BR transfer functions are dissipative. Therefore our problem is to obtain the model transfer function $\mathsf{H}(s, \boldsymbol{\vartheta})$ in such a way that it fulfills (2) and that is PR or BR, depending on the model representation.

2.2 Model Structure

Our approach performs model generation based on model structure [5]

$$\mathsf{H}(s,\boldsymbol{\vartheta}) = \frac{\mathsf{N}(s,\boldsymbol{\vartheta})}{\mathsf{D}(s,\boldsymbol{\vartheta})} = \frac{\sum_{i=0}^{n}\sum_{\boldsymbol{\ell}\in\mathscr{I}_{\overline{\boldsymbol{\ell}}}} R_{i,\boldsymbol{\ell}}\, b_{\boldsymbol{\ell}}^{\overline{\boldsymbol{\ell}}}(\boldsymbol{\vartheta})\,\varphi_i(s)}{\sum_{i=0}^{n}\sum_{\boldsymbol{\ell}\in\mathscr{I}_{\overline{\boldsymbol{\ell}}}} r_{i,\boldsymbol{\ell}}\, b_{\boldsymbol{\ell}}^{\overline{\boldsymbol{\ell}}}(\boldsymbol{\vartheta})\,\varphi_i(s)}. \tag{4}$$

In the above, the rational dependence on the variable s is induced by the basis functions $\varphi_i(s)$, constructed from a set of fixed poles $\{q_1,\dots,q_n\}$ with $\Re\{q_i\} < 0 \forall i > 0$ as

$$\begin{cases} \varphi_i(s) = (s-q_i)^{-1}, & q_i \in \mathbb{R} \\ \varphi_i(s) = [(s-q_i)^{-1} + (s-q_i^*)^{-1}] & q_i \in \mathbb{C} \\ \varphi_{i+1}(s) = j[(s-q_i)^{-1} - (s-q_i^*)^{-1}] & q_{i+1} = q_i^* \in \mathbb{C}, \end{cases} \tag{5}$$

and $\varphi_0(s) = 1$. Bases $b_{\boldsymbol{\ell}}^{\overline{\boldsymbol{\ell}}}(\boldsymbol{\vartheta})$ are multivariate Bernstein polynomials that parameterize the model with respect to $\boldsymbol{\vartheta}$. Finally, $r_{i,\boldsymbol{\ell}} \in \mathbb{R}$ and $R_{i,\boldsymbol{\ell}} \in \mathbb{R}^{\mathrm{P}\times\mathrm{P}}$ are the unknown model coefficients. We remark that N and D are rational transfer functions sharing the same set of common poles, but exhibiting different parameterized residues. Since the common poles simplify in (4), each element of $\mathsf{H}(s,\boldsymbol{\vartheta})$ is actually a ratio of n-degree polynomials of s, with parameterized coefficients. The poles of $\mathsf{H}(s,\boldsymbol{\vartheta})$ are the parameterized zeros of $\mathsf{D}(s,\boldsymbol{\vartheta})$.

We will make use of the following state space realizations associated to $\mathsf{D}(s,\boldsymbol{\vartheta})$

$$\mathsf{D}(s,\boldsymbol{\vartheta}) \leftrightarrow \Sigma_{\mathsf{D}} = \left(\begin{array}{c|c} A_1 & B_1 \\ \hline C_1(\boldsymbol{\vartheta}) & D_1(\boldsymbol{\vartheta}) \end{array}\right), \tag{6}$$

$$\mathbb{I}_{\mathrm{P}}\mathsf{D}(s,\boldsymbol{\vartheta}) \leftrightarrow \left(\begin{array}{c|c} \mathbb{I}_{\mathrm{P}}\otimes A_1 & \mathbb{I}_{\mathrm{P}}\otimes B_1 \\ \hline \mathbb{I}_{\mathrm{P}}\otimes C_1(\boldsymbol{\vartheta}) & \mathbb{I}_{\mathrm{P}}\otimes D_1(\boldsymbol{\vartheta}) \end{array}\right) = \left(\begin{array}{c|c} A & B \\ \hline C_{\otimes}(\boldsymbol{\vartheta}) & D_{\otimes}(\boldsymbol{\vartheta}) \end{array}\right), \tag{7}$$

where the constant matrices A_1, B_1 are

$$A_1 = \mathrm{blkdiag}\{A_{1,i}\} \in \mathbb{R}^{n\times n}, \quad B_1 = [\dots, B_{1,i}, \dots]^{\top} \in \mathbb{R}^n, \tag{8}$$

$$A_{1,i} = \begin{cases} q_i, & q_i \in \mathbb{R} \\ \begin{bmatrix} \sigma_i & \omega_i \\ -\omega_i & \sigma_i \end{bmatrix}, & q_i = \sigma_i \pm j\omega_i \in \mathbb{C} \end{cases}, \quad B_{1,i} = \begin{cases} 1, & q_i \in \mathbb{R} \\ \begin{bmatrix} 2 & 0 \end{bmatrix}, & q_i = \sigma_i \pm j\omega_i \in \mathbb{C} \end{cases} \tag{9}$$

Here, (A_1, B_1) is controllable and A_1 is Hurwitz. The parameterized output matrices are Bernstein polynomial expansions defined as

$$C_1(\boldsymbol{\vartheta}) = \sum_{\boldsymbol{\ell}\in\mathscr{I}_{\overline{\boldsymbol{\ell}}}} C_1^{\boldsymbol{\ell}}\, b_{\boldsymbol{\ell}}^{\overline{\boldsymbol{\ell}}}(\boldsymbol{\vartheta}), \quad C_1^{\boldsymbol{\ell}} = [r_{1,\boldsymbol{\ell}}, \dots, r_{n,\boldsymbol{\ell}}] \in \mathbb{R}^{1\times n}, \tag{10}$$

$$D_1(\boldsymbol{\vartheta}) = \sum_{\boldsymbol{\ell}\in\mathscr{I}_{\overline{\boldsymbol{\ell}}}} D_1^{\boldsymbol{\ell}}\, b_{\boldsymbol{\ell}}^{\overline{\boldsymbol{\ell}}}(\boldsymbol{\vartheta}), \quad D_1^{\boldsymbol{\ell}} = r_{0,\boldsymbol{\ell}} \in \mathbb{R}. \tag{11}$$

Defining $A = \mathbb{I}_{\mathrm{P}} \otimes A_1$ and $B = \mathbb{I}_{\mathrm{P}} \otimes B_1$, $\mathsf{N}(s,\boldsymbol{\vartheta})$ admits the realization

$$\mathsf{N}(s,\boldsymbol{\vartheta}) \leftrightarrow \Sigma_{\mathsf{N}} = \left(\begin{array}{c|c} A & B \\ \hline C_2(\boldsymbol{\vartheta}) & D_2(\boldsymbol{\vartheta}) \end{array}\right) \tag{12}$$

$$C_2(\boldsymbol{\vartheta}) = \sum_{\boldsymbol{\ell}\in\mathscr{I}_{\bar{\boldsymbol{\ell}}}} C_2^{\boldsymbol{\ell}} b_{\boldsymbol{\ell}}^{\bar{\boldsymbol{\ell}}}(\boldsymbol{\vartheta}) \qquad C_2^{\boldsymbol{\ell}} \in \mathbb{R}^{\mathrm{P}\times n\mathrm{P}}, \tag{13}$$

$$D_2(\boldsymbol{\vartheta}) = \sum_{\boldsymbol{\ell}\in\mathscr{I}_{\bar{\boldsymbol{\ell}}}} D_2^{\boldsymbol{\ell}} b_{\boldsymbol{\ell}}^{\bar{\boldsymbol{\ell}}}(\boldsymbol{\vartheta}) \qquad D_2^{\boldsymbol{\ell}} = R_{0,\boldsymbol{\ell}} \in \mathbb{R}^{\mathrm{P}\times\mathrm{P}}. \tag{14}$$

For any $\boldsymbol{\ell}$, $C_2^{\boldsymbol{\ell}}$ collects the entries of $R_{i,\boldsymbol{\ell}}$, $i > 0$ with suitable ordering. Being ν a placeholder for either 2 or $\otimes$, we define for brevity the matrices

$$X_\nu(\boldsymbol{\vartheta}) = \begin{bmatrix} C_\nu^\top(\boldsymbol{\vartheta}) \\ D_\nu^\top(\boldsymbol{\vartheta}) \end{bmatrix} \left[C_\nu(\boldsymbol{\vartheta})\; D_\nu(\boldsymbol{\vartheta})\right] = \sum_{\boldsymbol{m}\in\mathscr{I}_{\bar{\boldsymbol{m}}}} X_\nu^{\boldsymbol{m}} b_{\boldsymbol{m}}^{\bar{\boldsymbol{m}}}(\boldsymbol{\vartheta}), \tag{15}$$

with $\bar{\boldsymbol{m}} = 2\bar{\boldsymbol{\ell}}$. Similarly, we define the following augmented-degree representation for the output matrices of Σ_{N}

$$Y(\boldsymbol{\vartheta}) = \begin{bmatrix} C_2^\top(\boldsymbol{\vartheta}) \\ D_2^\top(\boldsymbol{\vartheta}) \end{bmatrix} = \sum_{\boldsymbol{\ell}\in\mathscr{I}_{\bar{\boldsymbol{\ell}}}} \begin{bmatrix} C_2^{\boldsymbol{\ell}\top} \\ D_2^{\boldsymbol{\ell}\top} \end{bmatrix} b_{\boldsymbol{\ell}}^{\bar{\boldsymbol{\ell}}}(\boldsymbol{\vartheta}) = \sum_{\boldsymbol{m}\in\mathscr{I}_{\bar{\boldsymbol{m}}}} Y^{\boldsymbol{m}} b_{\boldsymbol{m}}^{\bar{\boldsymbol{m}}}(\boldsymbol{\vartheta}), \tag{16}$$

which is always possible thanks to the degree elevation property of the Bernstein polynomials. In the above, each $Y^{\boldsymbol{m}}$ is obtained as a linear combination of the matrices $\left[C_2^{\boldsymbol{\ell}}\; D_2^{\boldsymbol{\ell}}\right]^\top$, with predefined coefficients. See [2] for further details.

3 Model Dissipativity Conditions

Conditions 1, 2, 3a, 3b depend continuously on the Laplace variable and on the parameter vector $\boldsymbol{\vartheta}$. Verifying numerically the dissipativity of a pROM based on these conditions is not feasible, as it would require checking an infinite number of constraints, one for each fixed frequency-parameter configuration. The following theorem represents our main result, providing sufficient conditions to assess the dissipativity of the pROM in terms of a finite number of semidefinite constraints on the model coefficients. Without loss of generality, we provide the statement for models in scattering representation. Similar results can be derived for the immitance case.

Theorem 1. *Let* $\Omega(P,Q,R) = \begin{bmatrix} P^\top R + RP & RQ \\ Q^\top R & 0 \end{bmatrix}$*. Then,*

a) $\mathsf{H}(s,\boldsymbol{\vartheta})$ *in (4) is uniformly asymptotically stable over* Θ *if*

$$\exists L^{\boldsymbol{\ell}} \in \mathscr{S}_n : K^{\boldsymbol{\ell}} = \Omega(A_1,B_1,L^{\boldsymbol{\ell}}) - \begin{bmatrix} 0 & C_1^{\boldsymbol{\ell}\top} \\ C_1^{\boldsymbol{\ell}} & 2D_1^{\boldsymbol{\ell}} \end{bmatrix} \prec 0 \quad \forall \boldsymbol{\ell} \in \mathscr{I}_{\bar{\boldsymbol{\ell}}} \tag{17}$$

b) $\mathsf{H}(s,\boldsymbol{\vartheta})$ *is uniformly Bounded Real over* Θ *if, additionally,*

$$\exists P^{\boldsymbol{m}} \in \mathscr{S}_{nP} : J^{\boldsymbol{m}} = \begin{bmatrix} \Omega(A,B,P^{\boldsymbol{m}}) - X_\otimes^{\boldsymbol{m}} & Y^{\boldsymbol{m}} \\ Y^{\boldsymbol{m}\top} & -\mathbb{I}_P \end{bmatrix} \preceq 0, \quad \forall \boldsymbol{m} \in \mathscr{I}_{\bar{\boldsymbol{m}}}. \tag{18}$$

We provide here a sketch of the proof for Theorem 1, more detailed derivations are available in [3]. The uniform model stability condition (a) stems from enforcing the denominator $\mathsf{D}(s, \boldsymbol{\vartheta})$ to be SPR. In fact, the zeros of a SPR function are guaranteed stable, and since the zeros of $\mathsf{D}(s, \boldsymbol{\vartheta})$ are the poles of $\mathsf{H}(s, \boldsymbol{\vartheta})$, uniform stability follows. The SPR conditions on $\mathsf{D}(s, \boldsymbol{\vartheta})$ are written in algebraic form using a (parameterized form of) the Kalman-Yakubovich-Popov (KYP) Lemma, which is then discretized in the parameter space by a Bernstein polynomial expansion. A straightforward derivation leads to the sufficient condition for uniform stability expressed by (17). A similar process is used to derive the uniform dissipativity condition in (b). The KYP lemma is again used to eliminate dependence on frequency in condition 3a, and a Bernstein polynomial expansion provides a discretized form of the corresponding parameterized LMI condition for BR-ness. The result is the sufficient condition in (18). The key enabling factors on which this proof is built are the model structure (4) with the associated parameterized state-space realizations of Sect. 2.2, and the properties of the Bernstein polynomials which allow proving positivity (negativity) of a parameter-dependent matrix by constraining the sign of its Bernstein coefficients.

4 Model Generation

In this section, we present our approach to generate dissipative pROMs in scattering representation via semidefinite programming, exploiting the theoretical results of Theorem 1. As in standard approaches based on model structure (4), we meet condition (2) iteratively, enforcing a sequence of linearized approximations

$$\frac{\mathsf{N}^{\mu}(j\omega_k, \boldsymbol{\vartheta}_m) - \mathsf{D}^{\mu}(j\omega_k, \boldsymbol{\vartheta}_m)\tilde{H}_{k,m}}{\mathsf{D}^{\mu-1}(j\omega_k, \boldsymbol{\vartheta}_m)} \approx 0, \quad k = 1, \ldots, \bar{k}, \quad m = 1, \ldots, \bar{m}, \tag{19}$$

where $\mu = 1, 2, \ldots$ is an index for the iteration. At iteration μ, $\mathsf{D}^{\mu-1}$ is numerically available[1] so that relation (19) can be recast in matrix form as

$$\begin{bmatrix} \Psi_x^{\mu} & \Psi_y^{\mu} \end{bmatrix} \begin{bmatrix} x^{\mu} \\ y^{\mu} \end{bmatrix} \approx 0 \tag{20}$$

where vectors x^{μ}, y^{μ} collect the current numerator and denominator coefficients $R_{i,\boldsymbol{\ell}}$ and $r_{i,\boldsymbol{\ell}}$, respectively, and Ψ_x^{μ} and Ψ_y^{μ} are known matrices. The approximation (20) is then enforced in a least-squares sense. The iteration stops whenever $\mathsf{D}^{\mu}(j\omega, \boldsymbol{\vartheta}) \simeq \mathsf{D}^{\mu-1}(j\omega, \boldsymbol{\vartheta})$, so that (19) becomes equivalent to (2).

Since only the denominator variables $y^{\mu-1}$ are required to set up problem (20), the iteration admits a fast implementation based on the elimination of the variables x^{μ}, that are computed only once convergence is met. The elimination procedure is based on computing the QR factorization of the matrix $\begin{bmatrix} \Psi_x^{\mu} & \Psi_y^{\mu} \end{bmatrix}$, as thoroughly discussed in [6]. After the variable elimination, (20) is replaced by the smaller denominator estimation problem

$$\Gamma_y^{\mu} y^{\mu} \approx 0, \tag{21}$$

[1] We set $\mathsf{D}^0(j\omega, \boldsymbol{\vartheta}) = 1$ at the first iteration to initialize the denominator estimate.

being Γ_y^{μ} a known matrix. We constrain the estimation with the stability conditions (17), by solving the following semi-definite program

$$\min_{y^{\mu}, L^{\boldsymbol{\ell}}} \|\Gamma^{\mu} y^{\mu}\|_2 \text{ subject to: } \quad \Omega(A_1, B_1, L^{\boldsymbol{\ell}}) - \begin{bmatrix} 0 & C_1^{\boldsymbol{\ell}\top} \\ C_1^{\boldsymbol{\ell}} & 2D_1^{\boldsymbol{\ell}} \end{bmatrix} \prec 0 \quad \forall \boldsymbol{\ell} \in \mathscr{I}_{\bar{\boldsymbol{\ell}}}, \tag{22}$$

so that the resulting y^{μ} guarantees a stable model by construction.

Problem (22) is solved repeatedly until convergence, that is practically met when the condition

$$\delta^{\mu} = \frac{\|y^{\mu} - y^{\mu-1}\|_2}{\|y^{\mu}\|_2} \leq \varepsilon \tag{23}$$

holds with a user-defined small threshold $\varepsilon > 0$. Supposing this condition is met at iteration $\bar{\mu}$, we complete the model generation by estimating the numerator unknowns $x^{\bar{\mu}}$. This can be done by substituting the available denominator coefficients $y^{\bar{\mu}}$ in (20) and enforcing the resulting condition

$$\Psi_x^{\bar{\mu}} x^{\bar{\mu}} \approx -\Psi_y^{\bar{\mu}} y^{\bar{\mu}}. \tag{24}$$

Since (17) holds by construction, we enforce (24) in such a way that the solution satisfies (18), so that $\mathsf{H}(s, \boldsymbol{\vartheta})$ is BR and the final model is dissipative. To this aim, we observe that the matrices $X_{\otimes}^{\boldsymbol{m}}$ in (18) are known, as they are defined upon the available denominator coefficients $y^{\bar{\mu}}$. Since the terms $Y^{\boldsymbol{m}}$ are obtained as linear combinations of the numerator unknowns $x^{\bar{\mu}}$ according to (16), we enforce (24) in a least-squares sense, by solving another semi-definite program

$$\min_{x^{\bar{\mu}}, P^{\boldsymbol{m}}} \left\| \Psi_x^{\bar{\mu}} x^{\bar{\mu}} + \Psi_y^{\bar{\mu}} y^{\bar{\mu}} \right\|_2 \quad \text{s.t.} \quad \begin{bmatrix} \Omega(A, B, P^{\boldsymbol{m}}) - X_{\otimes}^{\boldsymbol{m}} & Y^{\boldsymbol{m}} \\ Y^{\boldsymbol{m}\top} & -\mathbb{I}_{\mathrm{P}} \end{bmatrix} \preceq 0, \quad \forall \boldsymbol{m} \in \mathscr{I}_{\bar{\boldsymbol{m}}}, \tag{25}$$

which guarantees the bounded realness of $\mathsf{H}(s, \boldsymbol{\vartheta})$.

We remark that since Theorem 1 provides only sufficient conditions for the verification of model dissipativity, enforcing constraints (17), (18) may introduce some conservativity in the model generation process, by over-restricting the set of feasible model coefficients. A systematic approach based on the degree elevation property of the Bernstein polynomials can be used to arbitrarily reduce this conservativity, at the price of introducing additional instrumental variables. Full details about this procedure are available in [3].

5 A Test Case

The proposed strategy is applied to generate a dissiparive pROM of a high-speed interconnect link, designed as in [4]. We let the structure behavior depend on $d = 2$ geometrical parameters related to the vertical interconnect on a Printed Circuit Board, namely the pad radius $\vartheta_1 \in [100, 300]$ μm and the associated antipad radius $\vartheta_2 \in [400, 600]$ μm. The dataset (1) is retrieved by performing virtual measurements of the 2×2 scattering matrix of the structure, computed from a physics-based Maxwell equations solver in the bandwidth $[0, 5]$ GHz. Then we generate the pROM as described in Sect. 4, fixing the model order to $n = 25$. The remarkable accuracy of the resulting model is demonstrated in Fig. 1, through a comparison with a set of validation responses (not used for training) for different parameter configurations.

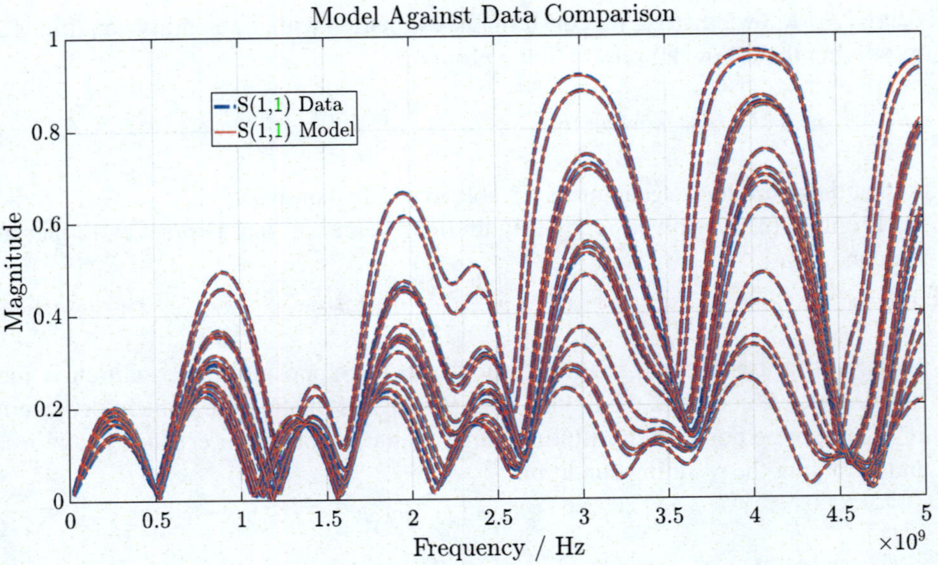

Fig. 1. Model-data comparison for different parameter configurations.

6 Conclusions

We presented a novel data-driven approach for generating pROMs with theoretical dissipativity certification. The method constrains the model training with finite dimensional linear matrix inequalities that are shown to guarantee the dissipativity of the model throughout the parameter space. Being based on convex programming, the approach is fully deterministic and returns accurate and compact parameterized models.

References

1. Benner, P., Goyal, P., Van Dooren, P.: Identification of port-Hamiltonian systems from frequency response data. Syst. Control Lett. **143**, 104741 (2020)
2. Berchtold, J., Bowyer, A.: Robust arithmetic for multivariate Bernstein-form polynomials. Comput. Aided Des. **32**(11), 681–689 (2000)
3. Bradde, T., Grivet-Talocia, S., Zanco, A., Calafiore, G.C.: Data-driven extraction of uniformly stable and passive parameterized macromodels. IEEE Access **10**, 15786–15804 (2022)
4. Preibisch, J.B., et al.: Exploring efficient variability-aware analysis method for high-speed digital link design using PCE. In: Proceedings of DesignCon, pp. 1–19 (2017)
5. Triverio, P., et al.: A parameterized macromodeling strategy with uniform stability test. IEEE Trans. Adv. Packag. **32**(1), 205–215 (2009)
6. Bradde, T., et al.: A scalable reduced-order modeling algorithm for the construction of parameterized interconnect macromodels from scattering responses. In: Proceedings of IEEE Symposium Electromagnetic Compatibility Signal Integrity Power Integrity (EMC, SI & PI), pp. 650–655. IEEE, Long Beach, CA, USA (2018)
7. Mehrmann, V., Unger, B.: Control of port-Hamiltonian differential-algebraic systems and applications. Acta Numer **32**, 395–515 (2023)
8. Ronald Wohlers, M.: Lumped and Distributed Passive Networks. Academic Press, New York (1969)

Splitting Methods for Linear Coupled Field-Circuit DAEs

Malak Diab(✉) and Caren Tischendorf

Humboldt Universität zu Berlin, Institute of Mathematics, Unter den Linden 6, 10099 Berlin, Germany
m.diab@stimulate-ejd.eu, tischendorf@math.hu-berlin.de

Abstract. The application of operator splitting methods to ordinary differential equations (ODEs) is well established. However, for differential-algebraic equations (DAEs) it is subjected to many restrictions due to the presence of (possibly hidden) constraints. In order to get convergence of the operator splitting for DAEs, it is important to have and exploit a suitable decoupled structure for the desired DAE system. Here we present a coupled field-circuit modeling via a loop-cutset analysis and the choice of a suitable tree that results in a port-Hamiltonian DAE system. Finally, we introduce an operator splitting approach of such linear coupled field-circuit DAEs and present convergence results for the proposed approach.

1 Introduction

In mathematical modeling, one often wishes to capture different aspects of a physical situation that are reflected in the model's system of equations as different operators. Operator splitting has been a successful strategy to deal with such complicated problems [1]. A straight forward transfer of operator splitting from ODEs to DAEs is not trivial. This is obviously because it is not applicable to algebraic equations. Accordingly, we will have to adapt the operator splitting for DAEs to the different nature of inherent DAE parts [2].

An important criterion that a splitting method must meet in order to solve physical and engineering problems is the conservation of physics, e.g., symplecticity, energy conservation, irreversibility, mass conservation, etc. The splitting approach proposed in this work for the linear coupled field-circuit DAE is designed to make use of the conservation of energy. In particular, for the index-1 DAE of consideration, a suitable decomposition of the matrices is achieved so that a natural port-Hamiltonian DAE structure is visible and can be exploited for a convergent splitting approach that is explicit and energy preserving in the dynamic part.

This paper is structured as follows. First, we describe the decoupling of the linear coupled field-circuit DAE to be exploited for a suitable operator splitting and we analyse its index. In Sect. 3, we introduce our operator splitting approach for the coupled DAE. It includes a convergence analysis and a discussion of some structural properties of the subsystems. Finally we demonstrate numerical results for a benchmark circuit in Sect. 4.

M. van Beurden et al. (Eds.): SCEE 2022, MI 43, pp. 159–166, 2024.
https://doi.org/10.1007/978-3-031-54517-7_18

2 Coupled Field-Circuit Modeling

We consider a model which couples partial differential equations for electromagnetic (EM) devices with linear DAEs describing the basic circuit elements. Since we are seeking a suitable modeling approach that fits the extension of the operator splitting method, we describe the electromagnetic field by the classical **E-H** formulation of Maxwell's equations to exploit the Hamiltonian structure. In an open bounded domain $\Omega \subset \mathbb{R}^3$ and $t \in \mathscr{I} = [t_0, T] \subset \mathbb{R}$ being the time interval, the evolution of electromagnetic fields on $\Omega \times \mathscr{I}$ is determined by Maxwell's equations. The **E-H** formulation of Maxwell's equations is given by

$$\nabla \times \mathbf{E} = -\frac{\partial}{\partial t}\mathbf{B}, \quad \nabla \cdot \mathbf{B} = 0, \quad \nabla \times \mathbf{H} = \mathbf{J} + \frac{\partial}{\partial t}\mathbf{D}, \quad \nabla \cdot \mathbf{D} = \rho \tag{1}$$

In these equations, **E** and **H** are the electric and magnetic fields,and **D** and **B** are the electric and magnetic flux densities, respectively. Further, ρ is the electric charge density and **J** is the electric current density.

The spatial discretization of Maxwell's equations (1) using the finite integration technique (FIT) [4] on a staggered grid pair with n primal grid points leads to the equations

$$C\mathbf{e} = -d_t\mathbf{b}, \quad \widetilde{C}\,\mathbf{h} = d_t\mathbf{d} + \mathbf{j}, \quad \widetilde{S}\,\mathbf{b} = 0, \quad S\mathbf{d} = \mathbf{q} \tag{2}$$

where $C, \widetilde{C} \in \mathbb{R}^{m\times m}$ are the discrete curl operators, $S, \widetilde{S} \in \mathbb{R}^{n\times m}$ the discrete divergence operators, which are defined on the primal and dual grid, respectively $(m \approx 3n)$. The fields $\mathbf{e}$, $\mathbf{h}$, $\mathbf{d}$, $\mathbf{b}$, $\mathbf{j} : \mathscr{I} \to \mathbb{R}^m$ and $\mathbf{q} : \mathscr{I} \to \mathbb{R}^n$ correspond to electric and magnetic voltages, electric and magnetic fluxes, electric currents and electric charges, respectively. Maxwell's grid equations (2) are linked with the material relations

$$\mathbf{d} = M_\varepsilon \mathbf{e}, \quad \mathbf{b} = M_\mu \mathbf{h}, \quad \mathbf{j} = M_\sigma \mathbf{e} + \mathbf{j}_s \tag{3}$$

where M_ε and M_μ are the material matrices of permittivities and permeabilities and M_σ is the matrix of conductivities. We consider an electrical circuit consisting of capacitors (C), resistors (R), inductors (L), voltage sources (V), current sources (I) and an electromagnetic device (E) satisfying well-posedness. And we denote by v_X and i_X the voltages and currents through a type-X element, $X \in \{C, R, L, V, I, E\}$.

We proceed -similar to the approach we followed in [2]- by modeling the electrical circuit using tree-based loop and cutset formulations [3]

$$i_t = -Q_l i_l, \qquad v_l = Q_l^\top v_t \tag{4}$$

where Q_l is a submatrix of the fundamental cutset matrix such that its columns represent the co-tree branches (links), see [2]. The subscripts $(\cdot)_l$ and $(\cdot)_t$ correspond to link and twig elements, respectively. In this work, we focus on topological conditions given in the following assumption.

Assumption 1. We assume that the circuit network has no CVE-loops, while CV-loops and LI-cutsets are allowed.

For the lumped circuit elements, we assume that all resistances, conductances, capacitances and inductances show a globally passive behavior. In addition, the independent functions v_s and i_s for voltage and current sources are assumed to be continuously differentiable. Notice that we used in our approach the conductive description for all resistances that belong to the tree and the resistive description for all resistances that does not belong to the tree, see below.

Based on the topological conditions of the Assumption 1, we can construct a tree as follows:

1. All voltage sources and the electromagnetic device belong to the tree.
2. All current sources do not belong to the tree.
3. Split resistors in such a way that all G-resistances belong to the tree and all R-resistances do not.
4. Split capacitors such that a capacitor is placed on a link if it belongs to a CV-loop.
5. Split inductors such that an inductor is placed on a twig if it belongs to an LI-cutset.

Consequently, the matrix Q_l of can be written

$$Q_l = \begin{bmatrix} Q_{VC} & Q_{VR} & Q_{VL} & Q_{VI} \\ Q_{CC} & Q_{CR} & Q_{CL} & Q_{CI} \\ 0 & Q_{ER} & Q_{EL} & Q_{EI} \\ 0 & Q_{GR} & Q_{GL} & Q_{GI} \\ 0 & 0 & Q_{LL} & Q_{LI} \end{bmatrix}$$

In the submatrices Q_{XY} with subscripts $X \in \{V,C,E,G,L\}$ and $Y \in \{C,R,L,I\}$, the subscript X refers to an X-type element on a twig, while the subscript Y refers to an Y-type element on a link.

We always have three zero submatrices in the first column since if there is a capacitor link, it will belong to a CV-loop, i.e., a loop that does not contain R, L, or E branches. Similarly, in the last row there are always two zero submatrices.

The circuit equations consist of the loop and cutset formulations (4) reflecting the Kirchhoff's laws together with elements constitutive equations

$$i_C = Cv'_C, \quad v_L = Li'_L, \quad i_G = Gv_G, \quad v_R = Ri_R, \quad i_I = i_s(t), \quad v_V = v_s(t). \tag{5}$$

In order to incorporate an electromagnetic device into the circut, the system needs to be completed with additional equations that distribute the circuit's voltages and currents to the field's domain. We assume that the source current density is given by

$$\mathbf{j}_s(t) = X_s i_E(t) \tag{6}$$

where $X_s \in \mathbb{R}^{m \times m_E}$ is a space-discretized function distributing the quantities, m_E is the number of contact parts on the boundary. Moreover, we have the expression

$$\frac{1}{\alpha} v_E = X_s^\top \mathbf{e} \tag{7}$$

with α being the number of mesh links in the direction between contacts.

The circuit's behaviour is then described by the formulations (4) together with the constitutive element relations (5) which are then coupled with Maxwell's grid equations (2) and the coupling Eqs. (6)-(7). The resultant coupled field-circuit system is an index-2 DAE [5] of the form

$$Dx'(t) + Jx(t) + My(t) = r_x(t) \tag{8a}$$

$$-M^\top x(t) + Sy(t) = r_y(t) \tag{8b}$$

$$z(t) + Kx'(t) + K_x x(t) + K_y y(t) = r_z(t) \tag{8c}$$

with $x = \begin{bmatrix} v_{Ct} \; i_{Ll} \; \mathbf{h} \; \mathbf{e} \end{bmatrix}^\top, y = \begin{bmatrix} i_{Rl} \; v_{Rt} \; v_E \; i_E \end{bmatrix}^\top, z = \begin{bmatrix} i_V \; v_I \end{bmatrix}^\top$

$$D = \begin{bmatrix} C_t + Q_{CC} C_l Q_{CC}^\top & 0 & 0 & 0 \\ 0 & L_l + Q_{LL}^\top L_t Q_{LL} & 0 & 0 \\ 0 & 0 & M_\mu & 0 \\ 0 & 0 & 0 & M_\varepsilon \end{bmatrix}, \quad J = \begin{bmatrix} 0 & Q_{CL} & 0 & 0 \\ -Q_{CL}^\top & 0 & 0 & 0 \\ 0 & 0 & 0 & \boldsymbol{C} \\ 0 & 0 & -\boldsymbol{C}^\top & M_\sigma \end{bmatrix},$$

$$M = \begin{bmatrix} Q_{CR} & 0 & 0 & 0 \\ 0 & -Q_{GL}^\top & -Q_{EL}^\top & 0 \\ 0 & 0 & 0 & 0 \\ 0 & 0 & 0 & X_s \end{bmatrix}, \quad S = \begin{bmatrix} R_l & -Q_{GR}^\top & -Q_{ER}^\top & 0 \\ Q_{GR} & G_t & 0 & 0 \\ Q_{ER} & 0 & 0 & I \\ 0 & 0 & -\frac{1}{\alpha} I & 0 \end{bmatrix}, \quad K = \begin{bmatrix} Q_{VC} C_l Q_{CC}^\top & 0 & 0 & 0 \\ 0 & Q_{LI}^\top L_t Q_{LL} & 0 & 0 \end{bmatrix},$$

$$K_x = \begin{bmatrix} 0 & Q_{VL} & 0 & 0 \\ -Q_{CI}^\top & 0 & 0 & 0 \end{bmatrix}, \quad K_y = \begin{bmatrix} Q_{VR} & 0 & 0 & 0 \\ 0 & -Q_{GI}^\top & -Q_{EI}^\top & 0 \end{bmatrix}, \quad r_z = \begin{bmatrix} -Q_{VI} i_s - Q_{VC} C_l Q_{VC}^\top v_s' \\ Q_{VI}^T v_s - Q_{LI}^\top L_t Q_{LI} i_s' \end{bmatrix}$$

$$r_x = \begin{bmatrix} -Q_{CI} i_s - Q_{CC} C_l Q_{VC}^\top v_s' \\ Q_{VL}^\top v_s - Q_{LL}^\top L_t Q_{LI} i_s' \\ 0 \\ 0 \end{bmatrix}, \quad r_y = \begin{bmatrix} Q_{VR}^\top v_s \\ -Q_{GI} i_s \\ -Q_{EI} i_s \\ 0 \end{bmatrix}$$

We note that the matrices D and S are nonsingular. Hence, Eqs. (8a) and (8b) are sufficient for the determination of the unknown vector variables x and y, while Eq. (8c) can be interpreted as an output equation for z. Further, the matrix $J = J_1 + J_2$ can be written as the sum of skew-symmetric and symmetric matrices. Notice that $r_z(t)$ in (8c) involves derivatives of the input functions v_s and i_s. Consequently, the DAE (8) has index 2 unless $Q_{VC} C_l Q_{VC}^\top = 0$ and $Q_{LI} L_t Q_{LI}^\top = 0$.

2.1 Coupled Index Analysis

As a part of the DAE analysis, we analyse the differentiation index of the coupled DAE (8). The result is given by the following theorem.

Theorem 1 (See [9]). *The DAE (8) yield by loop and cutset formulations is, for consistent initial conditions, uniquely solvable and*

- *is a DAE of index-1 if the following 2 statements hold*
  - *there is no CV-loop* **or** *any fundamental loop associated to a (link) capacitor does not contain other (twig) capacitors*
  - *there is no LI-cutset* **or** *any fundamental cutset associated to an inductor does not contain other (link) inductors*
- *Otherwise, the DAE is of differentiation index-2.*

Remark 1. For index-1 case, the coefficient matrix K is a zero matrix and DAE (8) fits into the port-Hamiltonian form, introduced by Mehrmann and Morandin [6].

3 Operator Splitting Approach

To this end, we have provided a new modeling approach for the coupled field-circuit system to which we are going to propose a splitting approach based on the inherent ODE. Therefore, we rewrite the DAE system (8a)-(8b) equivalently as

$$Dx' + (J_1 + J_2)x + MS^{-1}M^\top x = r_x(t) - MS^{-1}r_y(t) \tag{9a}$$

$$y = S^{-1}\big(r_y(t) + M^\top x\big). \tag{9b}$$

We split (9a), using Lie-Trotter splitting, into the subsystems $Dx' + J_1 x = 0$ and

$$Dx' + J_2 x + MS^{-1}M^T x = r_x(t) - MS^{-1}r_y(t). \tag{10}$$

Next, we reformulate (10) with (9b) back as DAE and obtain the following splitting approach (SADAE) for the coupled field-circuit DAEs.

1. Initialize $x_2(t_0) := x_0$ and $n = 0$.
2. Solve on $[t_n, t_{n+1}]$ the first subsystem

$$Dx_1' + J_1 x_1 = 0, \quad x_1(t_n) = x_2(t_n) \tag{splitDAE 1}$$

3. Solve on $[t_n, t_{n+1}]$ the second subsystem

$$Dx_2'(t) + J_2 x + My(t) = r_x(t), \quad x_2(t_n) = x_1(t_{n+1}) \tag{splitDAE 2a}$$

$$-M^T x_2(t) + Sy(t) = r_y(t). \tag{splitDAE 2b}$$

4. Set $n = n+1$ and go to 2. unless t_n is the final time point.

3.1 Subsystem Properties

We observe that the first subsystem (splitDAE 1) is a Hamiltonian ODE system with the Hamiltonian

$$\begin{aligned} H(x) = \frac{1}{2}x^\top Dx = \frac{1}{2}v_{Ct}^\top(C_t + Q_{CC}C_l Q_{CC}^\top)v_{Ct} + \frac{1}{2}i_{Ll}^\top(L_l + Q_{LL}^\top L_t Q_{LL})i_{Ll} \\ + \frac{1}{2}\mathbf{h}^\top M_\mu \mathbf{h} + \frac{1}{2}\mathbf{e}^\top M_\varepsilon \mathbf{e} =: H(v_{Ct}, i_{Ll}, \mathbf{h}, \mathbf{e}) \end{aligned} \tag{11}$$

where

$$\frac{d}{dt}H(x) = x^\top Dx' = -x^\top J_1 x = 0$$

since J_1 is skew-symmetric. If we additionally have no CVE-loops nor LI-cutsets then $Q_{VC} = Q_{CC} = 0$ and $Q_{LL} = Q_{LI} = 0$ and the Hamiltonian H becomes

$$H(x) = \frac{1}{2}x^\top Dx = \frac{1}{2}v_C^\top C v_C + \frac{1}{2}i_L^\top L i_L + \frac{1}{2}\mathbf{h}^\top M_\mu \mathbf{h} + \frac{1}{2}\mathbf{e}^\top M_\varepsilon \mathbf{e}$$

describing the total energy stored in the capacitors, inductors and the electromagnetic device. Convenient time integration methods to solve the first subsystem (splitDAE 1) are symplectic methods [7].

The second subsystem (splitDAE 2a)-(splitDAE 2b) is dissipative. It leads to non-symmetric but positive definite linear systems after time discretization that allows the exploitation of suitable iterative methods [8].

3.2 Convergence Analysis

In order to verify the convergence of DAE operator splitting method, one has to rely on the convergence of the ODE operator splitting method that results form consistency and stability [1].

Theorem 2 (Theorem 5.2.1 **in** [9]**).** *Let the time step size h be sufficiently small, the initial currents and voltages as well as the source functions of current and voltage sources be bounded. Let* $(x(t), y(t), z(t))$ *and* (x_n, y_n, z_n) *be the exact and the approximated solutions by (SADAE) of the DAE system* (8)*, respectively. Then*

1. *The approximated solution variables* x_n *and* y_n *converge to the exact solutions x and y of the reduced system* (8a)-(8b) *with order* 1.
2. *The order of convergence of* z_n *to z is* 1 *if the index of the DAE* (8) *is* 1. *Otherwise, the convergence of* z_n *is not guaranteed.*

Theorem 3 (Corollary 5.2.1 **in** [9]**).** *Under the same conditions of Theorem* 2*, if a higher order operator splitting method of order* $p \geq 2$ *is applied, then*

1. *The approximated solution variables* x_n *and* y_n *converge to the exact solutions x and y of the reduced system* (8a)-(8b) *with order p.*
2. *The order of convergence of* z_n *to z is p if the index of the DAE* (8) *is* 1. *Otherwise, the convergence order is* $p-1$.

4 Numerical Results

We aim in this section to check the convergence of operator splitting methods for coupled systems. We consider the coupled field-circuit problem in Fig. 1 operating in a GHz regime.

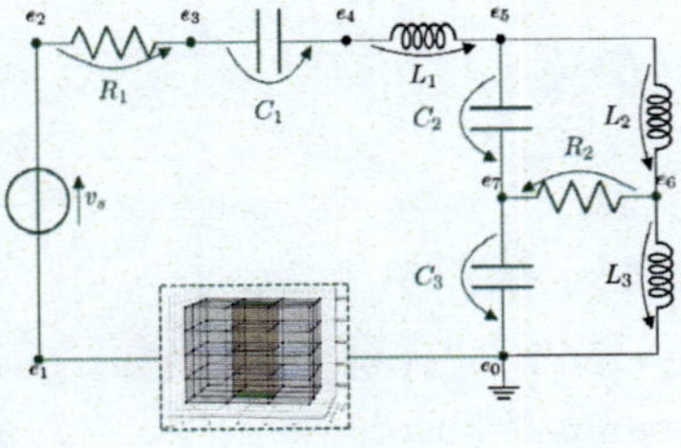

$C_1 = 10^{-12}$ F, $\quad C_2 = 5 \cdot 10^{-13}$ F, $\quad C_3 = 10^{-12}$ F

$L_1 = 5 \cdot 10^{-7}$ H, $\quad L_2 = 5 \cdot 10^{-7}$ H, $\quad L_3 = 5 \cdot 10^{-7}$ H

$G_1 = 50\ \Omega$, $\quad G_2 = 50\ \Omega$, $\quad v_s(t) = \sin(10^9 t)$

Fig. 1. RLC-circuit with EM device. The green branches form the tree considered for the model equations.

For comparison, we consider the following three variants of numerical simulation of the circuit:

1. Solve (8) by implicit Euler method.

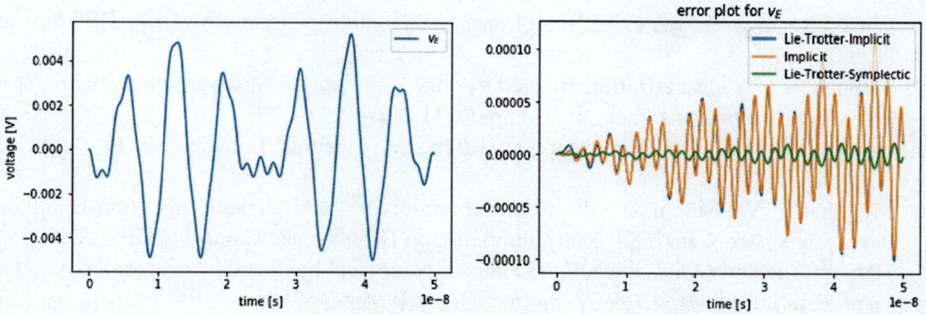

Fig. 2. Reference solution for EM device voltage v_E for circuit in Fig. 1 (left) and the error for numerical solution (right).

2. Solve (splitDAE 1) and (splitDAE 2a)-(splitDAE 2b) by implicit Euler method
3. Solve (splitDAE 1) by symplectic Euler and (splitDAE 2a)-(splitDAE 2b) by implicit Euler method.

In Fig. 2 we see the reference solution computed by time stepsize $h = 1e - 13$ and the error between the numerical solution for the three simulation variants with time stepsize $h = 1e - 12$ and the reference solution. The results show that the solution of the DAE splitting approach (variant 1) is almost the same as for the non-split solution (variant 2). The use of the DAE splitting approach with the symplectic Euler method (variant 3) gives the best results, of better accuracy. Further, we note that the symplectic Euler method for the first subsystem (8a) is an explicit method.

5 Conclusions and Outlook

The presented modelling and splitting approach for coupled field-circuit DAEs has the advantage that the skew symmetric dynamical part of high dimension (that preserves energy) can be solved efficiently with an explicit symplectic scheme and afterwards corrected by solving the symmetric dissipative part by an A-stable implicit scheme. The simulation results show that the splitting approach is faster and more accurate than a standard implicit (BDF) scheme for the coupled field-circuit system.

Acknowledgements. This project has received funding from the European Union's Horizon 2020 research and innovation program under grant agreement No 76504.

References

1. Bjórhus, M.: Operator splitting for abstract Cauchy problems. IMA J. Numer. Anal. **18**, 419–443 (1988)
2. Diab, M., Tischendorf, C.: Splitting methods for linear circuit DAEs of index 1 in port-Hamiltonian form. In: van Beurden, M., Budko, N., Schilders, W. (eds.) Scientific Computing in Electrical Engineering. Mathematics in Industry, vol. 36, pp. 211–219. Springer, Cham (2021). https://doi.org/10.1007/978-3-030-84238-3_21

3. Chua, L.O., Desoer, C.A., Kuh, E.S.: Linear and Nonlinear Circuits. McGraw-Hill, Singapore (1987)
4. Weiland, T.: A discretization method for the solution of Maxwell's equations for six-component fields. Int. J. Electron. Commun. **31**, 116–120 (1977)
5. Lamour, R., März, R., Tischendor, C.: Differential-Algebraic Equations: A Projector Based Analysis (2013)
6. Mehrmann, V., Morandin, R.: Structure-preserving discretization for port-Hamiltonian descriptor systems. In: IEEE 58th Conference on Decision and Control (CDC) (2019)
7. Hairer, E., Wanner, G., Lubich, C.: Geometric Numerical Integration. Springer Series in Computational Mathematics, vol. 31. Springer, Heidelberg (2002)
8. Chronopoulos, A.T.: s-step iterative methods for (non) symmetric (in) definite linear systems. SIAM J. Numer. Anal. **28**(6), 1776–1789 (1991)
9. Diab, M.: Splitting methods for partial differential-algebraic systems with application on coupled field-circuit DAEs Ph.D. Thesis, Humboldt-Universität zu Berlin, Mathematisch-Naturwissenschaftliche Fakultät (2023). https://doi.org/10.18452/25731

Structure-Preserving Identification of Port-Hamiltonian Systems—A Sensitivity-Based Approach

Michael Günther(✉), Birgit Jacob, and Claudia Totzeck

Bergische Universität Wuppertal, IMACM, Gaußstraße 20, 42119 Wuppertal, Germany
{guenther,bjacob,totzeck}@uni-wuppertal.de

Abstract. We present a gradient-based calibration algorithm to identify a port-Hamiltonian system from given time-domain input-output data. The gradient is computed with the help of sensitivities and the algorithm is tailored such that the structure of the system matrices of the port-Hamiltonian system (skew-symmetry and positive semi-definitness) is preserved in each iteration of the algorithm. As we only require input-output data, we need to calibrate the initial condition of the internal state of the port-Hamiltonian system as well. Numerical results with synthetic data show the feasibility of the approach.

1 Introduction

In structure-preserving modelling of coupled dynamical systems the port-Hamiltonian framework allows for constructing overall port-Hamiltonian systems (PHS) provided that (a) all subsystems are PHS and (b) a linear coupling between the input and outputs of the subsystems is provided [4,5,8,9]. In realistic applications this approach reaches its limits: for a specific subsystem, either no physics-based knowledge is available which allows for defining a physics-based PHS or (b) one is forced to use user-specified simulation packages with no information of the intrinsic dynamics, and thus only the input-output characteristics are available.

In both cases a remedy for such a subsystem is as follows: generate input-output data either by physical measurements or evaluation of the simulation package, and based on that derive a PHS surrogate that fits these input-output data best. This PHS surrogate can than be used to model the subsystem, and overall one gets a coupled PHS with structure-preserving properties.

Our approach aims at constructing a best-fit PHS model in one step, without the need of first deriving a best-fit linear state-space model and then, in a post-processing step, finding the nearest port-Hamiltonian realization, see, for example, [2,3]. The articles [1] and [10] consider port-Hamiltonian realizations in the frequency domain, using a Loewner framework or a parametrization of the system matrices, respectively. The identification of a PHS using parametrization leads to a high-dimensional problem as a dynamic with n states and k inputs and outputs is represented by $n(\frac{3n+1}{2}+2k)+k^2$ parameters. In contrast to these approaches we follow a time domain approach. In [6] an adjoint-based approach has been investigated to compute the gradient in order to derive

M. van Beurden et al. (Eds.): SCEE 2022, MI 43, pp. 167–174, 2024.
https://doi.org/10.1007/978-3-031-54517-7_19

a structure-preserving calibration algorithm for port-Hamiltonian input-output systems in the time domain. In this article, we develop a gradient-based calibration algorithm to identify a PHS from given time-domain input-output data. The gradient is computed with the help of sensitivities.

Consequently, we consider the surrogate PHS system given by

$$\frac{d}{dt}x = (J-R)Qx + Bu, \qquad x(0) = \hat{x}, \tag{1a}$$

$$y = B^\top Qx, \tag{1b}$$

where B, J, R, Q are matrices of suitable dimensions with $J^\top = -J$, $R \geq 0$ and $Q > 0$.

The task is to fit the system matrices and the initial conditions $v = (J, Q, R, \hat{x})$ to the data. We therefore define the cost functional

$$\mathscr{J}(x,v) = \frac{1}{2}\int_0^T |y(t) - y_{\text{data}}(t)|^2 dt = \frac{1}{2}\int_0^T |B^T Qx(t) - y_{\text{data}}(t)|^2 dt$$

leading us to the calibration problem

$$\min \mathscr{J}(x,v) \quad \text{subject to} \quad (1). \tag{P}$$

Note that well-posedness of P is a priori not guaranteed. For a detailed discussion we refer to [6].

As we are only interested in the input-output behaviour of the system, we can eliminate Q from the dynamics. In fact, by Cholesky decomposition we obtain V with $Q = VV^\top$.

$$w = V^\top x,\ \tilde{B} = V^\top B,\ \tilde{J} = V^\top JV,\ \tilde{R} = V^\top RV$$

yields the system

$$\frac{d}{dt}w = (\tilde{J} - \tilde{R})w + \tilde{B}u, \qquad w(0) = \hat{w}(= V^\top \hat{x}), \tag{2}$$

$$y = \tilde{B}^\top w. \tag{3}$$

For later use we define the state operator e corresponding to (2) as

$$e(w,v) = \begin{pmatrix} \frac{d}{dt}w - (\tilde{J} - \tilde{R})w - \tilde{B}u \\ w(0) - w_0 \end{pmatrix}.$$

Hence, (2) is equivalent to $e(w,v) = 0$.

The transformed cost functional is given by

$$\tilde{J}(w,v) = \frac{1}{2}\int_0^T |y(t) - y_{\text{data}}(t)|^2 dt = \frac{1}{2}\int_0^T |\tilde{B}^T w(t) - y_{\text{data}}(t)|^2 dt.$$

After the transformation we are left to identify the matrices $\tilde{J}, \tilde{R}$ and w_0. For notational convenience we define the space of admissible controls

$$\mathscr{V} = \{(\tilde{J}, \tilde{R}, w_0) \in \mathbb{R}^{n\times n} \times \mathbb{R}^{n\times n} \times \mathbb{R}^n : \tilde{J}^\top = -\tilde{J},\ \tilde{R} \geq 0\}.$$

Note that the system of differential equations admits a unique solution by standard ODE theory. This allows us to define the control to state map

$$S\colon \mathcal{V} \mapsto C([0,T],\mathbb{R}^n), \quad S(v) = w.$$

Moreover, we use S to define the reduced cost functional

$$\hat{J}(v) := \frac{1}{2}\int_0^T |\tilde{B}^T S(v)(t) - y_{\text{data}}(t)|^2 dt.$$

In the following we aim to derive an gradient-based algorithm that allows us to solve the calibration problem numerically. In particular, we require to compute the gradient of $\hat{J}$. Details are presented in the next section. From now on we only work with the transformed system and drop the $\sim$ for notational convenience.

2 Sensitivity Approach

We emphasize that the system matrices J, R as well as the initial condition $\hat{x}$ are finite dimensional. It is therefore feasible to employ an sensitivity approach [7] for the calibration problem.

To compute the sensitivities require admissible directions for the Gâteaux derivatives. Due to the structural restrictions, J can only be varied in direction h_J satisfying $h_J^\top = -h_J$ and R can only be varied by symmetric matrices.

The directional derivative of $\hat{J}$ in direction $h = (h_J, h_R, h_x)$ is given by

$$d\hat{J}(v)[h] = \langle \hat{J}'(v), h\rangle = \langle d_w J(w,v), S'(v)h\rangle + \langle d_v J(w,v), h\rangle$$

To evaluate this, we require $dw(v,h) = S'(v)h$ the so-called sensitivity. Here, we make use of the state equation $e(w,v) = 0$. In fact, it holds

$$e_w(w,v)dw(v,h) + e_v(w,v) = 0 \qquad \Leftrightarrow \qquad e_w(w,v)dw(v,h) = -e_v(w,v)h. \quad (4)$$

We emphasize that in order to identify the gradient $\hat{J}'(v)$ we need to compute the directional derivative w.r.t. all basis element of the tangent space of $\mathcal{V}$.

3 Gradient-Descent Algorithm

In the previous section we established the theoretical foundation of the gradient descent algorithm we present in the following.

Starting from an initial guess of system matrices and initial condition $v_0 = (J_0, R_0, \hat{x}_0)$ we compute the sensitivities $dw(v,h)$ for all basis elements of the tangent space of $\mathcal{V}$ by solving (4) and use the sensitivity information to evaluate the gradient $\hat{J}'(v_0)$. Then we seek for an admissible stepsize σ using Armijo-rule [7], see the pseudo code in Algorithm 2 and update the system matrices and the initial condition $v_0 \leftarrow v_0 - \sigma \hat{J}'(v_0)$. The calibration procedure is stopped when the cost functional value is sufficiently small. A pseudo code of the calibration algorithm can be found in Algorithm 3.

The presented algorithm can be used for numerical studies. In the following we discuss a proof of concept with states $x \in C([0,T],\mathbb{R}^2)$.

Algorithm 2. Armijo step size search

Require: gradient g, initial step size σ and safety parameter γ
Ensure: admissible step size σ, new parameter set v'
  $v' \leftarrow v + \sigma g$
  while $\hat{J}(v') - \hat{J}(v') > -\gamma\sigma\|g\|^2$ **do**
    $\sigma \leftarrow 0.5\sigma$
    $v' \leftarrow v - \sigma g$
  end while

Algorithm 3. Gradient-based calibration algorithm

Require: initial guess v_0 and additional parameters
Ensure: calibrated system matrices and initial condition $v = (J, R, \hat{x})$
  while $\hat{J}(v_0) > \varepsilon_{\text{stop}}$ **do**
    for all admissible directions h **do**
      compute $dw(v_0, h)$ by solving (4)
    end for
    identify $\hat{J}'(v_0)$
    find admissible step size σ by Armijo-rule, see Algorithm 2
    $v_0 \leftarrow v_0 - \sigma \hat{J}'(v_0)$
  end while

4 Proof of Concept

In the following we discuss a proof of concept with states $x \in C([0,T], \mathbb{R}^2)$ and output $y \in C([0,T], \mathbb{R})$. In the two dimensional setting the basis elements of the tangent space of $\mathscr{V}$ are manageable. Indeed, we have the basis elements

$$J_1 = \begin{pmatrix} 0 & -1 \\ 1 & 0 \end{pmatrix}, R_1 = \begin{pmatrix} 1 & 0 \\ 0 & 0 \end{pmatrix}, R_2 = \begin{pmatrix} 0 & 0 \\ 0 & 1 \end{pmatrix}, R_3 = \begin{pmatrix} 0 & 1 \\ 1 & 0 \end{pmatrix}, x_1 = \begin{pmatrix} 1 \\ 0 \end{pmatrix}, x_2 = \begin{pmatrix} 0 \\ 1 \end{pmatrix}.$$

We assume that $B = \begin{pmatrix} 1 & 1 \end{pmatrix}$ is known and that input signals at the time steps t_k are given as $u(t_k) = 1 + 0.1N(0,1)$ where $N(0,1)$ denotes a realization of a normally distributed random variable with mean 0 and standard deviation 1.

For simplicity, we assume that the time steps $t_k, k = 1, \ldots, K$ coincide with the time step of the Euler discretization that is implemented to solve the state ODE. Indeed, with the initial guess we solve (2) using the Euler scheme. Then we obtain the output y by (3), which we use to evaluate the cost functional for the initial guess. If the cost values is higher than the tolerance $\varepsilon_{\text{stop}}$ we start the calibration procedure.

For notational convenience we split the sensitivity $dw(v,h)$ into the parts h_J, h_R and h_x. The sensitivity w.r.t. J is computed by solving $e_w(w,v)dw(v,h_J) = -e_v(w,v)h_J$ which can be written explicitly as

$$\frac{d}{dt} dw(v,h_J) - (J-R)dw(v,h_J) = h_J w, \qquad dw(v,h_J)(0) = 0.$$

In the two dimensional case, there is only one admissible direction $h_J = J_1$. For the sensitivities w.r.t. R we solve

$$\frac{d}{dt}dw(v,h_R) - (J-R)dw(v,h_R) = -h_R w, \qquad dw(v,h_R)(0) = 0$$

for $h_R = \{R_1, R_2, R_3\}$. For the initial condition we solve

$$\frac{d}{dt}dw(v,h_x) - (J-R)dw(v,h_x) = 0, \qquad dw(v,h_x)(0) = h_x$$

for $h_x = \{x_1, x_2\}$.

The directional derivative of the cost functional reads

$$d\hat{J}(v)[h] = \langle B^\top S(v) - y_{\text{data}}, B^\top S'(v)h\rangle = \int_0^T B\big(B^\top S(v)(t) - y_{\text{data}}(t)\big), (S'(v)h)(t)dt,$$

which we can evaluate with the help of the sensitivities computed above. Note that $d\hat{J}(v)[h_\bullet] \in \mathbb{R}$ for all $h_\bullet$ discussed above. Hence, the gradient is assembled as follows

$$\hat{J}'(v) = \left[d\hat{J}(v)[J_1]J_1 \quad \sum_{\ell=1}^{3} d\hat{J}(v)[R_\ell]R_\ell \quad \sum_{\ell=1}^{2} d\hat{J}(v)[x_\ell]x_\ell \right]^\top$$

5 Numerical Results

For our proof of concept we generate synthetic data by solving the state system for fixed data matrices $J_{\text{data}}, R_{\text{data}}$ and initial condition $\hat{x}_{\text{data}}$. For the following results we choose

$$J_{\text{data}} = \begin{pmatrix} 0 & 1 \\ -1 & 0 \end{pmatrix}, \quad R_{\text{data}} = \begin{pmatrix} 0.5 & 0 \\ 0 & 0.3 \end{pmatrix}, \quad \hat{x}_{\text{data}} = \begin{pmatrix} 1 \\ 2 \end{pmatrix}. \tag{5}$$

The data yields the reference output y_{data} shown in Fig. 1 (left).

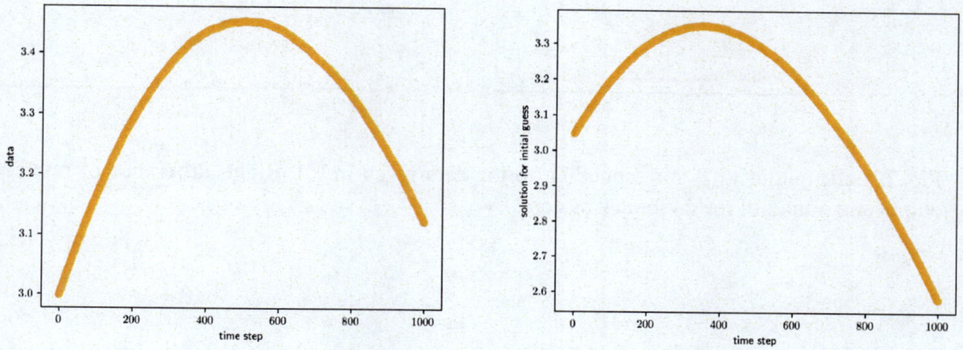

Fig. 1. Left: output y_{data} corresponding to the data given in (5). Right: output y_0 corresponding to the initial guess (6).

We start the proof of concept with the initial guess given by

$$J_0 = \begin{pmatrix} 0 & 1.2 \\ -1.2 & 0 \end{pmatrix}, \quad R_0 = \begin{pmatrix} 0.4 & 0 \\ 0 & 0.4 \end{pmatrix}, \quad \hat{x}_0 = \begin{pmatrix} 1.1 \\ 1.95 \end{pmatrix} \tag{6}$$

leading to the output in Fig. 1 (right). We set $T = 1$ and use 1000 time steps for the Euler discretization. The Armijo-search for an admissible step size is initialized with $\sigma = 10$ and the $\sigma \leftarrow \sigma/2$ if the current step size is not admissible.

Algorithm 3 is able to reproduce the output y_{data} with $\varepsilon_{\text{stop}} = 1e^{-4}$ in 22 gradient steps. The evolution of the cost function is shown in Fig. 2 (left) and the difference $y_{\text{data}} - y_{\text{opt}}$ is plotted in Fig. 2 (right). The calibrated matrices and initial data read

$$J_{\text{opt}} = \begin{pmatrix} 0 & 1.073 \\ -1.073 & 0 \end{pmatrix}, \quad R_{\text{opt}} = \begin{pmatrix} 0.379 & -0.080 \\ -0.080 & 0.367 \end{pmatrix}, \quad \hat{x}_{\text{opt}} = \begin{pmatrix} 1.039 \\ 1.929 \end{pmatrix},$$

where we rounded to precision $1e^{-3}$. It jumps to the eye that R_{opt} has nonzero off-diagonal entries. Out of curiosity we run the same toy problem with R restricted diagonal matrices. We obtain the calibrated matrices and initial data by

$$J_{\text{opt},2} = \begin{pmatrix} 0 & 1.016 \\ -1.016 & 0 \end{pmatrix}, \quad R_{\text{opt},2} = \begin{pmatrix} 0.351 & 0 \\ 0 & 0.335 \end{pmatrix}, \quad \hat{x}_{\text{opt},2} = \begin{pmatrix} 1.023 \\ 1.973 \end{pmatrix}, \tag{7}$$

again rounded to precision $1e^{-3}$. The additional structural information in R yields overall to better calibrated results. Compare Fig. 3 for the cost evolution and the difference of the outputs for the calibration with R restricted to diagonal matrices.

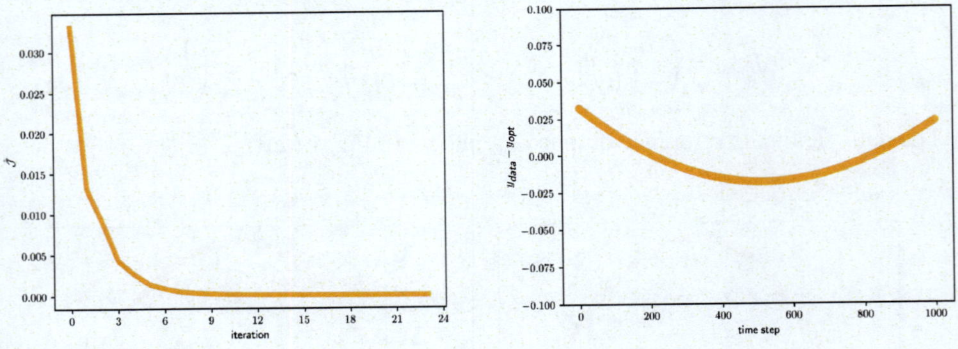

Fig. 2. Left: output y_{data} corresponding to the data given in (5). Right: difference of reference output and output of the calibrated system.

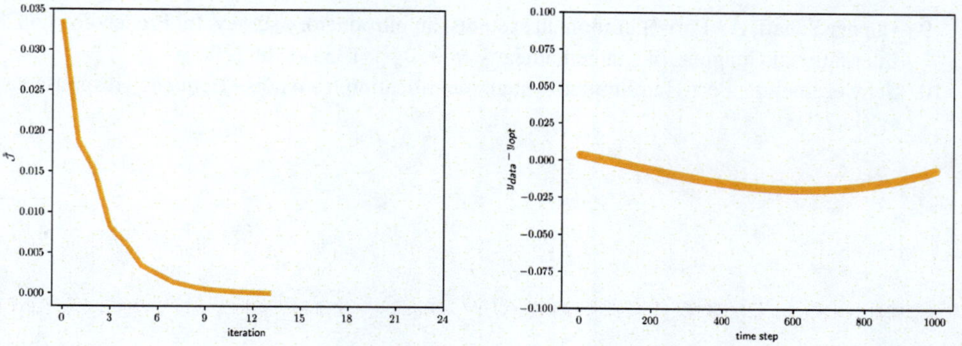

Fig. 3. Left: output y_{data} corresponding to the data given in (5). Right: difference of reference output and output of the calibrated system with R diagonal.

6 Conclusion and Outlook

We present a gradient-based algorithm to identify a port-Hamiltonian system consisting of ordinary differential equation to given input-output data. The gradient is computed with the help of a sensitivity approach. A proof of concept shows the feasibility of the approach.

As the effort of the sensitivity approach scales with the number of basis elements of the tangent space, the proposed calibration algorithm is only recommended for small systems. From an industrial standpoint the case of noisy data measurements is relevant. The performance of the proposed algorithm in the noisy setting is subject to future work.

References

1. Benner, P., Goyal, P., van Dooren, P.M.: Identification of port-Hamiltonian systems from frequency response data. Syst. Control Lett. **143**(4), 104741 (2020)
2. Cherifi, K., Goyal, P.K., Benner, P.: A non-intrusive method to inferring linear port-Hamiltonian realizations using time-domain data. Electron. Trans. Numer. Anal. Special Issue SciML **56**, 102–116 (2022)
3. Cherifi, K., Mehrmann, V., Hariche, K.: Numerical methods to compute a minimal realization of a port-Hamiltonian system. arXiv:1903.07042v1
4. Duindam, V., Macchelli, A., Stramigioli, S., Bruyninckx, H. (eds.): Modeling and Control of Complex Physical Systems. Springer, Germany (2009)
5. Eberard, D., Maschke, B.M., van der Schaft, A.J.: An extension of Hamiltonian systems to the thermodynamic phase space: towards a geometry of nonreversible processes. Rep. Math. Phys. **60**(2), 175–198 (2007)
6. Günther, M., Jacob, B., Totzeck, C.: Data-driven adjoint-based calibration of port-Hamiltonian systems in time domain. arXiv:2301.03924
7. Hinze, M., Pinnau, R., Ulbrich, M., Ulbrich, S.: Optimization with PDE Constraints. Springer, Berlin (2009)
8. Mehrmann, V., Morandin, R.: Structure-preserving discretization for port-Hamiltonian descriptor systems. In: 2019 IEEE 58th Conference on Decision and Control (CDC), pp. 6863-6868 (2019)

9. van der Schaft, A.: Port-Hamiltonian systems: an introductory survey. In: Proceedings on the International Congress of Mathematicians, vol. 3, pp. 1339–1366 (2006)
10. Schwerdtner, P.: Port-Hamiltonian system identification from noisy frequency response data. arXiv:2106.11355

BG Approximations of Multiphysics pH Distributed Systems with Finite Number of Ports

Daniel Ioan(✉) and Gabriela Ciuprina(✉)

Politehnica University of Bucharest, Spl. Independentei 313, 060042 Bucharest, Romania
{daniel,gabriela}@lmn.pub.ro

Abstract. This paper proposes a procedure for the modeling of linear passive devices with distributed parameters as Hamiltonian systems with a finite number of ports, in the view of their coupling with external systems with lumped parameters (circuits). To obtain this particular Dirac structure, appropriate boundary conditions (BC) are used for the PDEs of several physical fields. Originally, they are Electric Circuit Element BC, here generalized for multidisciplinary fields such as elastic solids, acoustic and thermal devices Their internal field is discretized by the Finite Element Method, thus obtaining the stiffness, damping and mass matrices of a second order ODEs system, transformed then into a first order pH canonical form, having as interaction variables the flow and effort of each terminal.

1 Introduction

The general class of abstract systems we consider are linear, port-Hamiltonian (pH), dissipative with distributed parameters and a finite number of ports. We will call them *pH systems* for short. Any device in this category occupies a 3D domain $\Omega \subset \mathbb{R}^3$, simple connected and smooth enough. On its boundary there are a set of disjoint surfaces S_k , $k = 1 : n+1$, called terminals (Fig. 1). The internal field is described by local, time dependent quantities $\varphi(M,t) : \Omega \times (0,T) \rightarrow \mathbb{R}^p$ which satisfy the evolution equations:

$$\mathscr{E}(\varphi) = 0. \tag{1}$$

Mathematically, they are PDE with Electric Circuit Element (ECE) boundary conditions, which will be defined below. This general class of systems is defined by the following three conditions [12]:

1. **Linearity:** $\forall\, \lambda_1, \lambda_2 \in \mathbb{R}$, $\forall\, \varphi^a$, φ^b

$$\mathscr{E}(\lambda_1 \varphi^a + \lambda_2 \varphi^b) = \lambda_1 \mathscr{E}(\varphi^a) + \lambda_2 \mathscr{E}(\varphi^b). \tag{2}$$

2. **Energy balance:**

$$I(t) = P(t) + \frac{\mathrm{d}W(t)}{\mathrm{d}t}, \tag{3}$$

a differential consequence of the evolution Eqs. (1), including three quantities: the energy, called also Hamiltonian (and denoted by $H(t)$ in mathematical contexts),

M. van Beurden et al. (Eds.): SCEE 2022, MI 43, pp. 175–183, 2024.
https://doi.org/10.1007/978-3-031-54517-7_20

a positive definite functional $W(t) = \mathscr{W}(\varphi(M,t))|_{M\in\Omega} \geq 0$; the dissipated power, a semi-definite functional $P(t) = \mathscr{P}(\varphi(M,t))|_{M\in\Omega_d} \geq 0$, $\Omega_d \subset \Omega$; the interaction power $I(t) = \mathscr{I}(i(M,t))|_{M\in\partial\Omega}$.

3. **Condition of finite interaction:**

$$I(t) = \sum_{k=1}^{n} u_k y_k. \tag{4}$$

The interaction power is a sum of n terms, one for each floating terminal, excepting for the last one (called reference). Each term is a product of two signals: the input u_k, and the output y_k.

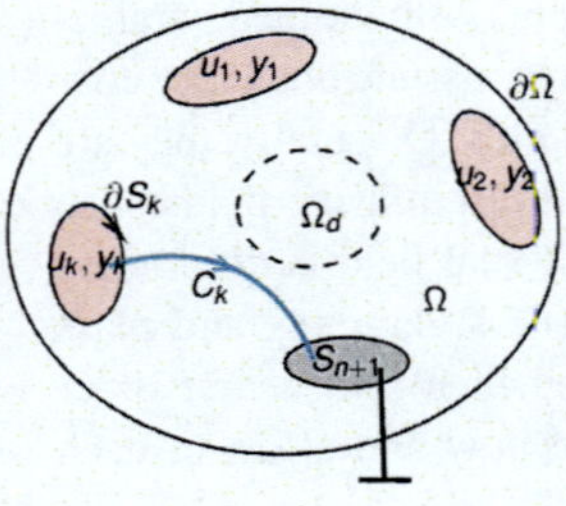

Fig. 1. A pH distributed system.

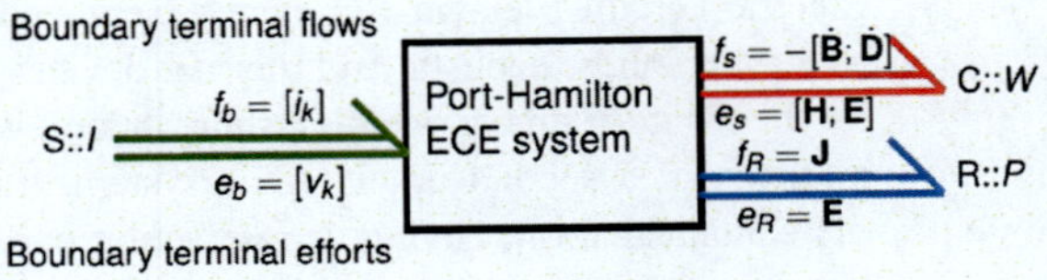

Fig. 2. BG of the EM device with ECE BC.

The first condition, expressed as (2), is an assumption. As we will see later, the second condition (3) is a consequence of the evolution equation (pH structure) given by (1). The third condition (4) is a consequence of the evolution equation (1), and the boundary conditions. For a particular physics, conditions (3) and (4) are thus theorems, but here, together with (2), they define the class of system we consider in this paper - the port-Hamiltonian (pH) linear dissipative/irreversible systems.

2 Devices with ECE BC

The first example of system in the class defined above is the **Electromagnetic** (EM) device [8, 11], described by the Maxwell Equations, linear constitutive relations and ECE boundary conditions (BC):

$$\begin{aligned}
&\text{(ece1)} \qquad && \mathbf{n}\cdot\frac{\partial \mathbf{B}(\mathbf{r},t)}{\partial t} = 0, \quad \forall \mathbf{r}\in\partial\Omega; \\
&\text{(ece2)} \qquad && \mathbf{n}\cdot(\nabla\times\mathbf{H}(\mathbf{r},t)) = 0 \quad \forall \mathbf{r}\in\partial\Omega - \cup_{k=1}^{n+1} S_k; \\
&\text{(ece3)} \qquad && \mathbf{n}\times\mathbf{E}(\mathbf{r},t) = \mathbf{0} \quad \forall \mathbf{r}\in S_k,\ k=1,\ldots,n+1.
\end{aligned} \tag{5}$$

According to these conditions, there is neither magnetic coupling through the boundary of the element $\partial\Omega$, nor electrical coupling through boundary excepting for the terminals S_k, which are equipotential. These BC allow the coupling of the device to an external electric circuit. Exterior calculus (differential forms) is the natural mathematical framework to describe these equations. The local field variables are $\varphi = [\mathbf{E};\mathbf{D};\mathbf{H};\mathbf{B};\mathbf{J}]$ (intensity and displacement of electric and magnetic fields, current density) out of which the state variables, which define the energy are $\mathbf{s} = [\mathbf{B};\mathbf{D}]$. The resulting energy balance $-\oint_{\partial\Omega}(\mathbf{E}_t \times \mathbf{H}_t)\cdot\mathbf{n}\,\mathrm{d}S = \int_{\Omega_d}\mathbf{J}\cdot\mathbf{E}\,\mathrm{d}x + \frac{\mathrm{d}}{\mathrm{d}t}\int_{\Omega} w\,\mathrm{d}x$, where $w = \mathbf{D}^2/(2\varepsilon) + \mathbf{B}^2/(2\mu)$ gives the expressions of energy W, dissipated power P and interaction power I, which is expressed in a finite manner as $I = \sum_{k=1}^{n} v_k i_k$, where v_k/i_k are the voltage/(total) current of the terminal k, properly defined as integrals of **E/H** on open/closed curves which belong to the boundary surface $\partial\Omega$: $i_k(t) = \oint_{\partial S_k}\mathbf{H}\cdot\mathbf{dl}$, C_k is an arbitrary curve, which links the terminal k to the reference $n+1$, as in Fig. 1. The equations and BC have important consequences, if the material constants are positive $\varepsilon, \mu, \sigma > 0$, both Kirchhoff current and voltage relations can be proven, as well as the passivity, reciprocity and linearity of the device. Thus, the impedance, admittance and in general, the hybrid matrix are properly defined, after the Laplace transform: $\mathbf{v}(s) = \mathbf{Z}(s)\mathbf{i}(s)$, $\mathbf{Y}(s) = \mathbf{Z}(s)^{-1} = \mathbf{Y}^T$, which are transcendental functions w.r.t. the complex variable s, having usually an infinite number of poles and zeros.

Being a pH system [13], the EM device with ECE BC has a Bond Graph (BG) [1] representation (Fig. 2), with n ports, each terminal characterized by its current and voltage, as flow and effort, which are conjugate variables since their product is the power transferred by the terminal. The storage port is infinite dimensional, with the flow $\mathbf{f}_S = [\dot{\mathbf{B}}, \dot{\mathbf{D}}]$ and effort $\mathbf{e}_S = [\mathbf{E}, \mathbf{H}]$. The dissipation port has the conjugate variables $\mathbf{f}_R = \mathbf{E}$ and $\mathbf{e}_R = \mathbf{J}$. Similar relations hold in other physical disciplines [7]:

Mechanical devices described by Cauchy-Newton relations with field variables $\varphi = \{\overline{\overline{\sigma}}, \overline{\overline{\varepsilon}}, \mathbf{j}, \mathbf{v}\}$ stress, strain, momentum density and local velocity, with ECE BC:

$$
\begin{aligned}
&(\text{ece1}) && S_k, k = 1,\ldots,n+1 \quad \text{are rigid (all their points have} \\
& && \text{the same velocities linear } \mathbf{v}_k/\text{angular } \omega_{\mathbf{k}} = \frac{1}{2}\mathrm{curl}\,\mathbf{v}); \\
&(\text{ece2}) && \mathbf{n}\cdot\overline{\overline{\sigma}} = 0, \quad \forall \mathbf{r} \in \partial\Omega - \cup_{k=1}^{n+1} S_k,
\end{aligned} \tag{6}
$$

with interaction variables $\mathbf{F}_k(t) = \int_{S_k}\mathbf{n}\cdot\overline{\overline{\sigma}}\,\mathrm{d}A$, $\mathbf{T}_k(t) = \int_{S_k}(\mathbf{r}\times\mathbf{n})\cdot\overline{\overline{\sigma}}\,\mathrm{d}A$ force and torque acting on the terminal k and $\mathbf{v}_k$, $\omega_{\mathbf{k}}$ - translation and rotation velocity, each terminal having as any rigid body up to six degrees of freedom.

Acoustic devices described by the linear equations of sound waves having field variables $\varphi = [\mathbf{j}; p; p_v; S; \mathbf{v}]$ pressure, compression, viscosity pressure, momentum density and velocity, with ECE BC:

$$
\begin{aligned}
&(\text{ece1}) && S_k, k = 1,\ldots,n+1 \text{ are rigid (all points have the same velocity } \mathbf{v}_k); \\
&(\text{ece2}) && p = 0 \text{ ("sound soft boundary")} \quad \forall \mathbf{r} \in \partial\Omega - \cup_{k=1}^{n} S_k.
\end{aligned} \tag{7}
$$

The interaction signals are $v_k(t) = \mathbf{v}_k \cdot \mathbf{n}$ the normal velocity [m/s] and $p_k(t) = \frac{1}{A_k}\int_{S_k} p\,\mathrm{d}A$ the average pressure [N/m^2], where A_k is the terminal aria.

Thermal systems described by the Fourier eqs. of heat transfer with ECE BC:

$$
\begin{aligned}
&(\text{ece1}) && S_k, k = 1,\ldots,n \text{ are isothermal (all points have the same temp.) } T_k; \\
&(\text{ece2}) && \mathbf{n}\cdot\mathbf{q} = 0 \text{ (adiabatic insulation, no heat transfer)} \forall \mathbf{r} \in \partial\Omega - \cup_{k=1}^{n} S_k.
\end{aligned} \tag{8}
$$

The local field variables are $\varphi = [u; \mathbf{q}; \mathbf{F}; T]$ density of the internal energy, temperature, density of the heat flux and temperature gradient, and the interaction variables are T_k - the temperature of the terminal k, and is its heat flux j_k.

All these devices can be represented as pH systems and BG. Table 1 holds the expressions of W, P, I for several physical domains. Note that the I/O signal of thermal devices are non-conjugate, and thus they are represented as pseudo-BG.

Table 1. W, P and I, for several devices. The formulas include the following material constants: ε-permittivity; μ-permeability; $\mathbf{S}$ - stiffness tensor; ρ - mass density; k - elasticity factor; γ - thermal capacity; τ - relaxation time; λ - thermal conductivity.

Device	W	P	I
EM	$\int_\Omega (\mathbf{D}^2/(2\varepsilon) + \mathbf{B}^2/(2\mu))\,\mathrm{d}x$	$\int_{\Omega_d} \mathbf{J}\cdot\mathbf{E}\,\mathrm{d}x$	$\sum_{k=1}^n v_k i_k$
Mechanical	$\int_\Omega \left(\overline{\overline{\overline{\mathbf{S}}}} : \overline{\overline{\varepsilon}} : \overline{\overline{\varepsilon}}/2 + \mathbf{j}\cdot\mathbf{v}/2\right)\mathrm{d}x$	$\int_\Omega \left(\mu \frac{\partial\overline{\overline{\varepsilon}}}{\partial t}\right) : \frac{\partial\overline{\overline{\varepsilon}}}{\partial t}\,\mathrm{d}x$	$\sum_{k=1}^n (\mathbf{F}_k\cdot\mathbf{v}_k + \mathbf{T}_k\cdot\omega_k)$
Acoustic	$\int_\Omega (\rho\mathbf{v}^2/2 + kp^2/2)\,\mathrm{d}x$	$\int_\Omega \left(\tau\mathbf{v}\cdot\mathrm{grad}\,\frac{\partial p}{\partial t}\right)\mathrm{d}x$	$\sum_{k=1}^n (p_k A_k) v_k$
Thermal	$\int_\Omega (\gamma T^2/2)\,\mathrm{d}x$	$\int_\Omega (\mathbf{q}^2/\lambda)\,\mathrm{d}x$	$\sum_{k=1}^n T_k j_k$

3 Interconnection of the pH Systems

An essential characteristic of pH systems is their ability to be interconnected. For this, it is enough to put their terminals in contact. For example, for two EM devices with the same number of terminals which are in contact, in each contact, the voltage will be the same and the currents will have zero sum. For several EM devices interconnected in an arbitrary manner, the terminals can be seen as nodes, and the obtained network satisfies Kirchhoff's relations, regardless how the components are: distributed or lumped. Each port has two interaction variables: input (excitation or controlled) u and output (response or observed) y, each of them can be a flow "through": f (e.g. currents i) or an effort "across": e (e.g. voltages v). Some authors prefer the inverse assignment of i/v to f/e. Therefore, to avoid confusions, this assignment should be specified explicitly. The transferred power is $I(t) = \mathbf{u}^T\mathbf{y} = \mathbf{v}^T\mathbf{i}$ in the case of EM devices and $I(t) = \mathbf{u}^T\mathbf{y} = \mathbf{e}^T\mathbf{f}$ in general. Due to the linearity, after the Laplace transform, $\mathbf{y}(s) = \mathbf{H}(s)\mathbf{u}(s)$, $\mathbf{v}(s) = \mathbf{Z}(s)\mathbf{i}(s)$, $\mathbf{i}(s) = \mathbf{Y}(s)\mathbf{u}(s)$, and in general, $I(t) = \mathbf{u}^T\mathbf{y} = \mathbf{e}^T\mathbf{f}$, $\mathbf{y}(s) = \mathbf{H}(s)\mathbf{u}(s)$, $\mathbf{e}(s) = \mathbf{Z}(s)\mathbf{f}(s)$, $\mathbf{f}(s) = \mathbf{Y}(s)\mathbf{e}(s)$. There are 2^n possibilities for assignments of the input/output signals to flows/efforts, and so as many ways to define the hybrid matrix.

It is important to note that the system obtained by interconnection of pH systems is also a pH system, since the interconnection is power conservative. We call the interconnection of lumped elements a Dirac structure. With Kirchhoff constrains, a circuit with b branches satisfies the Tellegen's theorem for pseudo-powers (in particular the power balance), regardless the type of elements, lumped or distributed: $\sum_{k=1}^b i'_k u''_k = (\mathbf{u}'')^T\mathbf{i}' \overset{\mathrm{KVL}}{=} (\mathbf{A}^T\mathbf{v})^T\mathbf{i}' = \mathbf{v}^T(\mathbf{A}\mathbf{i}') \overset{\mathrm{KCL}}{=} 0$, and thus the Dirac structure - an essential ingredient of pH systems is mathematically defined as

$$\mathscr{D} \subset \mathscr{E} \times \mathscr{F}, (\mathbf{e}', \mathbf{f}'') \in \mathscr{D} \quad \text{iff} \quad (\mathbf{e}', \mathbf{f}'') = 0. \tag{9}$$

Here $\mathscr{E}$, $\mathscr{F}$ are the spaces of efforts and flows, respectively, both of dimension n, since they are conjugate $\mathscr{F} = \mathscr{E}^*$. The Dirac structure [13] is a linear subspace of same dimension n. In the case of distributed systems of infinite dimension, a Stokes-Dirac structure defined by the inner product $< \mathbf{e}', \mathbf{f}'' >= 0$ (an integral over Ω or $\partial\Omega$ of the products of two fields, e.g. $\mathbf{E} \cdot \mathbf{J}$ or $\mathbf{E} \times \mathbf{H}$) is used instead of Dirac structure, defined by the dot product $(\mathbf{e}', \mathbf{f}'') = 0$, which is used in lumped systems or in distributed systems with ECE BC. The Dirac structures refers to finite number of variables, they describe interconnections, or junctions in BG terminology (i.e., wires in electric circuits, paths in PCB, or metallic traces in IC). A Dirac structure connects typically three kind of components: energy storage C (where the flow-effort relation is $\mathbf{f}_S = \mathbf{J}\mathbf{e}_S$ with $\mathbf{J}$ a skew-symmetric matrix), dissipative R (were $\mathbf{e}_R = \mathbf{R}\mathbf{f}_R$ with a symmetric and positive semi-definite matrix of resistances $\mathbf{R} \geq 0$) and the interactive port (of sources) P with total transferred power $\mathbf{u}^T\mathbf{y} = \mathbf{e}_p^T\mathbf{f}_p$. The fundamental energy balance is a consequence of the Dirac relations:

$$-\mathbf{e}_S^T\mathbf{f}_S = (\nabla H\mathbf{x})^T\frac{\mathrm{d}\mathbf{x}}{\mathrm{d}t} = \frac{\mathrm{d}H}{\mathrm{d}t}, \qquad \mathbf{e}_S^T\mathbf{f}_S + \mathbf{u}^T\mathbf{y} + \mathbf{e}_R^T\mathbf{f}_R = 0, \tag{10}$$

where H - the Hamiltonian is the same as W - the energy.[1] In (10) all three ports have the same orientation, not as in Fig. 2, where I is opposite. Thus, the canonical form of the state equations of the pH systems are:

- Structural equation: $-\mathbf{f}_s = (\mathbf{J} - \mathbf{R})\mathbf{e}_s + \mathbf{G}\mathbf{u}, \quad \mathbf{y} = \mathbf{G}^T\mathbf{e}_s$;
- Dynamic equation (flow f_s): $\frac{\mathrm{d}\mathbf{x}}{\mathrm{d}t} = -\mathbf{f}_s$;
- Constitutive equations (effort e_s): $\mathbf{e}_s = \nabla H\mathbf{x}$,

where $\mathbf{x}$ is the state vector, H is the Hamiltonian (stored energy), $\mathbf{J} = -\mathbf{J}^T$ is the structure matrix, which describes the interconnections of the storing components, $\mathbf{R} = \mathbf{R}^T \geq 0$ is the dissipation matrix and $\mathbf{G}$ is the port matrix. This state equation is a system of ODEs. The DAE form of the Dirac/structure equations is discussed in [10].

Let's consider an open circuit (or a Kirchhoffian network) excited by external voltage and current sources. The elements are linear, one-port or multi-port, lumped (ideal) or distributed, controlled in voltages. Let's split the nodes in: "a"- excited in known currents/flows $\mathbf{f} = \mathbf{i}_a$, "b"= excited in known voltages/efforts $\mathbf{e} = \mathbf{v}_b$ and "c"-internal nodes. In the case of one-port elements with admittance Y_k, the nodal admittance is $Y_{xy} = \mathbf{A}_x \operatorname{diag}(Y_1, Y_2, ..., Y_b)\mathbf{A}_y^T$ (where $\mathbf{A}_x$ is the incidence matrix between nodes "x" and branches), the nodal technique gives the equation:

$$\begin{bmatrix} \mathbf{Y}_{aa} & \mathbf{Y}_{ab} & \mathbf{Y}_{ac} \\ \mathbf{Y}_{ba} & \mathbf{Y}_{bb} & \mathbf{Y}_{bc} \\ \mathbf{Y}_{ca} & \mathbf{Y}_{cb} & \mathbf{Y}_{cc} \end{bmatrix} \begin{bmatrix} \mathbf{v}_a \\ \mathbf{v}_b \\ \mathbf{v}_c \end{bmatrix} = \begin{bmatrix} \mathbf{i}_a \\ \mathbf{i}_b \\ \mathbf{0} \end{bmatrix} \quad \Rightarrow \quad \mathbf{v}_c = -\mathbf{Y}_{cc}^{-1}(\mathbf{Y}_{ca}\mathbf{v}_a + \mathbf{Y}_{cb}\mathbf{v}_b)$$

It follows that:

$$\mathbf{v}_a = \underbrace{(\mathbf{Y}_{aa} - \mathbf{Y}_{ac}\mathbf{Y}_{cc}^{-1}\mathbf{Y}_{ca})^{-1}}_{\mathbf{H}_{aa}}\mathbf{i}_a \underbrace{-\mathbf{H}_{aa}(\mathbf{Y}_{ab} - \mathbf{Y}_{ac}\mathbf{Y}_{cc}^{-1}\mathbf{Y}_{cb})}_{\mathbf{H}_{ab}}\mathbf{v}_b. \tag{11}$$

[1] We use two notations for the same quantity so as to emphasize both interpretations: the physical and the mathematical one.

$$\mathbf{i}_b = \underbrace{(\mathbf{Y}_{ba} - \mathbf{Y}_{bc}\mathbf{Y}_{cc}^{-1}\mathbf{Y}_{ca})\mathbf{H}_{aa}}_{\mathbf{H}_{ba}}\mathbf{i}_a + \underbrace{[(\mathbf{Y}_{ba} - \mathbf{Y}_{bc}\mathbf{Y}_{cc}^{-1}\mathbf{Y}_{ca})\mathbf{H}_{ab} + (\mathbf{Y}_{bb} - \mathbf{Y}_{bc}\mathbf{Y}_{cc}^{-1}\mathbf{Y}_{cb})]}_{\mathbf{H}_{bb}}\mathbf{v}_b,$$

Based on these relations, an algorithm to interconnect several lumped or distributed pH models in an arbitrary network can be build. The circuit is reciprocal if $\mathbf{H}_{ab} = -\mathbf{H}_{ba}^T$ and it is passive with $\mathbf{H}$ real positive if Y_k are real positive. Equation (11) describes an arbitrary interconnection of circuit components. We can say that the linearity, passivity and reciprocity are preserved by interconnection. The global circuit has the BG representation in Fig. 3, if it is reciprocal and in Fig. 4, if not (e.g. with operational amplifiers or controlled sources).

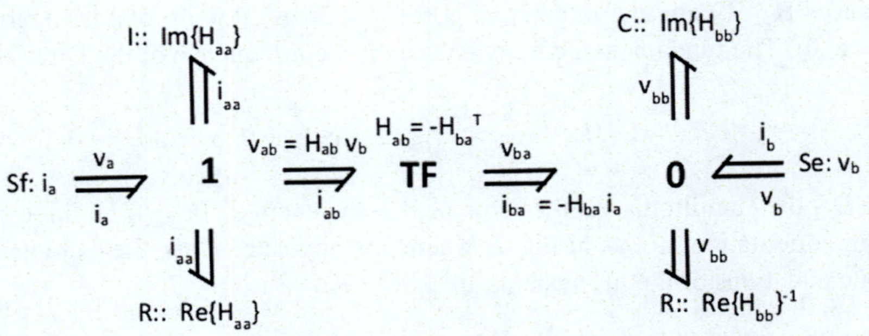

Fig. 3. BG of a reciprocal passive circuit.

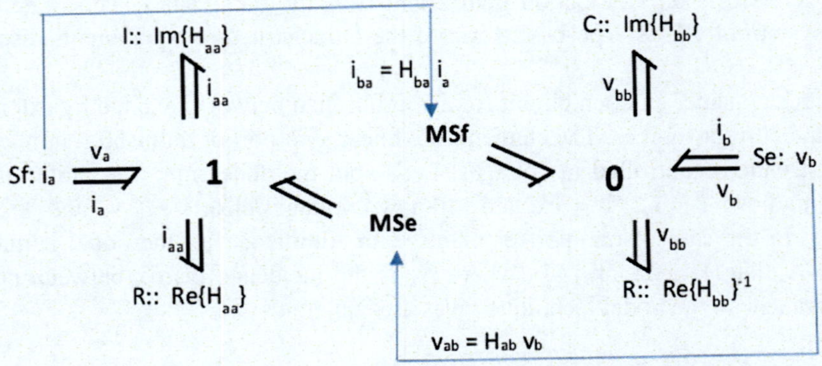

Fig. 4. BG of a non-reciprocal circuit (e.g. with op. amps).

In the particular cases of finite and infinite series or parallel connections, the impedances $\mathbf{Z}$ or admittances $\mathbf{Y}$ are added in sums and series [9].

4 Discrete Models of Distributed PH-ECE Systems

From Maxwell equations with ECE BC results in the frequency domain the 2nd order PDE for both electric and magnetic fields. Their strong form is a equation of complex curl-curl Helmholtz type [2]. Their weak forms are obtained by projection over a curl-conform space $\mathscr{H}$(curl) of the test functions. Since Hamiltonian plays an important role in the numerical variational methods, there is a close relation between weak form and the energy balance (Table 1). After solving this system, the vector of state variables $\mathbf{q} = [\underline{U}_j, \underline{V}_i, \underline{V}_k]$ (voltages of internal edges, node voltages on nonterminal boundary and terminal voltages) is obtained, which are the state variables of 2nd order ODE of FEM in time domain:

$$\mathbf{Kq} + \mathbf{D}\frac{\mathrm{d}\mathbf{q}}{\mathrm{d}t} + \mathbf{M}\frac{\mathrm{d}^2\mathbf{q}}{\mathrm{d}t^2} = \mathbf{B}'\mathbf{u} + \mathbf{B}''\frac{\mathrm{d}\mathbf{u}}{\mathrm{d}t}, \qquad \mathbf{y} = \mathbf{C}'\mathbf{q} + \mathbf{C}''\frac{\mathrm{d}\mathbf{q}}{\mathrm{d}t} + \mathbf{D}'\mathbf{u}, \tag{12}$$

with $\mathbf{K}$ the stiffness matrix, $\mathbf{D}$ the dissipation matrix; and $\mathbf{M}$ the mass matrix. all symmetric, positive definite and sparse matrices, and $\mathbf{u}$, $\mathbf{y}$ the vectors of input and output signals, currents or voltages, depending how terminals are excited. Similar equations are obtained after using other numeric methods, such as FIT (Finite Integrals Technique) or DEC (Discrete External Calculus), from which $\mathbf{q}$ can be extracted. By changing the variables, so as to use as states $\mathbf{r} = [\mathbf{q}; \mathbf{p}]$, where $\mathbf{q}$ is the "position" vector (states of 2nd order ODE), and $\mathbf{p} = \mathbf{Mq}$ is the momentum vector, the 1st order, canonical form of the pH equations is obtained:

$$\begin{cases} \frac{\mathrm{d}\mathbf{r}}{\mathrm{d}t} = (\mathbf{J} - \mathbf{R})\mathbf{Qr} + \mathbf{Bu} \\ \mathbf{y} = \mathbf{B}^T\mathbf{Qr} \end{cases} \qquad \begin{cases} \dot{\mathbf{r}} = \mathbf{Ar} + \mathbf{Bu} \\ \mathbf{y} = \mathbf{Cr} \end{cases}$$

$$\mathbf{J} = \begin{bmatrix} \mathbf{0} & \mathbf{I} \\ -\mathbf{I} & \mathbf{0} \end{bmatrix} \quad \mathbf{R} = \begin{bmatrix} \mathbf{0} & \mathbf{0} \\ \mathbf{0} & \mathbf{D} \end{bmatrix} \quad \mathbf{Q} = \nabla H = \begin{bmatrix} \mathbf{K} & \mathbf{0} \\ \mathbf{0} & \mathbf{M}^{-1} \end{bmatrix}$$

$$\mathbf{A} = (\mathbf{J} - \mathbf{R})\mathbf{Q} \qquad \mathbf{C} = \mathbf{B}^T\mathbf{Q} \tag{13}$$

with $\mathbf{J}$ a skew-symmetric (symplectic) matrix, $\mathbf{R}$ a symmetric positive semi-definite matrix, which describes the dissipation, and $\mathbf{Q}$, the gradient of the Hamiltonian, which describes the structure of energy storage. The port matrix $\mathbf{B}$ describes the interaction structure. After mass lumping, the inversion of $\mathbf{M}$ is not difficult.

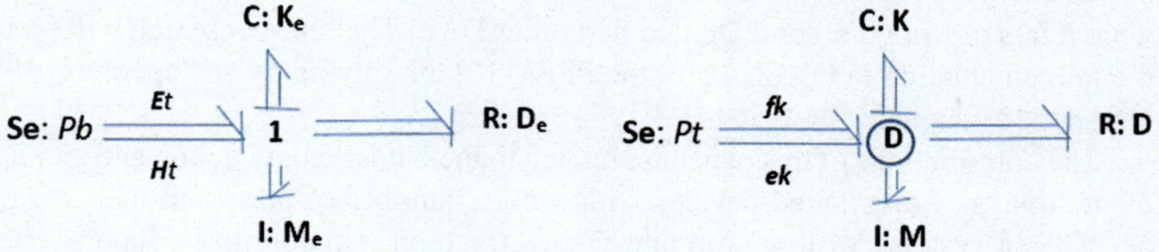

Fig. 5. BG for each element (left); BG for the global device (right).

Consequently, each finite element is a pH sub-system, and after their assembling, the global numerical model is also a pH systems (Fig. 5). The energy is balanced locally as well as globally, the mesh acting as a Dirac structure. The first order canonical form of equations can be obtained using an alternative way, called PFEM, as in [6], starting from the weak form of the Maxwell equation of 1st order and eventually obtained a canonical form of DAE type: $\mathbf{E}d\mathbf{x}/dt = (\mathbf{J}-\mathbf{R})\mathbf{Q}\mathbf{x}+\mathbf{B}\mathbf{u}$, with $\mathbf{x} = [\mathbf{E};\mathbf{D};\mathbf{B};\mathbf{H};\mathbf{J}]$. The advantage of 2nd order form is given by the Lax-Milgram theorem which guarantees the well formulation of both continuous and discrete problems, whereas PFEM preserves the conservative properties (divergence of flux densities **B**, **D**, and the energy conservation). However in DAE form of PFEM there is a waste of computing resources, since all 3D vectors **E**, **D**, **B**, **H** and **J** are stored.

The proposed method for field-circuits coupling is illustrated by simulation in a free software environment ONELAB (`onelab.info`) of a two-port patch GPS antenna (modeled FW with ECE BC+ABC) connected to a Bridge Line Connector (BLC), procedure described in [7]. The antenna's admittance was computed by FEM for several frequencies (only 25 samples were enough to obtain an error less than 1% in Adaptive Frequency Sampling). This frequency characteristic was approximated with VFIT (Vector Fitting), as a rational function of a reduced order $n = 9$, with an error less than 1e-4. Then an equivalent Spice circuit with this admittance was synthesized. The BLC modeled with transmission lines was coupled to this circuit and it was simulated in LTSpice.

5 Conclusions

The representation of distributed models by lumped parameter BG/pH has a practical relevance since in most technical devices the field effects are spatially distributed, but their interaction with the environment is often described only by a finite number of scalar quantities. In the literature, the multiphysics distributed devices are studied as infinite pH systems using a Stokes-Dirac structure. Our approach uses ECE BC for the field equations, thus generating pH systems with a finite number of ports, which are simpler finite Dirac structures instead of infinite Stokes-Dirac ones. The approach we propose is extremely efficient because the model order reduction is applied to the individual components, before their interconnection. Moreover, it can be applied for several physical fields such as: electro-magnetic, elasto-dynamic, thermal, acoustic. For all these domains, the PDE state equations are similar, both in strong and weak forms, and an energy balance. The FEM discretizations for all these domains are also similar in what refers to second order ODE and first order DAE. The future research will focus on the implementation in ONELAB of the PFEM [6], its solving by an "operator splitting method" [3,5] and MOR as in [4].

The contributions of this paper are related to the numerical modeling and simulation of multiphysics distributed devices with a finite number of ports and their treatment as BG, pH systems with appropriate Dirac structures, suitable to be incorporated in complex networks.

References

1. Borutzky, W.: Bond Graph Methodology. Springer, London (2010). https://doi.org/10.1007/978-1-84882-882-7
2. Ciuprina, G., Ioan, D., Sabariego, R.V.: Electric circuit element boundary conditions in the finite element method for full-wave passive electromagnetic devices. J. Math. Industry **12**, 7 (2022)
3. Günther, M.: Port-Hamiltonian systems: a useful approach in electrical engineering? Commun. SCEE 2022, Amsterdam (2022)
4. Günther, M., Jacob, B., Totzeck, C.: Structure-preserving identification of port-Hamiltonian systems. Commun. SCEE 2022, Amsterdam (2022)
5. Diab, M., Tischendorf, C.: Splitting methods for coupled field-circuit DAEs. Commun. SCEE 2022, Amsterdam (2022)
6. Payen, G., Matignon, D., Haine, G.: Modelling and structure-preserving discretization of Maxwell's equations as port-Hamiltonian system. IFAC-PapersOnLine **53**, 7581–7586 (2020)
7. Ioan, D., Ciuprina, G.: BG Approximation of multiphysics pH distributed systems with finite number of ports. In: Presentation at SCEE (2022). https://www.researchgate.net/publication/362432741
8. Ioan, D., Munteanu, I.: Missing link rediscovered: the electromagnetic circuit element concept. JSAEM Stud. Appl. Electromagn. Mech. **8**, 302–320 (1999)
9. Jacob, B., Zwart, H.J.: Linear Port-Hamiltonian Systems on Infinite-Dimensional Spaces. Springer, Basel (2012)
10. Mehrmann, V., Morandin, R.: Structure-preserving discretization for port-Hamiltonian descriptor systems. In: IEEE 58th Conference on Decision and Control, pp. 6863-6868. France (2019)
11. Răduleţ, R., Timotin, Al., Ţugulea, A.: Introducerea parametrilor tranzitorii în studiul circuitelor electrice lineare avînd elemente nefiliforme şi pierderi suplimentare. St. cerc. energ. electr. **16**(4), 857–929 (1966)
12. Răduleţ, R., Timotin, Al., Ţugulea, A.: O teorie de câmp structurală (a unei clase de sisteme lineare). Ed. Academiei, Bucureşti, România (1972)
13. Van Der Schaft, A., Jeltsema, D.: Port-Hamiltonian systems theory: an introductory overview. Found. Trends Syst. Control **1**(2–3), 173–378 (2014)

Bilinear Realization from I/O Data with NNs

D. S. Karachalios[1,2(✉)], I. V. Gosea[2], K. Kour[2], and A. C. Antoulas[2,3,4]

[1] Institute of Electrical Engineering in Medicine, University Luebeck, Luebeck, Germany
[2] Max Planck Institute for Dynamics of Complex Technical Systems, Magdeburg, Germany
dimitrios.karachalios@uni-luebeck.de,
{gosea,kour}@mpi-magdeburg.mpg.de
[3] Department of Electrical and Computer Engineering, Rice University, Houston, TX 77005, USA
aca@rice.edu
[4] Baylor College of Medicine, Houston, TX 77030, USA

Abstract. We present a method that connects a well-established nonlinear (bilinear) identification method from data in the time domain with the advantages of neural networks (NNs). The main challenge for fitting bilinear systems is the accurate recovery of the corresponding Markov parameters from the input and output measurements. Afterward, a realization algorithm similar to that proposed by Isidori can be employed. The novel step is that NNs are used here as a surrogate data simulator to construct input-output (i/o) data sequences from a single experiment. Then, classical realization theory is used to build an interpretable bilinear model that can further optimize engineering processes through robust simulations and control design.

1 Introduction

Evolutionary phenomena can be formally described as continuous dynamical models with partial differential equations (PDE)s. The continuous nature of these physical models is equipped with analytical results for an efficient discrete approximation in space and in time. In particular, methods such as finite elements or finite differences bridge the continuous analytical laws of the physical world with computational science [1]. On the other hand, data science allows model discovery when the identification feature is considered [2]. Quantification of these equivalences, in combination with the stochastic nature that governs real-world applications, aims to explain the digital twin [3]. Spatial discretization of PDEs, in many cases, results in a continuous in-time system of ordinary differential equations (ODE)s that is described by the operators $(\mathbf{F},\ \mathbf{G})$ and can be approximated with Carleman linearization (e.g., bilinear system form) [4] where we present the single-input single-output (SISO) case with the continuous operators to be denoted with the subscript "c."

$$\boldsymbol{\Sigma}:\begin{cases}\dot{\mathbf{x}}(t)=\mathbf{F}(\mathbf{x}(t))+\mathbf{G}(\mathbf{x}(t))u(t)\\ y(t)=\mathbf{H}\mathbf{x}(t),\ \mathbf{x}_0=\mathbf{0},\ t\geq 0.\end{cases}\ \xrightarrow[\boldsymbol{\Sigma}\approx\boldsymbol{\Sigma}_{bil}]{\text{Carleman}}\ \boldsymbol{\Sigma}_{bil}:\begin{cases}\dot{\mathbf{x}}(t)=\mathbf{A}_c\mathbf{x}(t)+\mathbf{N}_c\mathbf{x}(t)u(t)+\mathbf{B}_c u(t)\\ y(t)=\mathbf{C}_c\mathbf{x}(t),\ \mathbf{x}_0=\mathbf{0},\ t\geq 0.\end{cases} \quad (1)$$

If the original system has dimension n, since Carleman linearization [4] preserves up to the quadratic term $\mathbf{x}(t)\otimes\mathbf{x}(t)$[1], the dimension of the resulting bilinear system $(\mathbf{A}_c,\ \mathbf{N}_c,\ \mathbf{B}_c,\ \mathbf{C}_c)$ increases to $N=n^2+n$.

[1] $\otimes$: Kronecker product.

M. van Beurden et al. (Eds.): SCEE 2022, MI 43, pp. 184–192, 2024.
https://doi.org/10.1007/978-3-031-54517-7_21

Data-driven methods can be classified into two general classes. The first provides prediction through regression techniques such as neural networks (NN)s from machine learning (ML). At the same time, the second has its roots in system theory and allows model discovery [2,5]. Generally, NNs are sensitive to parameter tuning and lack model interpretability due to the inherent "black-box" structure [6], while the latter construct interpretable models and can explain the hidden dynamics. ML models learn the features by composing non-linear activation functions and utilizing mainly the backpropagation algorithm to adjust the network weights during training. Therefore, by using data points for training, the prediction would be expressed as a function of these data points (finite memory). Until recently, the ML and system identification (SI) techniques were developed independently. But in recent years, great effort has been invested into establishing a common ground [7].

The authors in [8] have extended the subspace realization theory from linear to bilinear systems. For example, in applications that concern chemical processes, the controls are flow rates, and, from the first principles, e.g., mass and heat balances, these will appear in the system equations as products with the state variables. Therefore, the bilinear equation has the physical form $M\dot{\mathbf{x}} = \sum_i \mathbf{q}_i\mathbf{x}_i - \sum_m \mathbf{q}_m\mathbf{x}_m$, $\mathbf{q}$(inputs), $\mathbf{x}$(state). The authors of [9] could construct bilinear systems with white noise input based on an iterative deterministic-stochastic subspace approach. The author in [10] uses the properties of the linear model of the bilinear system when subjected to a constant input. Constant inputs can transform the bilinear model to an equivalent linear model [11].

In Sect. (2), we introduce the theory of bilinear realization by explaining in detail the data acquisition procedure to compute the bilinear Markov parameters that will enter the bilinear Hankel matrix. Further, we present a concise algorithm that can achieve bilinear identification, detailed by two examples. In Sect. (3), we train a neural network with a single i/o data sequence to mimic the unknown simulator and combine it with the bilinear realization theory. As a result, we could construct a bilinear model from a single i/o data with slightly better fit performance compared with another state-of-the-art bilinear SI approach. Finally, we provide the conclusion and the outlook in Sect. (4).

2 The Bilinear Realization Framework

In the case of linear systems, Ho and Kalman [12] have provided the mathematical foundations for realizing linear systems from i/o data. In the nonlinear case and towards the exact scope of identifying nonlinear systems, Isidori in [13] has extended these results for the bilinear case, and Al Baiyat in [14] has provided an SVD-based algorithm.

Time discretization as in [15] of the single-input single-output (SISO) bilinear system Eq. (1) with sampling time Δt, results in fully discrete models defined at time instances given by $0 < \Delta t < 2\Delta t < \cdots < k\Delta t$, with $\mathbf{x}_c(k\Delta t) = \mathbf{x}_k$ and $u(k\Delta t) = u_k$ for $k = 0, \ldots, m-1$

$$\boldsymbol{\Sigma}_{\text{disc}} : \begin{cases} \mathbf{x}_{k+1} = \mathbf{A}\mathbf{x}_k + \mathbf{N}\mathbf{x}_k u_k + \mathbf{B}u_k, \\ y_k = \mathbf{C}\mathbf{x}_k,\ \mathbf{x}_0 = \mathbf{0}. \end{cases} \tag{2}$$

The discrete-time system in Eq. (2) has state dimension N, so, $\mathbf{x} \in \mathbb{R}^N$ and the operators have dimensions $\mathbf{A}, \mathbf{N} \in \mathbb{R}^{N\times N}$, $\mathbf{B}, \mathbf{C}^T \in \mathbb{R}^N$. We have assumed homogeneous initial conditions and a zero feed-forward term (e.g., $\mathbf{D} = 0$ term). As far as the authors are aware, the forward Euler scheme is the only numerical scheme that preserves the bilinear structure in a discrete set-up with the cost of conditional stability. Moreover, a more sophisticated scheme can exactly interpolate the continuous model at the sampling points in [16] but is restricted to only a subclass of bilinear systems. Therefore, a good choice in terms of stability is the backward Euler scheme from [15], which preserves the bilinear structure asymptotically, and the transformation in Eq. (3) that leads to the discrete system is

$$\begin{aligned}&\phi:\ \mathbf{A}=(\mathbf{I}-\Delta t\mathbf{A}_c)^{-1},\ \mathbf{N}=\Delta t(\mathbf{I}-\Delta t\mathbf{A}_c)^{-1}\mathbf{N}_c,\ \mathbf{B}=\Delta t(\mathbf{I}-\Delta t\mathbf{A}_c)^{-1}\mathbf{B}_c,\ \mathbf{C}=\mathbf{C}_c,\\ &\mathbf{\Sigma}_b^c:(\mathbf{A}_c,\mathbf{N}_c,\mathbf{B}_c,\mathbf{C}_c)\overset{\phi^{-1}}{\leftrightarrow}\mathbf{\Sigma}_b^d:(\mathbf{A},\mathbf{N},\mathbf{B},\mathbf{C})\end{aligned} \tag{3}$$

Definition 1. The reachability matrix $\mathscr{R}_n=\begin{bmatrix}\mathbf{R}_1 & \cdots & \mathbf{R}_n\end{bmatrix}$ is defined recursively from the following relation: $\mathbf{R}_j=\begin{bmatrix}\mathbf{A}\mathbf{R}_{j-1} & \mathbf{N}\mathbf{R}_{j-1}\end{bmatrix}$, $j=2,\ldots n$, $\mathbf{R}_1=\mathbf{B}$.

Then, the state space of the bilinear system is spanned by the states reachable from the origin if and only if $\mathrm{rank}(\mathscr{R}_n)=n$.

Definition 2. The observability matrix $\mathscr{O}_n=\begin{bmatrix}\mathbf{O}_1 & \cdots & \mathbf{O}_n\end{bmatrix}^T$ is defined recursively from the following relation: $\mathbf{O}_j^T=\begin{bmatrix}\mathbf{O}_{j-1}\mathbf{A} & \mathbf{O}_{j-1}\mathbf{N}\end{bmatrix}^T$, $j=2,\ldots n$, $\mathbf{O}_1=\mathbf{C}$.

Then the state space of the bilinear system is observable iff $\mathrm{rank}(\mathscr{O}_n)=n$. The following Def. (3) will allow a concise representation of the i/o relation.

Definition 3. $\mathbf{u}_j(h)=\begin{bmatrix}\mathbf{u}_{j-1}(h)\\ \mathbf{u}_{j-1}(h)u(h+j-1)\end{bmatrix}$, $j=2,\ldots,$ $\mathbf{u}_1(h)=u(h)$.

Let $\{\mathbf{w}_1,\mathbf{w}_2,\ldots,\mathbf{w}_j,\ldots\}$ be an infinite sequence of row vectors, in which $\mathbf{w}_j\in\mathbb{R}^{1\times 2^{j-1}}$ and is defined recursively as follows $\mathbf{w}_j=\mathbf{C}\mathbf{R}_j$, $j=1,2,\ldots;$

The state response of system Eq. (2) from the state $\mathbf{x}_0=\mathbf{0}$ at time $k=0$, under a given input function, can be expressed as:

$$\begin{aligned}&\mathbf{x}_1=\mathbf{B}u_0\triangleq\mathbf{R}_1\mathbf{u}_1(0),\\ &\mathbf{x}_2=\mathbf{A}\mathbf{R}_1\mathbf{u}_1(0)+\mathbf{N}\mathbf{R}_1\mathbf{u}_1(0)u(1)+\mathbf{B}u(1)\triangleq\mathbf{R}_2\mathbf{u}_2(0)+\mathbf{R}_1\mathbf{u}_1(1),\\ &\vdots\\ &\mathbf{x}_k=\sum_{j=1}^{k}\mathbf{R}_j\mathbf{u}_j(k-j),\ k=1,2,\ldots;\end{aligned} \tag{4}$$

Finally, the zero-state input-output map of system Eq. (2) after multiplication with the vector $\mathbf{C}$ from the left can be written as:

$$y_k=\sum_{j=1}^{k}\mathbf{w}_j\mathbf{u}_j(k-j),\ k=1,2,\ldots; \tag{5}$$

2.1 The Bilinear Markov Parameters

The bilinear Markov (invariant) parameters are encoded in the $\{\mathbf{w}_j\}$ vectors for $j \in \mathbb{Z}_+$. These are invariant quantities of the bilinear system in connection with the input-output relation. After making use of Def. (3), we can write

$$\underbrace{\begin{bmatrix} y_1 \\ y_2 \\ \vdots \\ y_k \end{bmatrix}}_{\mathbf{Y}} = \underbrace{\begin{bmatrix} \mathbf{u}_1^T(0) & 0 & \cdots & 0 \\ \mathbf{u}_1^T(1) & \mathbf{u}_2^T(0) & \cdots & 0 \\ \vdots & \vdots & \ddots & \vdots \\ \mathbf{u}_1^T(k-1) & \mathbf{u}_2^T(k-2) & \cdots & \mathbf{u}_k^T(0) \end{bmatrix}}_{\mathbf{U}} \cdot \underbrace{\begin{bmatrix} \mathbf{w}_1^T \\ \mathbf{w}_2^T \\ \vdots \\ \mathbf{w}_k^T \end{bmatrix}}_{\mathbf{W}}, \tag{6}$$

where the dimensions are: $\mathbf{Y} \in \mathbb{R}^{k\times 1}$, $\mathbf{U} \in \mathbb{R}^{k\times m}$, and $\mathbf{W} \in \mathbb{R}^{m\times 1}$.

The least squares problem filled out with k time steps will remain under-determined $\forall k \in \{2,3,\ldots\}$ as long as the $m = 2^k - 1$ bilinear Markov parameters are activated. Thus, we must deal with the k equations and the $2^k - 1$ unknowns. Solving an under-determined system is not impossible, but the solutions are infinite, and regularization schemes cannot easily lead to identification. Therefore, one way to uniquely identify bilinear Markov parameters and determine the solution vector $\mathbf{W}$ can be achieved by solving a coupled least squares system after applying several simulations to the original system.

To uniquely determine the $(2^k - 1)$ parameters, the column rank of the matrix $\mathbf{U}$ should be complete. This can be accomplished by adding rows with more experiments to the matrix $\mathbf{U}$ until the new augmented matrix $\hat{\mathbf{U}}$ has more rows than columns. Thus, we need at least 2^{k-1} independent simulations of the original system. That is exactly the bottleneck expected for nonlinear identification frameworks that deal with time-domain data. Later, we will relax this condition in a novel way using NNs. Equation (7) describes the coupled linear least squares system with $d = 2^{k-1}$ independent simulations that can provide the unique solution $\mathscr{W}$ with bilinear Markov parameters.

$$\underbrace{\begin{bmatrix} \mathbf{Y}_1 \cdots \mathbf{Y}_d \end{bmatrix}^T}_{\hat{\mathbf{Y}}} = \underbrace{\begin{bmatrix} \mathbf{U}_1 \cdots \mathbf{U}_d \end{bmatrix}^T}_{\hat{\mathbf{U}}} \cdot \mathscr{W} \tag{7}$$

Hence, we repeat the simulation d times, and each time we get k equations, with the ith simulation to be $\mathbf{Y}_i = \begin{bmatrix} y_1^{(i)} \; y_2^{(i)} \cdots y_k^{(i)} \end{bmatrix}^T$ and accordingly for the $\mathbf{U}_i$, the real matrix $\hat{\mathbf{U}}$ has dimension $2^k \times (2^k - 1)$. After concatenating all the lower triangular matrices with full column rank, the matrix $\hat{\mathbf{U}}$ results. To enforce that $\hat{\mathbf{U}}$ will also have full column rank, one choice is to use a white input (sampled from a Gaussian distribution) for the simulations. The use of a white input is widespread for SI. Still, in that case, a careful choice of deterministic inputs can make the inversion exact and recover the bilinear Markov parameters. The solution is as follows: $\text{rank}(\hat{\mathbf{U}}) = 2^k - 1$, so, the unique solution is: $\mathscr{W} = \hat{\mathbf{U}}^{-1}\hat{\mathbf{Y}} \in \mathbb{R}^{2^k - 1}$. The vector $\mathscr{W}$ contains the $2^k - 1$ bilinear Markov parameters. A generalized Hankel matrix can be computed from the bilinear Markov parameters.

2.2 The Bilinear Hankel Matrix

The bilinear Hankel matrix is the product of the observability and reachability matrices. The bilinear Hankel matrix is denoted with $\mathscr{H}_b$ and is defined as the product of the following two infinite matrices $\mathscr{O}$, $\mathscr{R}$,

$$\mathscr{H}_b = \mathscr{O}\mathscr{R} = \begin{bmatrix} \mathbf{C} \\ \mathbf{CA} \\ \mathbf{CN} \\ \vdots \end{bmatrix} \begin{bmatrix} \mathbf{B}\ \mathbf{AB}\ \mathbf{NB} \cdots \end{bmatrix} = \begin{bmatrix} \mathbf{CB} & \mathbf{CAB} & \mathbf{CNB} & \cdots \\ \mathbf{CAB} & \mathbf{CA^2B} & \mathbf{CANB} & \cdots \\ \mathbf{CNB} & \mathbf{CNAB} & \mathbf{CN^2B} & \cdots \\ \vdots & \vdots & \vdots & \ddots \end{bmatrix} \tag{8}$$

Equation (8) reveals the connection with the bilinear Markov parameters $\mathscr{W} = \mathbf{C}\mathscr{R}$ that appear in the first row of $\mathscr{H}_b$. In general, the construction of the bilinear Hankel matrix is described in [13] with the partial and completed realization theorems along with the partitions[2] $\mathscr{S}^A$, $\mathscr{S}^N$ [14].

2.3 Bilinear Realization Algorithm

Input: Input-output time-domain data from a system $u \rightarrow \boxed{\boldsymbol{\Sigma}?} \rightarrow y$.
Output: A minimal bilinear system $(\mathbf{A}_r, \mathbf{N}_r, \mathbf{B}_r, \mathbf{C}_r)$ of low dimension r that $\boldsymbol{\Sigma}_r \approx \boldsymbol{\Sigma}$.

1. Excite the system $\boldsymbol{\Sigma}$ k times with $\mathbf{u}_m \sim \mathbf{N}(\mu, \sigma)$ and collect $\mathbf{y}_m$, where $k = 2^{m-1}$.

1st simulation $[u_1(1)\cdots u_1(m)] \rightarrow \boxed{\boldsymbol{\Sigma}} \rightarrow [y_1(1)\cdots y_1(m)] = \mathbf{Y}_1$, and $\mathbf{U}_1$ as in Definition 3	
$\vdots$	$\vdots$
kth simulation $[u_k(1)\cdots u_k(m)] \rightarrow \boxed{\boldsymbol{\Sigma}} \rightarrow [y_k(1)\cdots y_k(m)] = \mathbf{Y}_k$, and $\mathbf{U}_k$ as in Definition 3.	

2. Identify the $(2^m - 1)$ bilinear Markov parameters by solving the system in (7).
3. Construct the bilinear Hankel matrix $\mathscr{H}_b$ and the sub-matrices $\mathscr{S}^{\mathbf{A}}$, $\mathscr{S}^{\mathbf{N}}$.
4. Compute $[\mathbf{U}, \boldsymbol{\Sigma}, \mathbf{V}] = \text{SVD}(\mathscr{H}_b)$ and truncate w.r.t the singular values decay ($r \ll n$) - the reduced/identified bilinear model $(\mathbf{A}_r, \mathbf{N}_r, \mathbf{B}_r, \mathbf{C}_r)$ is constructed

$$\mathbf{A}_r = \boldsymbol{\Sigma}^{-1/2}\mathbf{U}^T \mathscr{S}^{\mathbf{A}}\mathbf{V}\boldsymbol{\Sigma}^{-1/2} \tag{9}$$

$$\mathbf{N}_r = \boldsymbol{\Sigma}^{-1/2}\mathbf{U}^T \mathscr{S}^{\mathbf{N}}\mathbf{V}\boldsymbol{\Sigma}^{-1/2} \tag{10}$$

$$\mathbf{B}_r = \boldsymbol{\Sigma}^{1/2}\mathbf{V}^T \quad \rightarrow \quad \text{1st column} \tag{11}$$

$$\mathbf{C}_r = \mathbf{U}\boldsymbol{\Sigma}^{1/2} \quad \rightarrow \quad \text{1st row} \tag{12}$$

Example 1. (A toy system) Let the following bilinear system of order 2 be

$$\mathbf{A} = \begin{bmatrix} 0.9 & 0.0 \\ 0.0 & 0.8 \end{bmatrix}, \mathbf{N} = \begin{bmatrix} 0.1 & 0.2 \\ 0.3 & 0.4 \end{bmatrix}, \mathbf{B} = \begin{bmatrix} 1.0 \\ 0.0 \end{bmatrix}, \mathbf{C} = \begin{bmatrix} 1.0 \\ 1.0 \end{bmatrix}^T. \tag{13}$$

Applying the algorithm in Sect. (2.3), by choosing $m = 4$, we can recover $2^m - 1 = 15$ bilinear Markov parameters. The solution of the system in Eq. (7) is:

$$\mathbf{W} = \begin{bmatrix} 1.0\ 0.9\ 0.4\ 0.81\ 0.33\ 0.36\ 0.22\ 0.729\ 0.273\ 0.297\ 0.183\ 0.324\ 0.18\ 0.198\ 0.118 \end{bmatrix}$$

[2] $\mathscr{S}^A = \{\text{set of } \mathscr{H}_b \text{columns : from } 2^m \text{ to } (3 \cdot 2^{m-1} - 1),\ m = 1, 2, \ldots\}$,
$\mathscr{S}^N = \{\text{set of } \mathscr{H}_b \text{columns : from } 3 \cdot 2^{m-1} \text{ to } (2^{m+1} - 1),\ m = 1, 2, \ldots\}$.

By reshuffling the vector $\mathscr{W}$, we can form the $\mathscr{H}_b$ matrix and the shifted versions $\mathscr{S}^{\mathbf{A}}$, $\mathscr{S}^{\mathbf{N}}$ as described above. The Hankel matrix (3 rows & 7 columns are displayed) along with the shifted versions are $\left\{\mathscr{H}_b,\ \mathscr{S}^{A},\ \mathscr{S}^{N}\right\} :=$

$$\left\{\begin{bmatrix} 1.0 & 0.9 & 0.4 & 0.81 & 0.33 & 0.36 & 0.22 \\ 0.9 & 0.81 & 0.33 & 0.729 & 0.273 & 0.297 & 0.183 \\ 0.4 & 0.36 & 0.22 & 0.324 & 0.18 & 0.198 & 0.118 \end{bmatrix}, \begin{bmatrix} 0.9 & 0.81 & 0.33 \\ 0.81 & 0.729 & 0.273 \\ 0.36 & 0.324 & 0.18 \end{bmatrix}, \begin{bmatrix} 0.4 & 0.36 & 0.22 \\ 0.33 & 0.297 & 0.183 \\ 0.22 & 0.198 & 0.118 \end{bmatrix}\right\}.$$

In Fig. (1), the 3rd normalized singular value has reached machine precision $\sigma_3/\sigma_1 = 5.2501e-17$, that is the criterion for choosing the order of the fitted system (which is minimal, in this case) of the underlying bilinear system. Therefore, we construct a bilinear model of order $r = 2$, and the realization obtained is equivalent to the original (minimal) one, up to a coordinate (similarity) transformation. Other ways of constructing reduced models from Hankel⊂Loewner matrices can be obtained with the CUR (cross approximations based) decomposition scheme as in [17] (Fig. 1).

$$\mathbf{A}_r = \begin{bmatrix} 0.89394 & 0.11305 \\ 0.0050328 & 0.80606 \end{bmatrix}, \mathbf{N}_r = \begin{bmatrix} 0.41116 & -0.2281 \\ -0.24782 & 0.088841 \end{bmatrix}, \mathbf{B}_r = \begin{bmatrix} -1.0001 \\ -0.053577 \end{bmatrix}, \mathbf{C}_r = \begin{bmatrix} -1.0001 \\ 0.0040101 \end{bmatrix}^T. \quad (14)$$

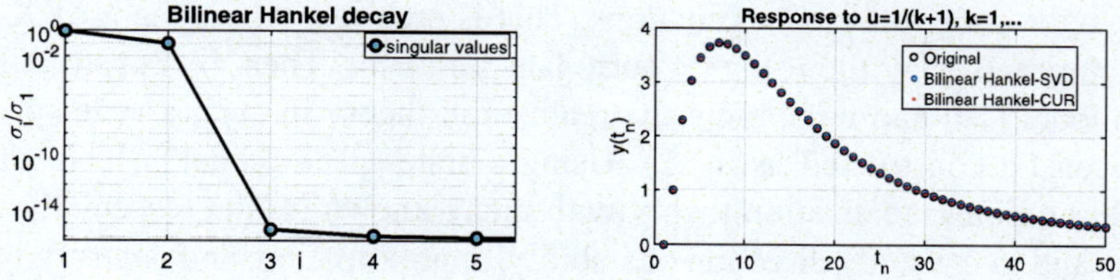

Fig. 1. On the left figure, the singular value decay of the bilinear Hankel matrix is depicted. On the right figure, the input response $u_k = 1/(k+1)$, $k = 0, 1, \ldots$ certifies that all models are equivalent.

Example 2. (The viscous Burgers' equation example) Following [15] after spatial semi-discretization and the Carleman linearization technique, yields a bilinear system of dimension $N = 30^2 + 30 = 930$. The viscosity parameter is $\nu = 0.1$; the sampling time is $\Delta t = 0.1$ and with $2^{m-1} = 512$ independent random inputs of length $m = 10$ each, we construct a database of $5,120$ points. Solving Eq. (7), we get the bilinear Markov parameters, and the bilinear Hankel matrix is constructed. On the left pane of Fig. (2), the decay of bilinear Hankel singular values captures the nonlinear nature of the Burgers' equation, while, on the other hand, the linear Hankel framework captures only the linear minimal response. It is evident in the right pane of Fig. (2) that after using the inverse transformation ϕ from (3), the reduced continuous-time bilinear model of order $r = 18$ performs well, producing an error $O(10^{-5})$ where at the same time the linear fit is off (Fig. 2).

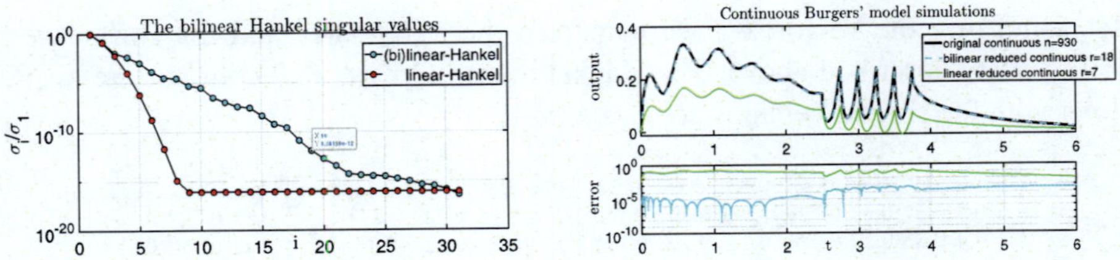

Fig. 2. Left pane: The recommended reduced bilinear model with $\Delta t = 0.1$ is of order $r = 18$ where $\sigma_{19}/\sigma_1 = 1.18 \cdot 10^{-12}$. **Right pane**: $u_1 = (1 + \cos(2\pi t))e^{-t}, t \in [0, 2.5], u_2 = 2\text{sawtooth}(8\pi t), t \in [2.5, 3.75], u_3 = 0$, it is compared with a continuous bilinear identification method based on the Loewner framework in both frequency and time domain approaches [5, 18].

3 From a Single Data Sequence to Bilinear Realization

A repetitive data assimilation simulation in the time domain is required to achieve bilinear realization as in [13]. In many cases, the data from a simulated system are available as a single i/o sequence [9]. Using the NARX-net-based model, in the case of a single experiment, the expensive, repetitive simulations can be avoided in a real engineering environment. These models learn from a unique data sequence and can predict the output behavior under different excitations. That is precisely where the NARX-net model architecture will play the role of a surrogate simulator. Then, by constructing an NN-based model [19] and combining the realization theory in [13], a state-space bilinear model can be constructed as in (2). Using a state-space model, which relies on the classical nonlinear realization theory with many known results (especially on bilinear systems and in the study direction of stability, approximation, and control), is beneficial compared to the NARX.

Example 3. (Heat exchanger) The process is a liquid-saturated steam heat exchanger, where water is heated by pressurized saturated steam through a copper tube. The input variable is the liquid flow rate, and the output variable is the outlet liquid temperature. The sampling time is 1(s), and the number of samples is 4,000. More details can be found in [20], and the data set can be downloaded from the database to identify systems (DaISy): https://homes.esat.kuleuven.be/~tokka/daisydata.html.

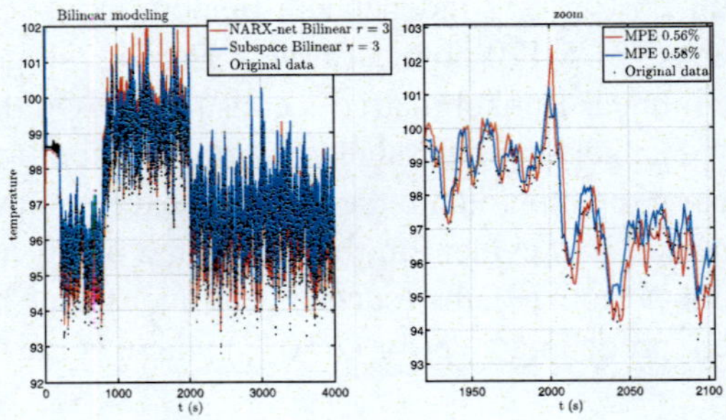

Fig. 3. Comparison and model fit of the proposed NARX-net bilinear model (15) with the subspace method from [9] for the same reduced order ($r = 3$).

$$\begin{cases} \dot{\mathbf{x}}(t) = \begin{bmatrix} 0.9164 & 0.09167 & -0.1847 \\ -0.2663 & -0.1515 & 0.1232 \\ -0.07227 & 0.4778 & 0.3571 \end{bmatrix} \mathbf{x}(t) + \begin{bmatrix} 0.02717 & 0.5169 & 0.5555 \\ -0.09674 & 0.5467 & 0.5696 \\ 0.1878 & -0.06846 & -1.981 \end{bmatrix} \mathbf{x}(t)u(t) + \\ \quad + \begin{bmatrix} 2.9063 \\ 2.909 \\ -0.16088 \end{bmatrix} u(t) + \begin{bmatrix} -1.073 \\ -1.074 \\ 0.05938 \end{bmatrix}, \\ y(t) = \begin{bmatrix} -0.7852 & 0.7794 & -0.05203 \end{bmatrix} \mathbf{x}(t) + 96.9358, \; \mathbf{x}(0) = \mathbf{0}, \; t \geq 0. \end{cases} \tag{15}$$

Figure 3 illustrates the superiority of the proposed method in terms of accuracy. From the single i/o data sequence, a neural network NN with 3-layers and 20-lags was trained using the same training data[3] as in [9] (1000 points). The trained NN was used in the bilinear realization algorithm to generate more data, and a stable reduced bilinear model of order $r = 3$ shown in Eq. (15) was successfully constructed. The original noisy data were explained with a lower mean percentage error MPE $= 0.56\%$ compared to the subspace method for the entire data set. Another NN architecture, s.a., the NARMAX[4] belongs to a subclass of bilinear systems and will filter some nonlinear features without achieving such a good MPE.

4 Conclusion

In conclusion, NN architectures are a superclass of NARMAX models used in the classical robust identification theory. Consequently, NN models share the same strong argument with the Carleman linearization scheme that can approximate general nonlinear systems. Finally, NN and realization theory successfully bridge data science with computational science to build reliable, interpretable nonlinear models. Different NN architectures (s.a., recurrent NNs, DeepOnets, etc.) in combination with other realization frameworks (s.a., the Loewner framework) and for other types of nonlinearities (s.a., quadratic-bilinear) are left for future research endeavors.

References

1. Antoulas, A.C.: Approximation of Large-Scale Dynamical Systems. SIAM, Philadelphia (2005)
2. Antoulas, A.C., Beattie, C.A., Gugercin, S.: Interpolatory Methods for Model Reduction. SIAM, Philadelphia (2020)
3. Kapteyn, M., Knezevic, D., Huynh, D., Tran, M., Willcox, K.: Data-driven physics-based digital twins via a library of component-based reduced-order models. Int. J. Numer. Meth. Eng. **123**(13), 2986–3003 (2022). https://doi.org/10.1002/nme.6423

[3] **Data detrend**: $u_n = (u - \bar{u})/\sigma_u$, $y_n = (y - \bar{y})/\sigma_y$; **zero-response**: data were doubled in size for learning the zero-response, i.e., $u_n = 0 \rightarrow \boxed{\Sigma} \rightarrow y_n = 0$.

[4] **NARMAX**: The nonlinear auto-regressive moving average model with exogenous input [14, 19].

4. Carleman, T.: Application de la théorie des équations intégrales linéaires aux systèmes d'équations différentielles non linéaires. Acta Math. **59**, 63–87 (1932)
5. Antoulas, A.C., Gosea, I.V., Ionita, A.C.: Model reduction of bilinear systems in the Loewner framework. SIAM J. Sci. Comput. **38**(5), B889–B916 (2016)
6. Schilders, W., Meijer, P., Ciggaar, E.: Behavioural modelling using the MOESP algorithm, dynamic neural networks and the Bartels-Stewart algorithm. Appl. Numer. Math. **58**(12), 1972–1993 (2008). https://doi.org/10.1016/j.apnum.2007.11.013
7. Ljung, L., Hjalmarsson, H., Ohlsson, H.: Four encounters with system identification. Eur. J. Control. **17**(5), 449–471 (2011). https://doi.org/10.3166/ejc.17.449-471
8. Favoreel, W., De Moor, B., Van Overschee, P.: Subspace identification of bilinear systems subject to white inputs. IEEE Trans. Autom. Control **44**(6), 1157–1165 (1999). https://doi.org/10.1109/9.769370
9. dos Santos, P.L., Ramos, J.A., de Carvalho, J.L.M.: Identification of bilinear systems with white noise inputs: an iterative deterministic-stochastic subspace approach. IEEE Trans. Control Syst. Technol. **17**(5), 1145–1153 (2009). https://doi.org/10.1109/TCST.2008.2002041
10. Juang, J.N.: Continuous-time bilinear system identification. Nonlinear Dyn. **39**(1), 79–94 (2005). https://doi.org/10.1007/s11071-005-1915-z
11. Gosea, I.V., Karachalios, D.S., Antoulas, A.C.: On computing reduced-order bilinear models from time-domain data. PAMM **21**(1), e202100,254 (2021). https://doi.org/10.1002/pamm.202100254
12. Ho, B.L., Kalman, R.E.: Editorial: effective construction of linear state-variable models from input/output functions. at - Automatisierungstechnik 14(1–12), 545–548 (1966). https://doi.org/10.1524/auto.1966.14.112.545
13. Isidori, A.: Direct construction of minimal bilinear realizations from nonlinear input-output maps. IEEE Trans. Autom. Control **18**(6), 626–631 (1973). https://doi.org/10.1109/TAC.1973.1100424
14. Al-Baiyat, S.A.: Model reduction of bilinear systems described by input-output difference equation. Int. J. Syst. Sci. **35**(9), 503–510 (2004). https://doi.org/10.1080/00207720410001734237
15. Benner, P., Breiten, T., Damm, T.: Generalised tangential interpolation for model reduction of discrete-time mimo bilinear systems. Int. J. Control **84**(8), 1398–1407 (2011). https://doi.org/10.1080/00207179.2011.601761
16. Dunoyer, A., Balmer, L., Burnham, K.J., James, D.J.G.: On the discretization of single-input single-output bilinear systems. Int. J. Control **68**(2), 361–372 (1997). https://doi.org/10.1080/002071797223668
17. Karachalios, D.S., Gosea, I.V., Antoulas, A.C.: The Loewner framework for system identification and reduction, pp. 181–228. De Gruyter (2021). https://doi.org/10.1515/9783110498967-006
18. Karachalios, D.S., Gosea, I.V., Antoulas, A.C.: On bilinear time-domain identification and reduction in the loewner framework. In: Benner, P., Breiten, T., Fabender, H., Hinze, M., Stykel, T., Zimmermann, R. (eds.) Model Reduction of Complex Dynamical Systems, vol. 171, pp. 3–30. Springer, Cham (2021). https://doi.org/10.1007/978-3-030-72983-7_1
19. Billings, S.A.: Neural Networks for Nonlinear System Identification, chap. 8, pp. 261–287. John Wiley & Sons, Ltd., Hoboken (2013). https://doi.org/10.1002/9781118535561.ch8
20. Bittanti, S., Piroddi, L.: Nonlinear identification and control of a heat exchanger: a neural network approach. J. Franklin Inst. **334**(1), 135–153 (1997). https://doi.org/10.1016/S0016-0032(96)00059-2

Coupling FMUs to Electric Circuits in Multiphysical System Simulation Software for the Development of Electric Vehicles

Michael Kolmbauer[1(✉)], Günter Offner[2], Ralf Uwe Pfau[2], and Bernhard Pöchtrager[1,3]

[1] MathConsult GmbH, Altenberger Street 69, 4040 Linz, Austria
michael.kolmbauer@mathconsult.co.at

[2] AVL List GmbH, Hans-List-Platz 1, 8020 Graz, Austria
{guenter.offner,ralf-uwe.pfau}@avl.com

[3] Radon Institute for Computational and Applied Mathematics (RICAM), Austrian Academy of Sciences, Altenberger Street 69, 4040 Linz, Austria
bernhard.poechtrager@ricam.oeaw.ac.at

Abstract. This work is devoted to the analysis of electric circuits stemming from automated modeling processes in system simulation software. Modern applications such as HEV (hybrid electric vehicle), BEV (battery electric vehicle) and FCEV (fuel cell electric vehicle) require not only couplings of electric networks with mechanical, thermal, fluid and gas systems. In many cases it is necessary to extend or control the physics with grey box models like FMUs (Functional Mock-up Units). In particular, the coupling of electric systems with FMUs can be done on various levels (model exchange, co-simulation) via different interfaces (controller, electric, electric-thermal) and is therefore a challenging task. In this work we concentrate on a co-simulation approch for an electric-thermal coupling in a BEV model.

1 Introduction

Multi-disciplinary modeling and simulation packages such as AVL CRUISE™M[1] provide concepts for the automatic generation and stable simulation of dynamic system models. The systems are built in a modular way. Standardized components can be coupled together to form physical networks. In turn, they can be coupled and form the overall system. For example, HEV, BEV and FCEV can be divided into electrical, fluid, gas, mechanical and thermal networks and their controllers.

A practical extension of this approach is the representation of dedicated subsystems as grey box models, i.e., physical or software components with unknown internal structures and principles but defined input and output interfaces. Those grey box models allow to incorporate external or customer-defined components, model parts or parts of the system into the software while guarding the intellectual property. Nowadays the FMI (Functional Mock-up Interface)[2] has been established as a standardized structured interface for those kind of grey box models, giving the input-output relation as differential

[1] https://www.avl.com/en/cruise-m.

[2] http://fmi-standard.org/.

M. van Beurden et al. (Eds.): SCEE 2022, MI 43, pp. 193–200, 2024.
https://doi.org/10.1007/978-3-031-54517-7_22

state and/or algebraic output equation. Depending on the characteristic of the FMI, the FMU can obtain model-exchange or co-simulation character. On the one hand, model exchange FMUs allow to describe the physical systems and grey box models by DAEs (differential algebraic equations), where solvability and index analysis are necessary to ensure a stable system simulation and the resulting system of DAEs is solved within one numerical scheme. On the other hand co-simulation FMUs can be incorporated within a multirate co-simulation framework as individual solver elements including individual integration and synchronisation time steps.

In this work we extend the results obtained in [4] for a multiphysical multi-rate framework to the case of an electric-thermal coupling via grey box FMUs. This approach leads to hybrid models, where some parts are described via multiphysical networks models and the FMU parts are described via input-output relations without any further physical interpretation.

2 Multiphysical System Formulation

The electrical network formulation is based on the well-known modified nodal analysis [3] and additionally includes the temperature of the network elements. In contrast to [7], where the temperature is modelled via temperature boundary elements, the formulation follows [3] and includes the temperature as given input as well as the heat flow as output. The electrical network can be described by the graph

$$\mathscr{N} = \{C, L, R, I, V, N, G\}, \tag{1}$$

which is composed of

$$\begin{aligned}
\text{capacitors } C &= \{C^1, \dots, C^{n_C}\}, n_C \in \mathbb{N}, \\
\text{inductors } L &= \{L^1, \dots, L^{n_L}\}, n_L \in \mathbb{N}, \\
\text{resistors } R &= \{R^1, \dots, R^{n_R}\}, n_R \in \mathbb{N}, \\
\text{current sources } I &= \{I^1, \dots, I^{n_I}\}, n_I \in \mathbb{N}, \\
\text{voltage sources } V &= \{V^1, \dots, V^{n_V}\}, n_V \in \mathbb{N}, \\
\text{nodes } N &= \{N^1, \dots, N^{n_N}\}, n_N \in \mathbb{N} \text{ and} \\
\text{grounds } G &= \{G^1, \dots, G^{n_G}\}, n_G \in \mathbb{N}.
\end{aligned}$$

With incidence matrix

$$A = \begin{pmatrix} A_C & A_L & A_R & A_I & A_V \end{pmatrix}$$

the system to be solved can be written as follows, cf. [3,4,7]:

For given inputs $u_E = (u_C^T, u_L^T, u_R^T)^T$, find currents $j = (j_C^T, j_L^T, j_R^T, j_I^T, j_V^T)^T$, node potentials $e = (e_N^T, e_G^T)^T$ and outputs $y_E = (y_C^T, y_L^T, y_R^T)^T$, such that

$$\begin{aligned}
0 &= A_C j_C + A_L j_L + A_R j_R + A_V j_V + A_I \bar{j}_I \\
0 &= j_C - \dot{q}(A_C^T e, u_C) \\
0 &= \dot{\phi}(j_L, u_L) - A_L^T e \\
0 &= j_R - g_R(A_R^T e, u_R) \\
\bar{v}_V &= A_V^T e \\
y_C &= |j_C \circ A_C^T e| \\
y_L &= |j_L \circ A_L^T e| \\
y_R &= |j_R \circ A_R^T e|
\end{aligned} \tag{2}$$

for given boundary conditions $e_G = 0$, given characteristic functions g, q, ϕ, prescribed currents $\bar{j}_I$ and prescribed voltages $\bar{v}_V$ as well as $\circ$ denoting the element-wise (Hadamard) product. For the electrical system, solvability and index results are available in [3].

In case of a multiphysical system, similar results are also available for fluid, gas and thermal solid systems, cf. [7]. A rigorous analysis of the full multiphysical equation system is required in order to ensure, that the resulting system of DAEs is of (differential) index 1 again. On an abstract level, every physical subsystem of interest $i = 1, \ldots, n$ yields a semi-explicit DAE with index 1:

$$\begin{aligned}
\dot{x}_i &= f_i(x_i, a_i, u_i, t) \\
0 &= r_i(x_i, a_i, u_i, t), \\
y_i &= g_i(x_i, a_i, u_i, t)
\end{aligned} \tag{3}$$

with states x_i, algebraic variables a_i, inputs u_i, outputs y_i, time t and functions f_i, r_i, g_i. Those physical subsystems can be coupled with each other by certain quantities, cf. [7]. For example the electrical and thermal solid system can be coupled by temperature and heat flow. The input output relation can be represented using a skew-symmetric coupling matrix C, cf. [7],

$$u = Cy. \tag{4}$$

Combining the multiphysical subsystems (3) as well as the coupling equation (4) yields the system to be solved. Imposing certain conditions on the individual subsystems as well as the coupling matrix C, it can be shown that the coupled multiphysical system is feasible and yields a DAE of (differential) index 1, cf. [6].

So far, we are still operating in the multiphysical work. As a next step, grey box FMUs are incorporated into the multiphysical framework.

3 Functional Mock-Up Units (FMUs)

In addition to the physical systems, grey box models can be incoporated in the multiphysical system modeling approach. In this work we restrict our use cases to FMUs which are conform with the *FMI 2.0* standard. FMUs may have *model exchange* or *co-simulation* character.

In a simplified approach a FMU for *model exchange* represents a DAE as an input-output system of the form

$$\begin{aligned} \dot{x} &= f(t,x,u), \\ y &= g(t,x,u), \end{aligned} \tag{5}$$

where t is the time, x is the differential state, u are inputs and y are outputs.

Furthermore a FMU for *co-simulation* represents an input-output system of the form

$$y = g(t, \Delta t, u), \tag{6}$$

where t is the time, Δt is a timestep, u are inputs and y are outputs.

The choice for model exchange or co-simulation FMUs is determined by the underlying problem formulation and the features of the generating platform. For example, multiphysical subsystems can be extracted as co-simulation FMUs in AVL CRUISE^{TM}M. Inputs and outputs of the FMU can be specified via a predefined interface. The cooling system of Fig. 2 can be generated by adding the corresponding interface and defining inputs and outputs, cf. Fig. 1. Co-simulation FMUs can be incorporated in various simulation software.

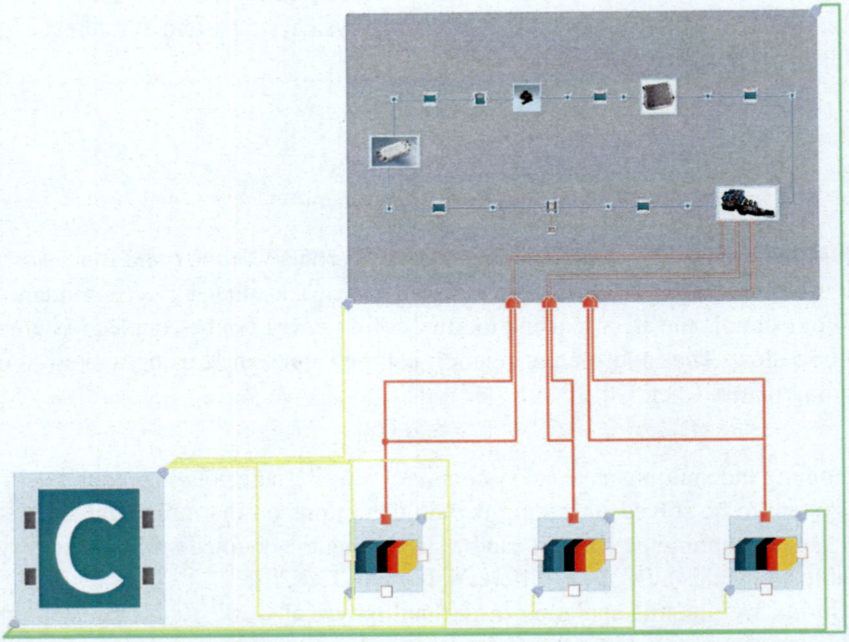

Fig. 1. Schematic representation of a cooling system extracted as FMU.

4 Co-simulation for Coupled Network DAEs

The full DAE is partitioned due to the physical background to n subsystems (3) (typically $n \gg 2$). The incorporation of model exchange FMUs (5) can be done on component level, i.e. a physical component or subsystem can be replaced via a FMU. Applying Runge-Kutta methods with micro-stepsizes h_i to each subsystem (3), a representation in co-simulation form is obtained. Coupling the physical subsystems with $k \in \mathbb{N}$ co-simulation FMUs (6), $m = n + k$, the overall coupled co-simulation system reads as

$$y_i = c_i(u_i, t, \Delta t), \qquad i = 1, \ldots, m. \tag{7}$$

Again the interaction of the individual systems is described via Eq. (4). Each subsystem (7) has its own internal solver and step size h_i. Since a synchronous communication approach is applied, the whole system exchanges its data with frequency Δt. The input values u_i are handled with appropriate interpolation, extrapolation or filtering techniques, depending on the slow or active characteristics of the interacting subsystems. At this communication level no time integration method is involved. The system is solved by applying a sequential non-iterative Gauss-Seidel type approach, cf. [2]. If all integration methods are of order p, the interpolation methods are of order $p-1$ and a contraction condition is fulfilled, the co-simulation approach is convergent of order p, cf. [1].

5 Numerical Example

We consider a BEV that demonstrates the modeling of an electrical system coupled to the required cooling system, cf. Fig. 2. The model consists of an electrical propulsion and two cooling circuits. This model has already been used and analysed in [4] and is now extended to the case of grey box models in terms of FMUs. An oil circuit is used for cooling of the electric machine and a coolant circuit is used for cooling of the battery pack, inverter and low voltage DC-DC converter. In model 1 the coolant circuit is described by a multiphysical system, cf. Fig. 2, while in model 2 the coolant circuit is modelled by a co-simulation FMU, cf. Fig. 3. The FMU in model 2 is exported from AVL CRUISE^{TM}M using the same coolant circuit as in model 1, cf. Fig. 1. The corresponding input-output relations are established in order to mimic the behavior of the original model, i.e. the electric FMU communications with the cooling circuits via input and output channels and not via physical connections. Both models are implemented and simulated using again AVL CRUISE^{TM}M. Although the importing and exporting tool is the same, information about the equation system is lost due to the specific characterization of the FMU interface. Model 1 and model 2 can be integrated with the co-simulation approach using different solver parametrizations, while providing reliable results. For example, the results of the temperature in a single battery pack of a multirate co-simulation case of model 2 (in blue) is compared to the reference solution of model 1 using the same multirate co-simulation parametrization (in red) in Fig. 4. In the standard multirate approach of model 1, model based inter- and extrapolation techniques of the invelved physical quantities (temperature, heat flux) yield a accurate result. Since in the multirate co-simulation FMU case of model 2 the corresponding models of the coupling variables are hidden in the FMU, only standard inter- and

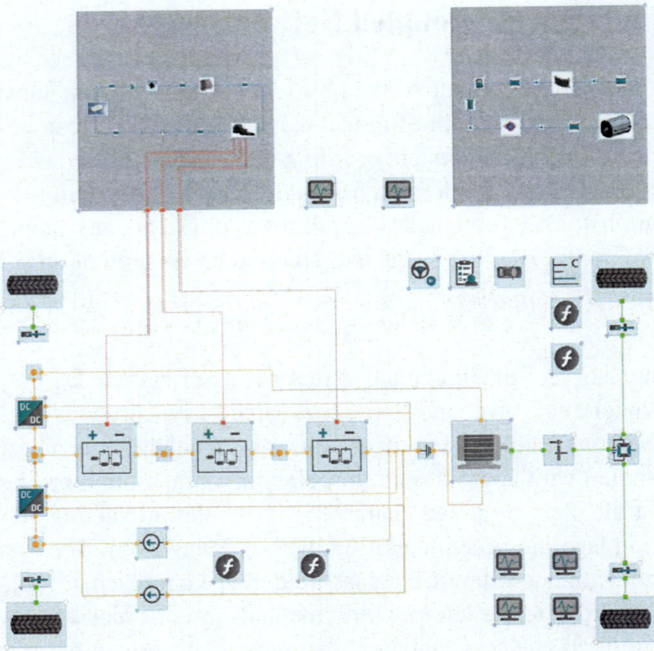

Fig. 2. Model 1: Schematic representation of a BEV with cooling system.

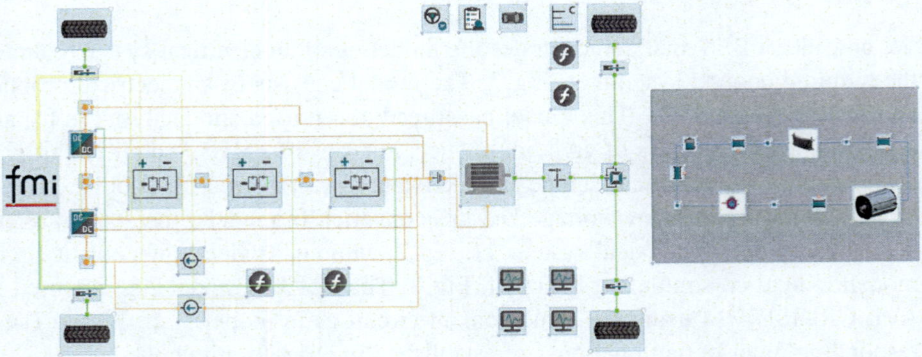

Fig. 3. Model 2: Schematic representation of a BEV with FMU cooling system.

extrapolation methods can be applied. Since the applied interpolation schemes differ for those two cases, differences in the simulation results occur. In this case the contraction condition, cf. [1], is fulfilled, as it is coupled by a differential state. System modeling using FMUs is expected to require increased computational effort. Performance losses have to be accepted when calling the external FMU code. In addition, one is limited in the optimization of the equation systems based on the given FMU structure. Due to the computation overhead the FMU co-simulation approach in model 2 is about 10% slower compared to the standard multirate approach in model 1.

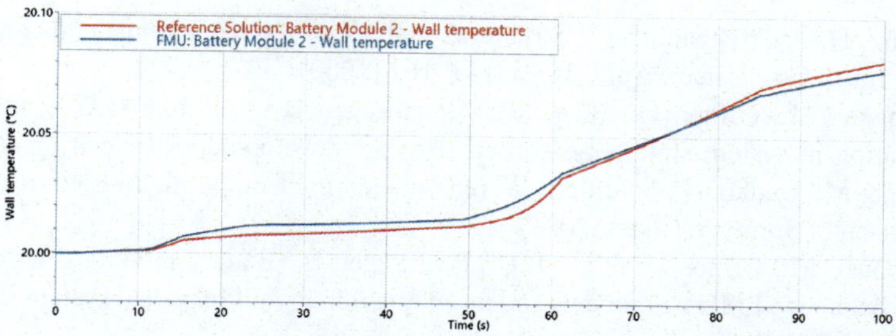

Fig. 4. Wall temperature within a battery pack for a reference solver parametrization (in red) as well as a parametrization for the co-simulation FMU case (in blue). (Color figure online)

6 Conclusion and Outlook

For the integration of FMUs in multiphysical systems, solvability and index statements for the overall system are essential for a stable system simulation. In the given thermal-electrical FMU coupling example above, the solvability and index statements follows directly from the results of the subsystems, since they were coupled via differential states. Since information is contained in the physical model, which can be exploited by the solver framework, it is always preferable to provide the networks as a physical system instead of a grey box model. Nevertheless the presented framework is not restricted to tool in tool simulation, but is open to handle any feasible co-simulation from any other tool. Here the hybrid multi-domain - grey box approach closes the bridge between multiphysical single-tool applications and signal based multi-tool co-simulation applications.

The presented framework is not restricted to the co-simulation case, but is also valid for the model-exchange case. Although model-exchange FMUs are also grey box models, some more information can be retrieved from the interface, since the corresponding coupling is equation based and not solver based. For the interested reader, we refer to [5], where a bunch of model exchange FMUs are coupled directly into the electrical system.

Furthermore, taking advantage of the possible port-Hamiltonian structure in the coupled DAE system (3) is under investigation for the follow-up steps.

Acknowledgements. Part of this work has been supported by the government of Upper Austria within the program "FTI Struktur" and by the FFG within the project "FFG DigiLab".

References

1. Bartel, A., Günther, M.: Inter/extrapolation-based multirate schemes: a dynamic-iteration perspective. In: Reis, T., Grundel, S., Schöps, S. (eds.) Progress in Differential-Algebraic Equations II, pp. 73–90. Springer, Cham (2020). https://doi.org/10.1007/978-3-030-53905-4_3
2. Bartel, A., Günther, M.: Multirate schemes—an answer of numerical analysis to a demand from applications. In: In: Gunther, M., Schilders, W. (eds.) Novel Mathematics Inspired by Industrial Challenges pp. 5–27. Springer, Cham (2022). https://doi.org/10.1007/978-3-030-96173-2_1

3. Schwarz, D.E., Tischendorf, C.: Structural analysis of electric circuits and consequences for MNA. Int. J. Circ. Theor. Appl. **28**(2), 131–162 (2000)
4. Kolmbauer, M., Offner, G., Pfau, R.U., Pöchtrager, B.: Multirate DAE-simulation and its application in system simulation software for the development of electric vehicles. In: van Beurden, M., Budko, N., Schilders, W. (eds.) Scientific Computing in Electrical Engineering, pp. 231–240. Springer, Cham (2021)
5. Kolmbauer, M., Offner, G., Pfau, R.U., Pöchtrager, B.: Battery module simulation based on model exchange FMU cell models and its application in multiphysical system simulation software. In: SCEE 2022 - Book of Abstracts (2022)
6. Kolmbauer, M., Offner, G., Pöchtrager, B.: Topological index analysis and its application to multi-physical systems in system simulation software. In: Cruz, M., Parés, C., Quintela, P. (eds.) Progress in Industrial Mathematics: Success Stories, pp. 171–191. Springer, Cham (2021). https://doi.org/10.1007/978-3-030-61844-5_10
7. Pöchtrager, B.: Coupling Multiphysical Systems in Automotive Simulation Software. PhD thesis, JKU Linz (2022)

Battery Module Simulation Based on Model Exchange FMU Cell Models and Its Application in Multi-physical System Simulation Software

Michael Kolmbauer[1(✉)], Günter Offner[2], Ralf Uwe Pfau[2], and Bernhard Pöchtrager[1,3]

[1] MathConsult GmbH, Altenberger Street 69, 4040 Linz, Austria
michael.kolmbauer@mathconsult.co.at
[2] AVL List GmbH, Hans-List-Platz 1, 8020 Graz, Austria
{guenter.offner,ralf-uwe.pfau}@avl.com
[3] Radon Institute for Computational and Applied Mathematics (RICAM), Austrian Academy of Sciences, Altenberger Street 69, 4040 Linz, Austria
bernhard.poechtrager@ricam.oeaw.ac.at

Abstract. Over the past years, the importance of battery development has increased significantly and will increase further. Consequently, the requirements for the system simulation of electric networks are rising. In modern simulation applications, the battery setup and battery modules can be simulated with different levels of complexity and can be coupled with detailed powertrain and cooling, yielding a multi-physcial problem. For some detailed investigations, e.g. about the thermal behavior, it is needed that in the electrical subsystem one can model and simulate detailed cells. The physical models of the cells can be given via grey-box models such as FMUs (functional mock-up units) (https://fmi-standard.org/). Simulating then combined battery cells increases the complexity of the model. In this case, care must be taken to ensure that the simulation delivers stable results and that the computing time is kept as short as possible. Here we use a hierarchical approach to solve the resulting equations for the electrical system.

1 Introduction

In today's vehicle development, BEV (battery electric vehicles) and HEV (hybrid electric vehicles) are playing an increasingly important role. Simulation packages like AVL CRUISE™M[1] offer the possibility of generic modeling of such problems, see [2]. Modular components can be coupled together to form physical subsystems, such as electrical as well as mechanical, thermal, gas and fluid circuits. These in turn can be linked together via predefined coupling points.

In the development of such vehicles, the simulation of batteries and the interaction with the other parts, especially the cooling, is crucial. For detailed investigations about the temperature distribution within the battery, cooling, and ageing effects, one needs a detailed simulation of the battery cells including electrical and thermal behavior. The cells are then grouped serial and parallel electrically into modules and the modules are

[1] https://www.avl.com/en/cruise-m.

M. van Beurden et al. (Eds.): SCEE 2022, MI 43, pp. 201–207, 2024.
https://doi.org/10.1007/978-3-031-54517-7_23

coupled to battery packs. Cells, modules, and packs then interact on different levels with controllers to achieve efficiency and improve the lifetime of the batteries. For the development and testing of control strategies, detailed simulations are needed. The system simulation approach allows a smooth transition from fine granular to coarse models.

The starting point is a cell model which includes the needed physical details like electrical characteristic, thermal behavior and in some cases mechanical properties. The cells are connected in series and parallel for a module, see Fig. 4. Nowadays, typical cell types are pouch cells, prismatic cells or cylindrical cells, see [5]. The cell type determines how they can be grouped and connected to the cooling. In Fig. 1 a schematic setup of pouch cells is given with intermediate cooling fins and the connection to the housing.

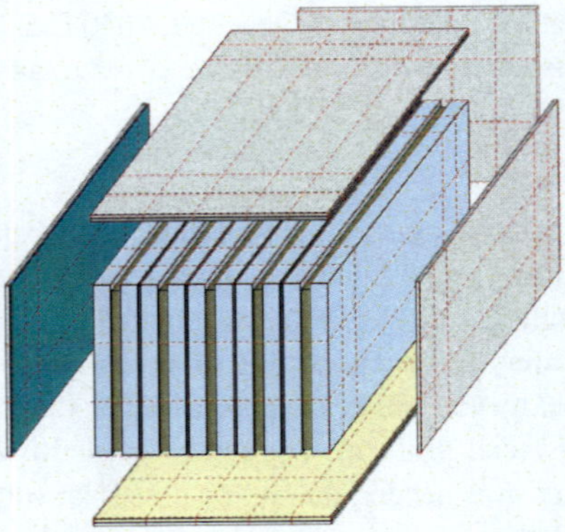

Fig. 1. Schematic representation of a battery module with cells, cooling fins and housing plates.

From the electrical point of view, a cell is often a voltage boundary depending on current. For detailed cell models, the actual state depends strongly on the temperature of the cell. Due to the complexity of the internal physics, the cell models are often done by specialists. For the simulation, cell models are provided as FMUs to encapsulate physics and to secure intellectual property. We focus on the case that the cells are grey-box voltage boundaries, coupled with the electrical system and the cooling network. The challenge in modeling and simulating this constellation is to properly integrate the FMUs of the cells in the electrical system and to ensure a high-performance simulation.

2 Mathematical Model

The simulation of the electrical system is based on the modified nodal analysis, cf. [1,3]. An electrical network $\mathscr{N} = \{R, C, L, V, I, N, G, B\}$ is composed of resistors R, capacitors C, inductors L, voltage sources V, current sources I, nodes N, grounds G and batteries B.

The electrical system is modelled using a directed graph $\mathscr{G} = \{\mathscr{V}, \mathscr{E}\}$.

Nodes and grounds are vertex elements, i.e. $\mathscr{V} = \{N, G\}$, while resistors, capacitors, inductors, voltage sources, current sources, and batteries are edge elements, i.e. $\mathscr{E} = \{R, C, L, V, I, B\}$.

Every vertex of the network graph $v \in \mathscr{V}$ has a corresponding potential e_v, where each ground has constant potential zero, i.e. $e_G = 0$. Furthermore each edge element of the network $e \in \mathscr{E}$ has an assigned current i_e. The DAE for the electrical network $\mathscr{N}$ is given by, cf. [4]: For given continuous temperature inputs T_R, T_C, T_B, find the potentials $e = (e_N^T, e_G^T)^T$, the currents $j = (j_R^T, j_C^T, j_L^T, j_V^T, j_B^T)^T$, such that

$$
\begin{aligned}
A_R j_R + A_C j_C + A_L j_L + A_V j_V + A_B j_B + A_I \bar{j}_I &= 0 \\
r(T_R) j_R - A_R^T e &= 0 \\
j_C - \frac{d(c(T_C) A_C^T e)}{dt} &= 0 \\
l \frac{d j_L}{dt} - A_L^T e &= 0 \\
A_V^T e &= \bar{v}_V \\
A_B^T e &= \bar{v}_B(j_B, T_B)
\end{aligned}
\tag{1}
$$

for given boundary conditions $e_G = 0$ and given resistance r, capacitance c and inductance l as well as prescribed currents $\bar{j}_I$ and prescribed voltages $\bar{v}_V$ and $\bar{v}_B$.

Since the battery cells are combined within the battery module with each other in a specific structure, we have voltage boundary loops within the electrical network, cf. Fig. 4. Hence the DAE has (differential) index 2, cf. [1,3].

In system simulation software individual elements are often provided by more complex systems in the form of model exchange or co-simulation FMUs. Typically, such FMUs are computationally expensive, hence one wants to avoid too many calls to the FMUs. As most of the time is within the evaluation of the FMUs, solving the DAE directly might not be efficient. The coupling of the cells leads to a regular structure in those equations. Due to this, the equations can be arranged to solve for the inner current distribution and the outside effect in the electrical network split, see Sect. 3.

Concerning the cell behavior, it is advantageous to use model exchange FMUs as those allow stable simulation with larger time steps. Often the simulated time goes in the range of days or weeks which should be simulated within hours. The FMU provides states x which have to be integrated with the electrical system, z are (continuous) inputs and y are the outputs of the FMU. The inputs into the FMU are for example the current and the heat flux into the cell, the outputs are then the cell voltage, the cell temperature and similar.

$$
\begin{aligned}
\dot{x} &= f(t, x, z), \\
y &= g(t, x, z).
\end{aligned}
$$

In order to solve the index 2 equation, it is advantageous that the FMU provides the current to voltage sensitivity information. Those are the needed parts in the chain rule for the index reductoin to 1 of the constraints. In contrast to the electrical equation, the change of the thermal behavior system happens slower. Still temperatures and heat fluxes need to be calculated sufficiently accurate due to the dependency of the cell behavior on its temperature.

AVL CRUISE^{TM}M allows here a multirate approach to integrate the electrical and thermal part with different time scales, cf. [4]. As the states of the FMU need to be integrated with the fastest scale those are included in the electrical domain and the thermal coupling is handled with the multirate framework. The thermal behavior of the cooling fins and the housing plates is simulated within the thermal domain.

3 Hierarchical Approach

For a more efficient solution of the electrical system we can utilize the special setup within the battery module as seen in Fig. 4b:

- For the equilibrium in the electrical system, the inner structure of the battery module is not important. Only the current to voltage behavior is relevant.
- The voltage of the battery module is given by the sum of voltage for each row.
- Each row can be solved for the voltage independently of the others. The coupling is mainly on the thermal level. So instead of a larger problem, we can solve a sequence of smaller problems of the same structure: with the current input I and the voltage output u_{level}.

Assuming, we have n rows and in each row m cells in parallel. If I is the input current to the battery module, $u_{i,j}$ is the voltage of the cell at row i and column j, we have with currents $i_{i,j}$ into the cell at position (i, j) for each row $i = 1,..,n$:

$$I = \sum_{j=1}^{m} i_{i,j} \tag{2}$$

$$u_{i,1} - u_{i,k} = 0, \forall k \in \{2,..,m\} \tag{3}$$

Note that we put a special role on the first column, but all other equalities $u_{i,j} - u_{i,k} = 0$, for all $j,k \in \{1,..,m\}$ follow therefrom. Additionally we have for the output voltage of the module $u_{module} = \sum_{i=1}^{n} u_{i,1}$. Using the FMU provided sensitivity information $\frac{\partial u_{i,j}}{\partial i_{i,j}}$, we can assemble the needed Jacobian analytically and the Jacobian is sparse. Hence, we can solve the nonlinear problem more efficiently with fewer right hand side evaluations which include the often expensive update calls to the FMUs.

4 Numerical Example

We focus here on the simulation of the battery module connected with an electrical current boundary condition. The current prescription is calculated from a BEV vehicle, see Fig. 2, with a NEDC (New European Driving Cycle) velocity boundary condition, see Fig. 3. The vehicle is simulated with a simpler battery model to generate the current prescription for the detailed model. One could couple the detailed battery model with a detailed powertrain and cooling model. We focus here on the two step approach to better see the performance improvement.

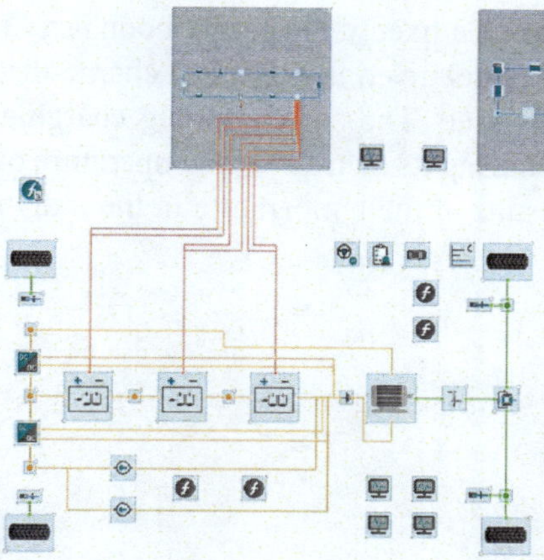

Fig. 2. Example model for a BEV vehicle.

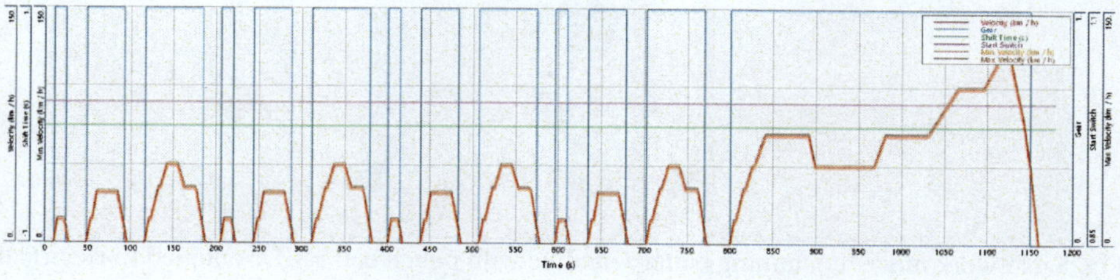

Fig. 3. NEDC cycle.

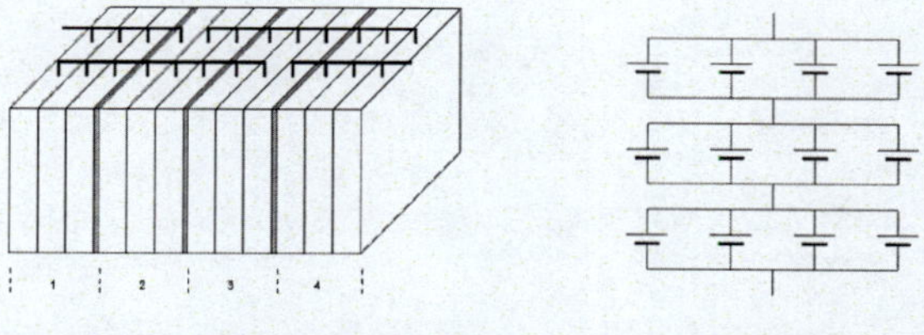

(a) Three dimensional representation.

(b) Network structure.

Fig. 4. Schematic picture of serial arrangement and electrical connections.

In the battery module, we consider the battery pouch cell *LG Chem E66A*[2], cf. [5], from Batemo[3] with twelve cells, three serial and four parallel layers, depicted in Fig. 4 with the electrical connections. Geometrically the cells are arranged sequentially with cooling fins in-between, see Fig. 4. This setup containing twelve model exchange FMUs is incorporated into an electrical system containing a liquid flow system for cooling, cf. Fig. 5. The influence of different setups, coolings and connections is the objective of the simulations and the comparisons. In our case, the bottom plate is connected to a cool-

2 https://www.batemo.de/products/batemo-cell-library/e66a/.

3 https://www.batemo.de.

ing network, here simplified as a fixed temperature boundary. The electrical boundary conditions for the network model given in Fig. 5 is a charge-discharge cycle with short resting period in-between is used. The corresponding charging/discharging current is given in Fig. 6. Each cell is having its own surface temperature over the simulation time. For particular times, the results of the temperature in the individual cells are shown in Fig. 7.

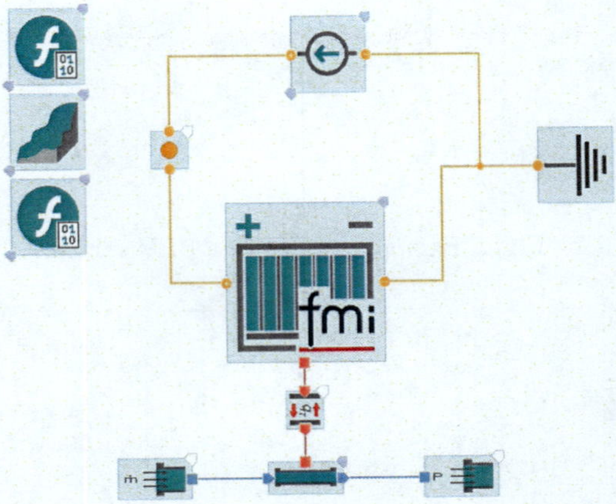

Fig. 5. Network model containing battery module with pouch cell FMUs coupled to electrical and cooling system.

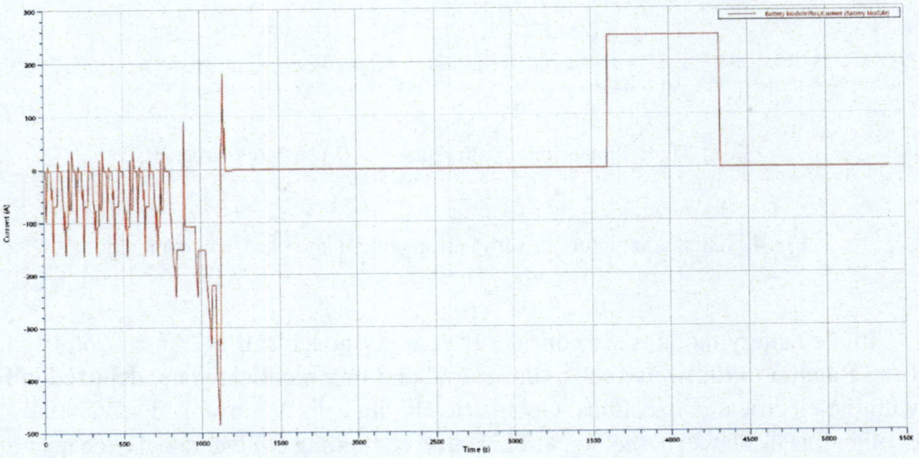

Fig. 6. Charging/discharging current for the battery module in Fig. 5.

We compare the solution of the hierarchical approach versus a model setup compared to a model where this is not utilized, i.e. in the electrical system all is solved

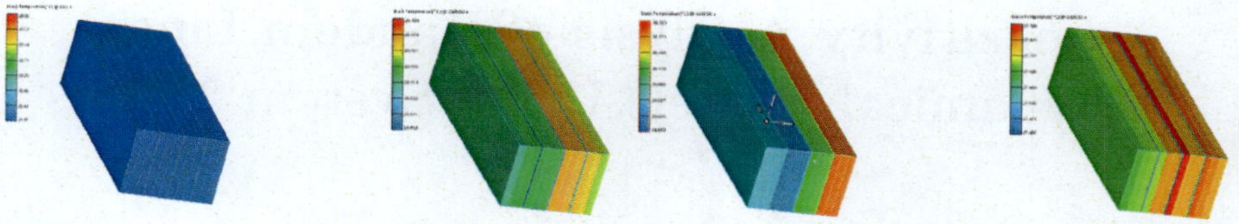

(a) At $t = 0s$. (b) At $t = 1200s$. (c) At $t = 3600s$. (d) At $t = 3900s$.

Fig. 7. Temperature distribution within stack over time t: at the beginning, after the driving cycle, before the charging phase, after the charging phase.

together, resulting in a larger set of unknowns. The hierarchical approach to solve the equation system brings a computational benefit of 10% and 20% speedup. For example in our test setup used here, the traditional simulation run requires 10721 s, while the hierarchical approach requires only 8856 s in order to simulate the 7200 s lasting decharging and charging cycle.

5 Conclusion

The presented hierarchical approach allows the efficient integration of model exchange FMUs representing the cell physics within a system simulation model within AVL CRUISE^{TM}M's multi-rate and multi-domain framework. In the structured setup, the resulting index 2 DAE can be efficiently solved by reducing the constraints to an index 1 system with the sensitivities provided by the FMUs. The resulting system can be simulated avoiding too many expensive FMU evaluations. This is then the base for coupling such systems with more detailed cooling networks and powertrain systems. Using FMUs allows to encapsulate the cell physics and secure the intellectual property as well as with the additional sensitivity information to solve the systems with less effort.

Acknowledgements. Part of this work has been supported by the government of Upper Austria within the program "FTI Struktur" and by the FFG within the project "FFG DigiLab".

References

1. Schwarz, D.E., Tischendorf, C.: Structural analysis of electric circuits and consequences for MNA. Int. J. Circ. Theor. Appl. **28**(2), 131–162 (2000)
2. Knaus, O., Wurzenberger, J.C.: System simulation in automotive industry. In: Hick, H., Kupper, K., Sorger, H. (eds.) Systems Engineering for Automotive Powertrain Development, pp. 499–532. Springer, Cham (2021)
3. März, R., Tischendorf, C.: Recent results in solving index 2 differential algebraic equations in circuit simulation. SIAM J. Sci. Stat. Comput. **18**(1), 135–139 (1997)
4. Kolmbauer, M., Offner, G., Pfau, R.U., Pöchtrager, B.: Multirate DAE-simulation and its application in system simulation software for the development of electric vehicles. In: van Beurden, M., Budko, N., Schilders, W. (eds.) Scientific Computing in Electrical Engineering, pp. 231–240. Springer, Cham (2021)
5. Battery University. Bu-301a: Types of battery cells (2019)

Sensitivity Analysis of Random Linear Dynamical Models Using System Norms

Roland Pulch(✉)

Institute of Mathematics and Computer Science, Universität Greifswald, Walther-Rathenau-Str. 47, 17489 Greifswald, Germany
roland.pulch@uni-greifswald.de

Abstract. We consider linear dynamical systems with a single output, where the systems include random parameters to perform an uncertainty quantification. Using the concept of polynomial chaos, a linear stochastic Galerkin system of higher dimension with multiple outputs is arranged. Quadratic combinations of the outputs yield approximations of time-dependent indices in global sensitivity analysis, which indicate the influence of each random parameter. We investigate system norms for the quadratic outputs, because these norms generate time-independent sensitivity measures. Numerical results are presented for a model of an electric circuit.

1 Introduction

Mathematical modelling often yields systems of differential equations including physical parameters. Uncertainty quantification (UQ) is required to investigate an output of the model with respect to a variability in the parameters. A common approach consists in the substitution of the parameters by random variables, see [10]. In addition, a global sensitivity analysis of the random-dependent model can be performed to characterise the importance of each random parameter. There are variance-based indicators (first-order indices and total-effect indices) as well as derivative-based indicators for global sensitivity analysis, see [4,5,8,9]. The resulting numerical values allow for a ranking of the parameters.

We examine linear dynamical systems composed of ordinary differential equations (ODEs) or differential-algebraic equations (DAEs). A single input or multiple inputs are induced, while a single output represents a quantity of interest (QoI). In the random-dependent system, we expand the state/inner variables as well as the output in the polynomial chaos (PC), see [10]. The stochastic Galerkin method yields a larger deterministic linear dynamical system with multiple outputs, which represent an approximation of coefficient functions in the expansion of the QoI. Quadratic combinations of the outputs produce approximations of three types of indices in global sensitivity analysis: first-order, total-effect, and derivative-based. Since the outputs depend on time, the sensitivity indices also vary in time.

M. van Beurden et al. (Eds.): SCEE 2022, MI 43, pp. 208–216, 2024.
https://doi.org/10.1007/978-3-031-54517-7_24

Alternatively, we derive system norms of the stochastic Galerkin system for each non-negative quadratic output. These system norms provide sensitivity measures, which are independent of both time and the input. In [6,7], this strategy was applied to investigate system norms associated to total-effect indices. Now we extend the approach to first-order indices as well as derivative-based indices. Therein, the derivative-based concept uses the $\mathcal{L}^2$-norm of the QoI's partial derivatives with respect to the parameters. All system norms are computable as $\mathcal{H}_\infty$-norms of corresponding transfer functions in frequency domain. Finally, we illustrate results of numerical computations employing the electric circuit of the Miller integrator.

2 Random Linear Dynamical Systems

Let a linear dynamical system be given in the form

$$\begin{aligned} \mathbf{E}(\mathbf{p})\tfrac{\mathrm{d}}{\mathrm{d}t}\mathbf{x}(t,\mathbf{p}) &= \mathbf{A}(\mathbf{p})\mathbf{x}(t) + \mathbf{B}(\mathbf{p})\mathbf{u}(t) \\ y(t,\mathbf{p}) &= \mathbf{c}(\mathbf{p})^\top \mathbf{x}(t,\mathbf{p}) \end{aligned} \tag{1}$$

with time $t \in I = [0,\infty)$. Single or multiple inputs $\mathbf{u} : I \to \mathbb{R}^{n_{\mathrm{in}}}$ are supplied. The matrices $\mathbf{A},\mathbf{E} \in \mathbb{R}^{n\times n}$, $\mathbf{B} \in \mathbb{R}^{n\times n_{\mathrm{in}}}$, and the vector $\mathbf{c} \in \mathbb{R}^n$ depend on physical parameters $\mathbf{p} \in \Pi \subseteq \mathbb{R}^q$. The variables $\mathbf{x} : I \times \Pi \to \mathbb{R}^n$ depend on time as well as the parameters. A single output $y : I \times \Pi \to \mathbb{R}$ is observed as a QoI. If the mass matrix $\mathbf{E}(\mathbf{p})$ is non-singular, then the system (1) consists of ODEs. Alternatively, a singular mass matrix implies a system of DAEs. A linear DAE system exhibits a (nilpotency) index $\nu \geq 1$, see [3]. We assume that the systems (1) are asymptotically stable for all $\mathbf{p} \in \Pi$. An initial value problem is specified by $\mathbf{x}(0,\mathbf{p}) = \mathbf{x}_0(\mathbf{p})$ with a predetermined function $\mathbf{x}_0 : \Pi \to \mathbb{R}^n$.

The parameters are often affected by uncertainties due to modelling errors or measurement errors, for example. A common approach to model their variability consists in replacing the parameters by independent random variables, see [10]. Consequently, the parameters become measurable functions $\mathbf{p} : \Omega \to \Pi$ on a probability space $(\Omega,\mathcal{A},P)$. We assume that there is a joint probability density function $\rho : \Pi \to \mathbb{R}$. Hence the expected value of a measurable function $f : \Pi \to \mathbb{R}$ reads as

$$\mathbb{E}[f] = \int_\Omega f(\mathbf{p}(\omega))\,\mathrm{d}P(\omega) = \int_\Pi f(\mathbf{p})\,\rho(\mathbf{p})\,\mathrm{d}\mathbf{p}. \tag{2}$$

We consider the Hilbert space

$$\mathcal{L}^2(\Pi,\rho) = \left\{ f : \Pi \to \mathbb{R} \;:\; f \text{ measurable and } \mathbb{E}[f^2] < \infty \right\}, \tag{3}$$

which is equipped with the inner product $\langle f,g\rangle = \mathbb{E}[fg]$ for two functions $f,g \in \mathcal{L}^2(\Pi,\rho)$ using the expected value (2). Its norm is $\|f\|_{\mathcal{L}^2(\Pi,\rho)} = \sqrt{\langle f,f\rangle}$.

Let an orthonormal basis $(\Phi_i)_{i\in\mathbb{N}}$ be given, which consists of multivariate polynomials $\Phi_i : \Pi \rightarrow \mathbb{R}$. Without loss of generality, $\Phi_1 \equiv 1$ is the unique polynomial of degree zero. We assume that the variables $\mathbf{x}(t,\cdot)$ as well as the output $y(t,\cdot)$ are (component-wise) functions in the space $\mathcal{L}^2(\Pi,\rho)$ for each $t \geq 0$. It follows that the functions can be expanded in the *polynomial chaos* (PC), see [10],

$$\mathbf{x}(t,\mathbf{p}) = \sum_{i=1}^{\infty} \mathbf{v}_i(t)\Phi_i(\mathbf{p}) \qquad \text{and} \qquad y(t,\mathbf{p}) = \sum_{i=1}^{\infty} w_i(t)\Phi_i(\mathbf{p}) \tag{4}$$

with time-dependent coefficient functions $\mathbf{v}_i : I \rightarrow \mathbb{R}^n$ and $w_i : I \rightarrow \mathbb{R}$. A truncation of the series (4) to $i = 1,\ldots,m$ with some integer $m \geq 1$ yields a finite approximation. Typically, all basis polynomials up to some total degree d are included. Hence the number of basis functions results to $m = (d+q)!/(d!q!)$.

3 Stochastic Galerkin Systems and Norms

The *stochastic Galerkin method* changes the random-dependent linear dynamical system (1) into the larger deterministic linear dynamical system

$$\hat{\mathbf{E}}\tfrac{\mathrm{d}}{\mathrm{d}t}\hat{\mathbf{v}}(t) = \hat{\mathbf{A}}\hat{\mathbf{v}}(t) + \hat{\mathbf{B}}\mathbf{u}(t) \tag{5}$$
$$\hat{\mathbf{w}}(t) = \hat{\mathbf{C}}\hat{\mathbf{v}}(t). \tag{6}$$

The constant matrices $\hat{\mathbf{A}}, \hat{\mathbf{E}} \in \mathbb{R}^{mn\times mn}$, $\hat{\mathbf{B}} \in \mathbb{R}^{mn\times n_{\mathrm{in}}}$, $\hat{\mathbf{C}} \in \mathbb{R}^{m\times mn}$ are derived from $\mathbf{A}$, $\mathbf{E}$, $\mathbf{B}$, $\mathbf{c}$, respectively. The definition of the matrices can be found in [6], for example. The state/inner variables are $\hat{\mathbf{v}} = (\hat{\mathbf{v}}_1^\top,\ldots,\hat{\mathbf{v}}_m^\top)^\top$. Now the system produces multiple outputs $\hat{\mathbf{w}} = (\hat{w}_1,\ldots,\hat{w}_m)^\top$ by (6). Both $\hat{\mathbf{v}}$ and $\hat{\mathbf{w}}$ include approximations of the exact coefficients in the PC expansions (4). The induced approximation of the random QoI in (1) becomes

$$\hat{y}^{(m)}(t,\mathbf{p}) = \sum_{i=1}^{m} \hat{w}_i(t)\Phi_i(\mathbf{p}). \tag{7}$$

We assume that the stochastic Galerkin system (5) is asymptotically stable. In the following, we always predetermine initial values $\hat{\mathbf{v}}(0) = \mathbf{0}$.

A non-negative quadratic output of the system (5) reads as

$$g(t) = \hat{\mathbf{w}}(t)^\top\hat{\mathbf{M}}\hat{\mathbf{w}}(t) = \hat{\mathbf{v}}(t)^\top\hat{\mathbf{C}}^\top\hat{\mathbf{M}}\hat{\mathbf{C}}\hat{\mathbf{v}}(t) \tag{8}$$

with a symmetric positive semi-definite matrix $\hat{\mathbf{M}} \in \mathbb{R}^{m\times m}$. Let $\Sigma(\hat{\mathbf{M}})$ be the system consisting of the dynamical part (5) and the quadratic output (8). Now we employ a symmetric decomposition

$$\hat{\mathbf{M}} = \hat{\mathbf{F}}^\top\hat{\mathbf{F}} \tag{9}$$

with a matrix $\hat{\mathbf{F}} \in \mathbb{R}^{m \times m}$, for example, using a pivoted Cholesky decomposition. We arrange a stochastic Galerkin system $\Sigma(\hat{\mathbf{F}})$ consisting of (5) and the linear outputs

$$\hat{\mathbf{z}}(t) = \hat{\mathbf{F}}\hat{\mathbf{C}}\hat{\mathbf{v}}(t). \tag{10}$$

We define the system norm belonging to (5), (8) as

$$\left\| \Sigma(\hat{\mathbf{M}}) \right\|_{\mathcal{L}^2} = \sup_{\mathbf{u} \neq \mathbf{0}} \frac{\left\| \sqrt{g} \right\|_{\mathcal{L}^2[0,\infty)}}{\|\mathbf{u}\|_{\mathcal{L}^2[0,\infty)}} \tag{11}$$

including the $\mathcal{L}^2$-norm in the time domain $[0, \infty)$. This norm involves the supremum of the set of all inputs $\mathbf{u} \in \mathcal{L}^2[0, \infty)\backslash\{\mathbf{0}\}$. In view of (10), it follows that

$$\left\| \Sigma(\hat{\mathbf{M}}) \right\|_{\mathcal{L}^2} = \sup_{\mathbf{u} \neq \mathbf{0}} \frac{\left\| \hat{\mathbf{F}}\hat{\mathbf{w}} \right\|_{\mathcal{L}^2[0,\infty)}}{\|\mathbf{u}\|_{\mathcal{L}^2[0,\infty)}} = \sup_{\mathbf{u} \neq \mathbf{0}} \frac{\left\| \hat{\mathbf{F}}\hat{\mathbf{C}}\hat{\mathbf{v}} \right\|_{\mathcal{L}^2[0,\infty)}}{\|\mathbf{u}\|_{\mathcal{L}^2[0,\infty)}} = \left\| \Sigma(\hat{\mathbf{F}}) \right\|_{\mathcal{L}^2}. \tag{12}$$

This system norm is independent of the chosen decomposition (9).

The input-output behaviour of a linear dynamical system can be described by a transfer function in frequency domain, see [1]. The matrix-valued transfer function of the system (5), (10) reads as $\hat{\mathbf{H}}(s) = \hat{\mathbf{F}}\hat{\mathbf{C}}(s\hat{\mathbf{E}} - \hat{\mathbf{A}})^{-1}\hat{\mathbf{B}}$ for almost all $s \in \mathbb{C}$. Now the system norm of $\Sigma(\hat{\mathbf{F}})$ in (12) coincides with the $\mathcal{H}_\infty$-norm of this transfer function. The $\mathcal{H}_\infty$-norm is computable by methods of numerical linear algebra, see [2]. If the linear stochastic Galerkin system (5) consists of ODEs or DAEs with index $\nu = 1$, then the finiteness of the $\mathcal{H}_\infty$-norm is guaranteed. Yet the $\mathcal{H}_\infty$-norm may still be finite for DAEs of index $\nu \geq 2$ depending on the definition of inputs and outputs. The following sensitivity analysis applies to any DAE system (5), (6) with finite $\mathcal{H}_\infty$-norm, because a system (5), (10) with modified output inherits a finite $\mathcal{H}_\infty$-norm.

4 Sensitivity Measures

Our aim is a global sensitivity analysis of the stochastic model (1) with respect to the influence of the individual random parameters. In general, there are variance-based sensitivity measures and derivative-based sensitivity measures, see [8,9]. Although the variance-based sensitivity indices originally were defined different, we use an equivalent specification by the PC expansion as in [5].

Let $V(t)$ be the variance of the random QoI $y(t, \cdot)$ for $t \geq 0$. We define the index sets $\mathbb{I}_j, \mathbb{I}'_j \subset \mathbb{N}$ for $j = 1, \ldots, q$ using the family of basis polynomials $(\Phi_i)_{i \in \mathbb{N}}$

$$\mathbb{I}_j = \{i \ : \ \Phi_i \text{ depends only on } p_j\},$$
$$\mathbb{I}'_j = \{i \ : \ \Phi_i \text{ depends (also) on } p_j\}.$$

It holds that $\mathbb{I}_j \subset \mathbb{I}'_j$ and $1 \notin \mathbb{I}_j$ due to $\Phi_1 \equiv 1$. Variance-based sensitivity measures are the *first-order indices*

$$\overline{S}_j^{\text{FO}}(t) = \frac{S_j^{\text{FO}}(t)}{V(t)} \quad \text{with} \quad S_j^{\text{FO}}(t) = \sum_{i \in \mathbb{I}_j} w_i(t)^2 \tag{13}$$

and the *total-effect indices*

$$\overline{S}_j^{\mathrm{TE}}(t) = \frac{S_j^{\mathrm{TE}}(t)}{V(t)} \quad \text{with} \quad S_j^{\mathrm{TE}}(t) = \sum_{i \in \mathbb{I}'_j} w_i(t)^2. \tag{14}$$

These real numbers satisfy $0 \leq \overline{S}_j^{\mathrm{FO}}(t) \leq \overline{S}_j^{\mathrm{TE}}(t) \leq 1$ for each $t \geq 0$ and $j = 1, \ldots, q$. If it holds that $V(t) = 0$ for some t, then the variance-based sensitivity indices are not defined. However, an UQ is obsolete in this case as there is no variability.

Furthermore, we examine *derivative-based indices* with respect to the norm of (3), i.e.,

$$S_j^{\mathrm{DB}}(t) = \left\| \frac{\partial y}{\partial p_j}(t, \cdot) \right\|^2_{\mathcal{L}^2(\Pi,\rho)} = \int_{\Pi} \left(\frac{\partial y}{\partial p_j}(t, \mathbf{p}) \right)^2 \rho(\mathbf{p}) \, \mathrm{d}\mathbf{p}. \tag{15}$$

Applying the PC expansion (4), it follows that

$$S_j^{\mathrm{DB}}(t) = \sum_{k,\ell=1}^{\infty} \eta_{jk\ell} w_k(t) w_\ell(t) \quad \text{with} \quad \eta_{jk\ell} = \int_{\Pi} \frac{\partial \Phi_k}{\partial p_j}(\mathbf{p}) \frac{\partial \Phi_\ell}{\partial p_j}(\mathbf{p}) \rho(\mathbf{p}) \, \mathrm{d}\mathbf{p} \tag{16}$$

assuming that infinite summations and integration can be interchanged.

Now the sensitivity indices S_j^{FO} in (13), S_j^{TE} in (14), and S_j^{DB} in (15) can be approximated by quadratic outputs (8) of the stochastic Galerkin system (5). Let $\mathbb{K} \subset \mathbb{N}$ be an index set. We define the diagonal matrix

$$\hat{\mathbf{D}}(\mathbb{K}) = \mathrm{diag}(d_1, \ldots, d_m) \quad \text{with} \quad d_k = \begin{cases} 1, \text{ if } k \in \mathbb{K}, \\ 0, \text{ if } k \notin \mathbb{K}. \end{cases}$$

This matrix owns the trivial symmetric factorisation $\hat{\mathbf{D}}(\mathbb{K}) = \hat{\mathbf{D}}(\mathbb{K})^\top \hat{\mathbf{D}}(\mathbb{K})$. On the one hand, we obtain the variance-based indices by

$$S_j^{\mathrm{FO}}(t) \approx \hat{S}_j^{\mathrm{FO}}(t) = \hat{\mathbf{w}}(t)^\top \hat{\mathbf{D}}(\mathbb{I}_j) \hat{\mathbf{w}}(t) \quad \text{and} \quad S_j^{\mathrm{TE}}(t) \approx \hat{S}_j^{\mathrm{TE}}(t) = \hat{\mathbf{w}}(t)^\top \hat{\mathbf{D}}(\mathbb{I}'_j) \hat{\mathbf{w}}(t).$$

On the other hand, we arrange the symmetric matrices $\hat{\mathbf{N}}_j = (\eta_{jk\ell})_{k,\ell=1,\ldots,m}$ including the coefficients from (16). It can be shown that the matrices are also positive semi-definite. Although the integrals $\eta_{jk\ell}$ include the derivatives of the basis polynomials, the majority of these integrals are zero, i.e., the matrices $\hat{\mathbf{N}}_j$ are sparse. Consequently, the approximation of the derivative-based indices reads as

$$S_j^{\mathrm{DB}}(t) \approx \hat{S}_j^{\mathrm{DB}}(t) = \hat{\mathbf{w}}(t)^\top \hat{\mathbf{N}}_j \hat{\mathbf{w}}(t).$$

We consider the system norms (11) with $\hat{\mathbf{M}}_j \in \{\hat{\mathbf{D}}(\mathbb{I}_j), \hat{\mathbf{D}}(\mathbb{I}'_j), \hat{\mathbf{N}}_j\}$ for $j = 1, \ldots, q$. These system norms represent sensitivity measures $\mu_1^\star, \ldots, \mu_q^\star$ for $\star \in$

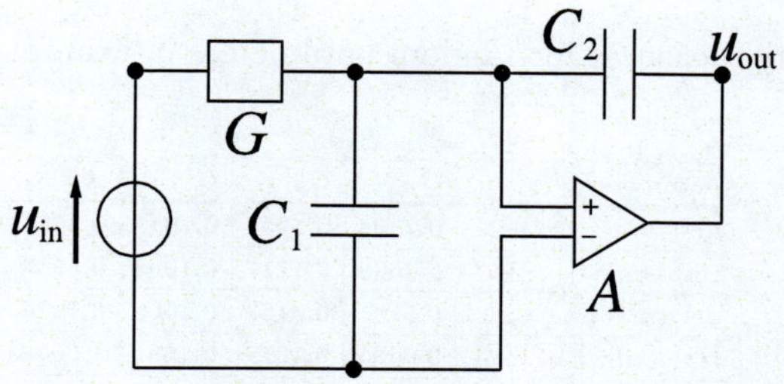

Fig. 1. Circuit diagram of Miller integrator.

$\{\mathrm{FO}, \mathrm{TE}, \mathrm{DB}\}$, which are independent of time t as well as the selection of inputs $\mathbf{u}$. A standardisation yields coefficients $0 \leq \tilde{\mu}_1^\star, \ldots, \tilde{\mu}_q^\star \leq 1$ with

$$\tilde{\mu}_1^\star + \cdots + \tilde{\mu}_q^\star = 1 \tag{17}$$

to investigate the relative differences for the random parameters.

5 Illustrative Example

We examine the electric circuit of a Miller integrator shown in Fig. 1. A mathematical model of the Miller integrator was presented in [3], which consists of a system of $n = 5$ linear DAEs. The system involves $q = 4$ physical parameters: two capacitances C_1, C_2, a conductance G, and an amplification factor A. The index of this DAE system is $\nu = 2$ for all (positive) parameters. Furthermore, the DAE systems are asymptotically stable. An input voltage u_{in} is supplied as single input. The output voltage $y = u_{\mathrm{out}}$ represents the QoI. The DAE system can be written in the form (1). Although the system of DAEs exhibits index two, the $\mathcal{H}_\infty$-norm of the associated transfer function is finite.

We replace the physical parameters by independent random variables with uniform distributions. The mean values are chosen as $\bar{C}_1 = 10^{-10}$, $\bar{C}_2 = 5 \cdot 10^{-11}$, $\bar{G} = 0.001$, $\bar{A} = 2$, whereas each random variable varies 20% around its mean value. We standardise the resulting hypercuboid Π to $[-1, 1]^4$, which changes only the magnitude of the derivative-based indices (15), since the derivatives are multiplied by constants in the transformation. The PC expansions (4) include basis polynomials, which are the products of univariate Legendre polynomials. We truncate the expansions such that all polynomials up to total degree $d = 4$ are included, i.e., $m = 70$ basis polynomials. The stochastic Galerkin method produces a system of DAEs (5) with dimension $mn = 350$. This linear dynamical system is asymptotically stable.

In [7], the Miller integrator was also used as a test example, where only the system norms associated to the total-effect sensitivity indices were investigated. Now we examine all cases of system norms introduced in Sect. 4. Table 1 depicts the computed system norms with respect to the three types of sensitivity indices. The standardised sensitivity measures satisfying (17) are illustrated in Fig. 2.

Table 1. Sensitivity measures for random parameters in example of Miller integrator.

parameter	C_1	C_2	G	A
first-order	0.2338	0.1257	0.1015	0.3564
total-effect	0.2699	0.1717	0.1088	0.3803
derivative-based	0.5187	0.3168	0.1906	0.7074
total-effect in time	0.6485	0.1879	0.1681	1.0000

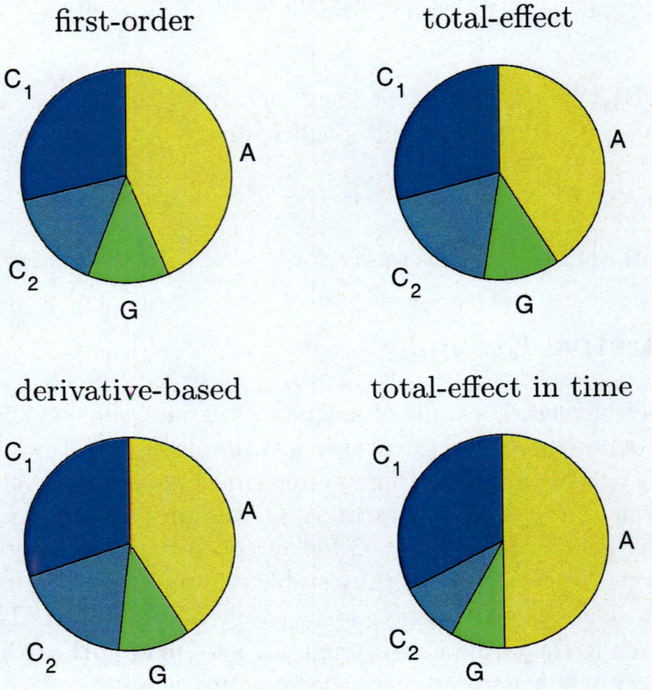

Fig. 2. Standardised sensitivity measures from system norms for first-order indices, total-effect indices, derivative-based indices, and from maxima of total-effect indices in time.

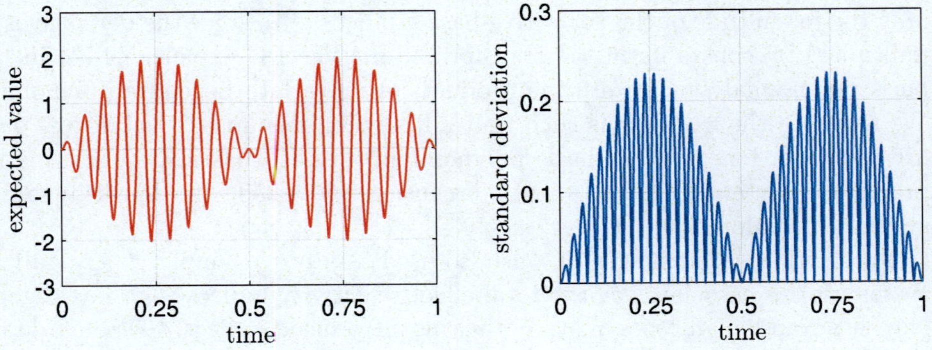

Fig. 3. Expected value (left) and standard deviation (right) of random output voltage in electric circuit of Miller integrator.

We perform a transient simulation of the stochastic Galerkin system for comparison. The two-tone signal

$$u_{\rm in}(t) = \sin(2\pi\, t)\sin(40\pi\, t)$$

is supplied as input voltage. Initial values are zero and the time interval $[0, 1]$ is considered. The backward differentiation formula (BDF) of order two yields a numerical solution of this initial value problem. The outcome (7) also provides approximations for the expected value as well as the standard deviation of the random output voltage, demonstrated in Fig. 3. Using the approximation from the stochastic Galerkin system, we calculate the maxima in time with respect to (14), i.e.,

$$\max\left\{\bar{S}_j^{\rm TE}(t) = S_j^{\rm TE}(t)/V(t) \ : \ t \in [\delta, 1]\right\} \qquad \text{for } j = 1, 2, 3, 4. \tag{18}$$

The threshold $\delta = 10^{-5} > 0$ is introduced, because the initial conditions imply $V(0) = 0$ and thus the variance exhibits tiny values at the beginning. Table 1 and Fig. 2 also show the sensitivity measures (18).

We observe that the ranking of the random parameters agrees in all four concepts: the amplification factor A is the most influential parameter, followed by the two capacitances C_1, C_2, and the conductance G is of least importance.

6 Summary

We investigated a sensitivity analysis for linear systems of ODEs or DAEs with respect to the influence of random variables. Three concepts were considered: two variance-based approaches and a derivative-based approach. Each concept yields sensitivity indices, which represent quadratic outputs of a stochastic Galerkin system. It follows that associated sensitivity measures are computable as $\mathcal{H}_\infty$-norms of the system. A test example demonstrates that this sensitivity analysis identifies a correct ranking of the random variables.

References

1. Antoulas, A.C.: Approximation of Large-Scale Dynamical Systems. SIAM, Philadelphia (2005)
2. Boyd, S., Balakrishnan, V., Kabamba, P.: A bisection method for computing the H_∞ norm of a transfer matrix and related problems. Math. Control Signals Syst. **2**, 207–219 (1989)
3. Günther, M., Feldmann, U., ter Maten, J.: Modelling and discretization of circuit problems. In: Ciarlet, P.G. (ed.) Handbook of Numerical Analysis, vol. 13, pp. 523–659. Elsevier, North-Holland (2005)
4. Liu, Q., Pulch, R.: Numerical methods for derivative-based global sensitivity analysis in high dimensions. In: Langer, U., Amrhein, W., Zulehner, W. (eds.) Scientific Computing in Electrical Engineering. MI, vol. 28, pp. 157–167. Springer, Cham (2018). https://doi.org/10.1007/978-3-319-75538-0_15

5. Mara, T., Becker, W.: Polynomial chaos expansions for sensitivity analysis of model output with dependent inputs. Reliab. Eng. Syst. Saf. **214**, 107795 (2021)
6. Pulch, R., Narayan, A.: Sensitivity analysis of random linear dynamical systems using quadratic outputs. J. Comput. Appl. Math. **387**, 112491 (2021)
7. Pulch, R., Narayan, A., Stykel, T.: Sensitivity analysis of random linear differential-algebraic equations using system norms. J. Comput. Appl. Math. **397**, 113666 (2021)
8. Sobol, I.M.: Sensitivity estimates for nonlinear mathematical models. Math. Model. Comput. Exp. **1**, 407–414 (1993)
9. Sobol, I.M., Kucherenko, S.: Derivative based global sensitivity measures and their link with global sensitivity indices. Math. Comput. Simul. **79**, 3009–3017 (2009)
10. Sullivan, T.J.: Introduction to Uncertainty Quantification. Springer, Cham (2015). https://doi.org/10.1007/978-3-319-23395-6

Compact Modelling of Wafer Level Chip-Scale Package via Parametric Model Order Reduction

Ibrahim Zawra[1,2](✉), Jeroen Zaal[3], Michiel van Soestbergen[3], Torsten Hauck[4], Evgeny Rudnyi[5], and Tamara Bechtold[1,2]

[1] Department of Engineering, Jade University of Applied Sciences, Wilhelmshaven, Germany
ibrahim.zawra@jade-hs.de
[2] Institute for Electronic Appliances and Circuits, University of Rostock, Rostock, Germany
[3] NXP Semiconductors, Nijmegen, The Netherlands
[4] NXP Semiconductors, Munich, Germany
[5] Cadfem GmbH, Munich, Germany

Abstract. The interconnect reliability of a packaged chip on the printed circuit board is a major requirement that should be met for assembling microelectronics. The solder connection fatigue is one of the main failure modes. It is caused by the mechanical stress due to thermal expansion. In this work, the finite element model of a package on the printed circuit board is built and the solder joint analysis is performed within a thermo-mechanical simulation. For efficient studies of the temperature impact on the solder joints, we present a successful application of parametric model order reduction for constructing a compact model starting from the full order finite element model. Temperature dependent Young's modulus, a parameter, which appears on both the left-hand and the right-hand side of the spatially discretized model, is preserved in the symbolic form within this compact model.

1 Introduction

For high-tech systems, such as motor control units for automated factories, smart infrastructures (streetlights, power grids), or autonomous vehicles, the requirements relating to quality and mechanical reliability should be met. With the development of modelling and simulation techniques in the last decades, the simulation software enables unprecedented reliability analysis and design optimization with reduced experimental cycles and costs. However, the faster growing computational demands of the industry exceed the power of desktop.

To solve this issue, the methodology of model order reduction (MOR) has been introduced [1–5,9]. Starting from a high-dimensional finite element model, MOR enables automatic generation of a lower dimensional but still accurate surrogate, which significantly reduces the computational cost and enables the system-level simulation. Conventional MOR methods already proved successful for linear single-physical-domain models [10]. However, microelectronic components require coupled domain thermo-mechanical simulations and exhibit temperature dependent material proprieties.

M. van Beurden et al. (Eds.): SCEE 2022, MI 43, pp. 217–228, 2024.
https://doi.org/10.1007/978-3-031-54517-7_25

To deal with that, parametric model order reduction (pMOR) methods have been developed, which enable to preserve the parameters in the symbolic form within the reduced order model [6–8, 11, 13, 14].

The European project COMPAS [17] aims to develop novel compact models and ultra-compact digital twins for predicting the thermo-mechanical reliability issues in high-tech systems, which integrate numerous highly complex components. The project starts with a test model of a wafer level chip-scale package provided by NXP Semiconductors. One major failure mode in such hardware is the solder connection fatigue (see Fig. 1). The mismatch between the coefficients of thermal expansion (CTE) of the package and of the printed circuited board (PCB) causes mechanical stress within the solder connection and leads to the solder fatigue and ultimate failure.

In this work, we successfully apply pMOR to the wafer level chip-scale package model for constructing a parametric reduced order model (pROM). The temperature dependent Young's modulus in the solder connection is defined as a parameter and preserved in the symbolic form within the compact model. This enables efficient reliability analysis.

The paper is organized as follows. In Sect. 2, the setup of the mechanical model of the wafer level chip-scale package under the thermal load is introduced. Then, the pMOR process is presented in Sect. 3 and the numerical simulation results of the parametric ROM are illustrated in Sect. 4. Section 5 concludes the contributions of this work and gives an outlook to future research.

Fig. 1. Crack in the solder ball due to thermal loading.[15]

2 Case Study: Wafer Level Chip-Scale Package

Figure 2 displays the model assembly that contains the PCB, solders, copper, passivation, chip and coating. The model consists of six elastic material domains and the parameter of interest is the Young's modulus of the solder domain. It is to be preserved in symbolic form within the reduced order model. Furthermore, Young's modulus of the silicon chip domain can also be preserved in the symbolic form. The three points demarked in Fig. 2 with $\{0, 1, 2\}$ on the border $\partial\Omega$ of computational domain Ω are subjected to the following mechanical Dirichlet boundary conditions, where point one is totally fixed, point two is free only in x direction and point three is only fixed in z direction:

$$u_0(x,y,z) = [0,0,0], \quad u_1(x,y,z) = [x,0,0], \quad u_2(x,y,z) = [x,y,0] \quad on \quad \partial\Omega \tag{1}$$

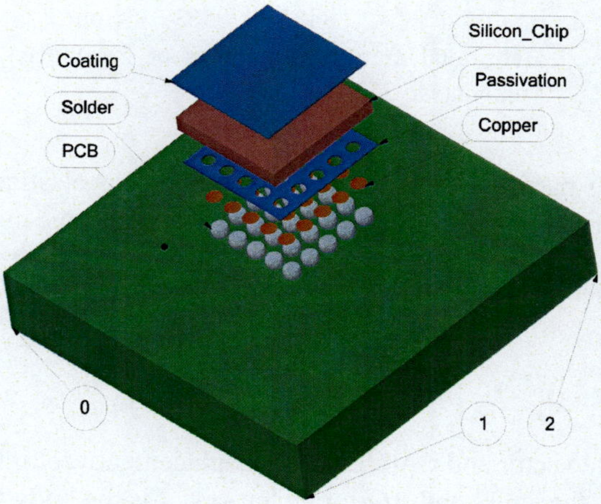

Fig. 2. An exploded view of the waver chip-scale package.

The reliability tests for these devices are performed inside an oven under homogeneous temperature cycles. This temperature cycling leads to mechanical deformations and stresses. Usually, such tests are performed in passive regime, that is, without turning on the Chip. This means that one can assume the homogeneous temperature distribution across the chip (corresponding to the temperature cycling) and describe it with a static finite element model. We use ANSYS®R 21.2. In Fig. 3 and Table 1 the finite element mesh and its statistics are displayed.

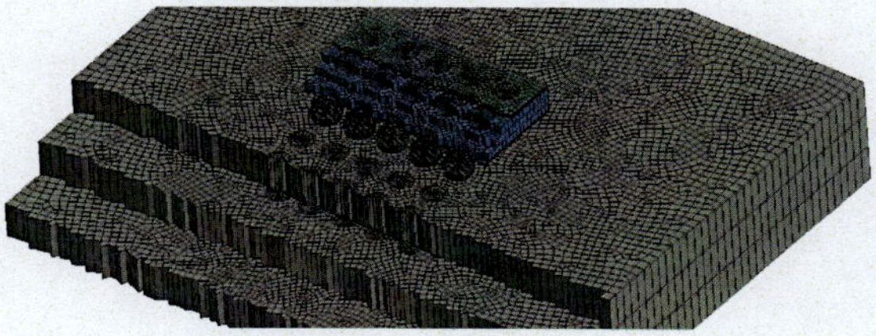

Fig. 3. Volumetric mesh sectioned through the whole model

Table 1. Mesh statistics of the model

Body name	Nodes	Elements	Type
Coating	12272	9123	Solid
Silicon Chip	12272	9123	Solid
Passivation	9172	6048	Solid
Copper	10375	7500	Solid
Solder	11725	8850	Solid
PCB	40152	29919	Solid
Total	84232	70563	Solid

The total number of degrees of freedoms amounts to 252690

The governing partial differential equations of linear elasticity over a continuous domain Ω, considering infinitesimal strain theory and isotropic materials can be written as follow:

$$\begin{aligned} -\nabla \boldsymbol{\sigma}(u) &= f \quad \text{in } \Omega \\ \boldsymbol{\sigma}(u) &= \lambda \operatorname{tr}(\boldsymbol{\varepsilon}(u))\mathbf{I} + 2\mu\boldsymbol{\varepsilon}(u) \\ \boldsymbol{\varepsilon}(u) &= \frac{1}{2}\left(\nabla u + (\nabla u)^T\right) \end{aligned} \tag{2}$$

where u is the state vector and represents the displacement vector field in the domain Ω, $\sigma(u)$ and $\varepsilon(u)$ are the stress and strain-rate tensors, f is the body force per unit volume, λ and μ are elasticity parameters of materials in Ω, I is the identity tensor, tr is the trace operator on a tensor.

Finite elements based spatial discretization of Eq. (2) leads to the following element matrices and element load vectors:

$$\begin{aligned} &\{\sigma\} = [D]\left\{\varepsilon^{el}\right\} \\ &\left\{\varepsilon^{el}\right\} = \{\varepsilon\} - \left\{\varepsilon^{th}\right\}, \quad \left\{\varepsilon^{th}\right\} = \Delta T\left[\alpha_x^{se}\,\alpha_y^{se}\,\alpha_z^{se} 0 \quad 0 \quad 0\right]^T \end{aligned} \tag{3}$$

$$\begin{aligned} \left([K_e] + \left[K_e^f\right]\right)\{u\} &= \left\{F_e^{th}\right\} \\ [K_e] &= \int_{Vol} [B]^T [D][B] d(\text{ vol }) \\ \left[K_e^f\right] &= \text{k} \int_{\text{area}} [N_n]^T [N_n] d\left(\text{ area }_f\right) \\ \left\{F_e^{th}\right\} &= \int_{vol} [B]^T [D]\left\{\varepsilon^{th}\right\} d(\text{ vol }) \end{aligned} \tag{4}$$

$$[D] = \frac{\text{E}}{(1+v)(1-2v)} \begin{bmatrix} 1-v & v & v & 0 & 0 & 0 \\ v & 1-v & v & 0 & 0 & 0 \\ v & v & 1-v & 0 & 0 & 0 \\ 0 & 0 & 0 & \frac{1-2v}{2} & 0 & 0 \\ 0 & 0 & 0 & 0 & \frac{1-2v}{2} & 0 \\ 0 & 0 & 0 & 0 & 0 & \frac{1-2v}{2} \end{bmatrix} \tag{5}$$

where D is the generalized Hook's low fourth order tensor, material property, that relates stress and strain. E and v are Young's modulus and Poisson's ratio. ε^{el} is the elastic strain, α_x^{se} is the first component of secant coefficient of thermal expansion vector, $\Delta T = T - T_{ref}$ while T_{ref} is the strain free temperature, $[B]$ strain-displacement matrix, based on the element shape functions, $\{u\}$ nodal displacement vector, $[K_e]$ is element stiffness matrix, $[K_e^f]$ is the element foundation stiffness matrix, $[N_n]$ is the matrix of shape functions for normal motions at the surface and $\{F_e^{th}\}$ is the element thermal load vector.

From Eq. (4) the system stiffness matrix K is assembled and the parameter of interest (E) can be factorized see Eq. (6). Note that it would be not so simple to factorize the Poisson's ratio as a parameter, because it enters the system matrix in a non-linear way. The same holds true for geometrical parameters.

3 Parametric Model Order Reduction

In this chapter we will define the linear parametric system arising from the finite element model defined in Sect. 2. Furthermore, the parametric reduced order model, which preserves inputs, outputs and the Young's modulus in symbolic form is defined. Many studies in the field of parametric model order reduction focus on treating dynamical systems, in which solely the left hand side is parameter-independent. However, the parametric system arising in this work, contains parameters also at the right hand side. Finally, we will describe the multi-point moment matching property of the MOR algorithm [6]. In this case, we can say that the reduced model is a partial realization or Padé-type approximation of the full order model.

3.1 Arising Parametric System

As discussed in Sect. 2 we can write the parametric full order model for a single material parameter as follows:

$$\Sigma_N : \begin{cases} \underbrace{(K_0 + E \cdot K_1)}_{=K(E)} \cdot x = \underbrace{(B_0 + E \cdot B_1)}_{:=B(E)} \cdot u(t) \\ y = C \cdot x \end{cases} \tag{6}$$

where N is the dimension of the full order model and is equivalent to the number of the degrees of freedoms defined in Table reftable:mesh (N= 252690), $K \in \mathbb{R}^{N \times N}$ is the system's stiffness matrix with factorized Young's modulus and $K_0, K_1 \in \mathbb{R}^{N \times N}$ are its parameter-independent and the parameter-dependent parts respectively. $u \in \mathbb{R}^m, y \in \mathbb{R}^o$ are the input and output vectors. $B \in \mathbb{R}^{N \times m}, C \in \mathbb{R}^{o \times N}$ are the input and output matrices, respectively. m, o are the number of inputs and user defined outputs. $x \in \mathbb{R}^N$ is the state vector of unknown displacements and E is the Young's modulus of the specified material domain.

In general multi-parameter case, the parametric system can be written as follows:

$$\Sigma_N : \begin{cases} \underbrace{(K_0 + E_1 \cdot K_1 + +E_2 \cdot K_2 + \ldots + E_p \cdot K_p)}_{=K(E)} \cdot x = \underbrace{(B_0 + E_1 \cdot B_1 + E_2 \cdot B_2 + \ldots + E_p \cdot B_p)}_{:=B(E)} \cdot u(t) \\ y = C \cdot x \end{cases} \tag{7}$$

where the subscript p denotes the total number of parameters. Physically each parameter can describe material property of a certain material domain, which enters the system matrices linearly and hence, can be factorized. The goal is to reduce such parameterized system to a compact form, which can be employed within a system level simulation. Single-parameter system Eq. (6) can be reduced by Galerkin approximation as follows:

$$\Sigma_r : \begin{cases} \underbrace{V^T(K_0 + E \cdot K_1)V}_{K_r(E)} \cdot x_r = \underbrace{V^T(B_0 + EB_1)}_{B_r} \cdot u(t) \\ y_r = \underbrace{CV}_{C_r} \cdot x_r \end{cases} \tag{8}$$

where $V \in \mathbb{R}^{N \times r}$, $K_r \in \mathbb{R}^{r \times r}$, $B_r \in \mathbb{R}^{r \times m}$, $C_r \in \mathbb{R}^{o \times r}$ and $r << N$ is the dimension of the reduced order model Note that, m, o are the same numbers of inputs and user defined outputs, as in the original system Eq. (6). $x \in \mathbb{R}^r$ is the reduced state vector and E is the Young's modulus of the specified material domain, which now preserved in the reduced space and can be changed at the system level simulation. The remaining question is how to define the projection subspace $\mathscr{K}_r$ with minimal approximation error as it will be demonstrated in the next section.

3.2 Moment Matching and Subspace Definition

The transfer function of the parametric system defined in Eq. (6) reads:

$$G(E) = Y(s)/U(s) = C[K(E)]^{-1} \cdot B(E) \tag{9}$$

This transfer function can be rewritten as follows:

$$G(E) = C\left[I - [-(E)K_1]K(E)^{-1}\right]^{-1} K(E)^{-1}[B(E) + (E) \cdot B_1] \tag{10}$$

Then, we apply the Taylor expansion and observe its coefficients (moments) around a chosen expansion point E_0:

$$\begin{aligned} G(E) = &\underbrace{CK(E)^{-1}B(E)}_{M_0^E} \\ &+ \underbrace{\sum_{i=1}^{\infty}\left\{C\left[-K(E)^{-1}K_1\right]^i K(E)^{-1}B(E) + C\left[-K(E)^{-1}K_1\right]^{i-1} K(E)^{-1}B_1\right\}(E)^i}_{M_i^E, i=1,2,\ldots} \end{aligned} \tag{11}$$

Based on these moments we can generate the Krylov subspace as follows:

$$\begin{aligned} &\text{colspan}\{V_1\} = \mathscr{K}_{r_1}\left\{-K(E)^{-1}K_1, K(E)^{-1}[B(E), B_1]\right\} \\ &B(E) = B_0 + E \cdot B_1 \end{aligned} \tag{12}$$

$$\begin{aligned} \text{colspan}\{V\} &= \mathscr{K}_{r_2}\left\{-K(E)^{-1}K_1, K(E)^{-1}[B_0, B_1]\right\} \\ &= \mathscr{K}_{r_1}\left\{-K(E)^{-1}K_1, K(E)^{-1}[B(E), B_1]\right\} \end{aligned} \tag{13}$$

The derivatives included in V can be matched by the reduced system such that: $M_i^E = V\hat{M}_i^E$, where $\hat{M}_i^E$ are the moments of the reduced system. Thus, we have moments of y and y_r are identical [6].

$$
\begin{aligned}
G(E_1,E_2) = \underline{G(0,0)} + \underline{\frac{\partial G}{\partial E_1}(0,0)\cdot E_1} + \underline{\frac{\partial G}{\partial E_2}(0,0)\cdot E_2} + \underline{\frac{1}{2!}\frac{\partial^2 G}{\partial E_1^2}(0,0)\cdot E_1^2} \\
+\frac{\partial^2 G}{\partial E_1 \partial E_2}(0,0)\cdot E_1\cdot E_2 + \frac{\partial^2 G}{\partial E_2 \partial E_1}(0,0)\cdot E_1\cdot E_2 + \underline{\frac{1}{2!}\frac{\partial^2 G}{\partial {E_2}^2}(0,0)\cdot {E_2}^2} + \ldots
\end{aligned}
\tag{14}
$$

$$
\begin{aligned}
G_r(E_1,E_2) = \underline{G_r(0,0)} + \underline{\frac{\partial G_r}{\partial E_1}(0,0)\cdot E_1} + \underline{\frac{\partial G_r}{\partial E_2}(0,0)\cdot E_2} + \underline{\frac{1}{2!}\frac{\partial^2 G_r}{\partial E_1^2}(0,0)\cdot E_1^2} \\
+\frac{\partial^2 G_r}{\partial E_1 \partial E_2}(0,0)\cdot E_1\cdot E_2 + \frac{\partial^2 G_r}{\partial E_2 \partial E_1}(0,0)\cdot E_1\cdot E_2 + \underline{\frac{1}{2!}\frac{\partial^2 G_r}{\partial {E_2}^2}(0,0)\cdot {E_2}^2} + \ldots
\end{aligned}
\tag{15}
$$

For multi-parameter system like in Eq. (7) building the reduced space is more complicated. As studied in [8] a comparison between three different algorithms, we here stick to the second proposed method, where building the reduced space is more efficient and robust. However, we apply a correction to deal with parametric right hand side. In this method, the derivatives are computed separately. For example, Eqs. (14) and (15) show two parameters expansion derivatives for both full and reduced, only the underlined moments are matched. For generalized case, we can define the subspace that preserve moment matching for each parameter p_i as follows:

$$
\begin{aligned}
\text{colspan}\left\{V_{p_i}\right\} = \mathscr{K}_{r_i}\left\{-K(E)^{-1}K_{p_i}, K(E)^{-1}\left[B(E), B_1\right]\right\} \\
V = \text{span}\left(V_{p_1},\ldots,V_{p_i}\right)
\end{aligned}
\tag{16}
$$

4 Numerical Results

In this chapter, we will demonstrate the efficiency and accuracy of our approach.

Table 2 shows the time comparison between the full finite element model and the reduced parameterized model with single material domain parameter. A speed up by a factor of 63 could be reached with maximal relative error of $0.2E-5$. Note that the speed up would be much larger if a transient simulation of the full model is required. A great time reduction in simulating the parametric reduced model over the full finite element model and keeping almost a negligible error. The full model runs in almost half an hour, while the reduced model do the job in a fraction of a second.

Figure 4 displays a schematic for the usage of pROM in the system level simulation. Engineers can define the Young's modulus of different material domains as an arbitrary function of temperature. In our case study temperature cycles are defined from $-40°$C to 125°C. Corresponding mechanical response of the full- and reduced-order model is shown in Fig. 5.

Table 2. Time comparison between reduced and full order models at Intel Core Processor (Broadwell) @3.0 GHz, 64 GB.

Model	DOF	Time[s]
Finite element model	252690	2058.9
pROM generation (offline)	86	32.208
pROM (online)	86	0.1600

Here, we have to define our error criteria, as the input $u(t)$ is time dependent and each output node in y is defined by a row-vector in C. Relative error is defined as:

$$e = \frac{|y - y_r|}{y} \qquad e_{vec} = \begin{bmatrix} \|e_{0j}\|_2 \\ \|e_{1j}\|_2 \\ \cdots \\ \|e_{oj}\|_2 \end{bmatrix} \qquad e_{rel} = \|e_{vec}\|_\infty \tag{17}$$

First, we calculate e which is the defined as the error above, while y and y_r are FOM and ROM outputs respectively, then we take the second norm for each row in e, e.g. e_{0j} is the first row of e and e_{oj} is the last row, which can represent the average error over time. Secondly, we have a vector of these averaged errors, e_{vec}, then we compute the infinite norm of it, which can be considered as the maximum relative error, e_{rel}, among the selected output nodes.

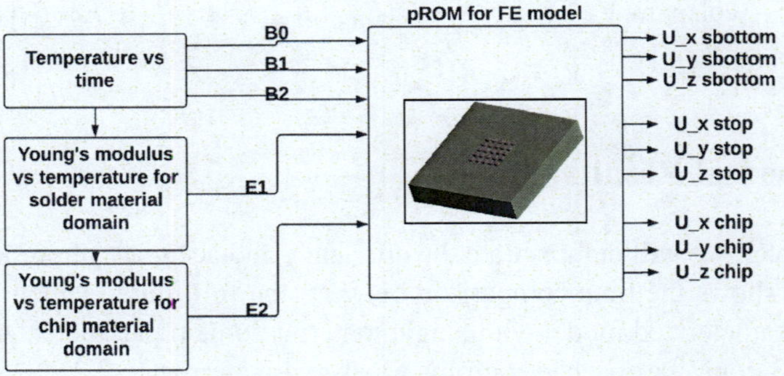

Fig. 4. The schematic of in the reduced space, while B0, B1, B2, E1 and E2 are consistent with the definitions in the equations in the previous sections. The outputs on the right hand side are arbitrary three points directional displacements. sbottom and stop are two points chosen arbitrarily in the solder material domain, while chip is in the silicon material domain.

Figure 6 shows the relative error between the parametric reduced order model and the full order model over the range of values for Young's modulus. In this case, the single material parameter is observed. As expected, the minimum error is in the vicinity of the chosen expansion point $E_0 = 2.9E10$ Pa. In the case of Multi material domain it shows the same conclusion.

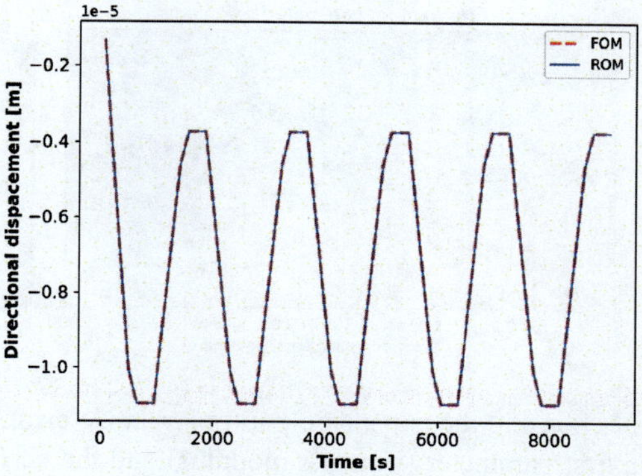

Fig. 5. A comparison between the full order model and parametric reduced model in response to temperature cycling. The response is the displacement in x-direction at the node sbottom defined in Fig. 4

Figure 7 shows the effect of using the expansion (extraction) point in chip material domain on the multi material parametric reduced model. The plot is generated by producing a reduced model with K_0 shown by X and Y axis of the plots, with the difference to Fig. 6 each point on Fig. 7 is a new reduced model. Then the relative error for a cyclic simulation (see Fig. 5) is evaluated and plotted. We can clearly observe that error drastically go up when we choose an expansion (extraction) point below the lower bound of the used curve which describes the young's modulus temperature dependency. Despite the fact that mathematically there should be no influence in selecting an extraction point far from the expansion point, Industrial models show many numerical problems here. Also, Optimality algorithm should be applied to identify the optimum choice for the expansion point in each material domain, maybe 'Iterative Rational Krylov Algorithm' [12]. We used the expansion and the extraction point interchangeably. Figure 8 shows how the choice of the subspace can influence the results in multi material domain study. In contrast to singe material parametric case, subspace building has a great influence on the error. As far as we know, the mixing moments absence in building the subspace can be one reason for that.

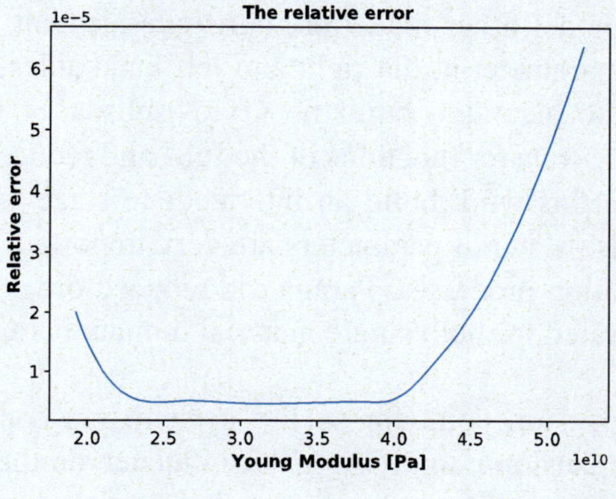

Fig. 6. Validation for a single domain parametric reduced order model. The chosen expansion point is 2.9E10 Pa.

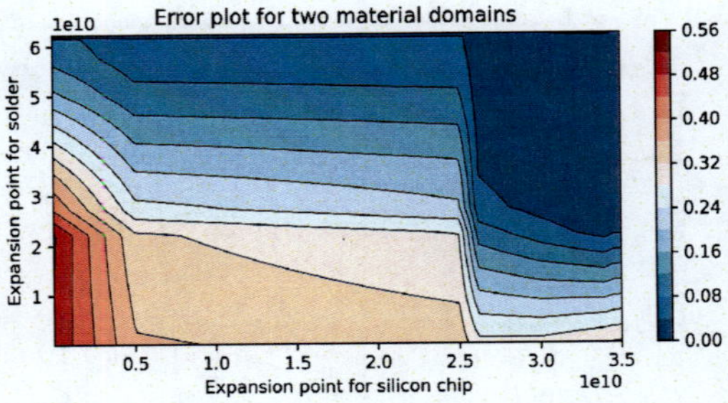

Fig. 7. The influence of choosing the expansion point on relative error, the horizontal axis is the expansion point of the first parameter (Young's modulus) and the vertical axis is the expansion point of the second parameter, while the color represents the error.

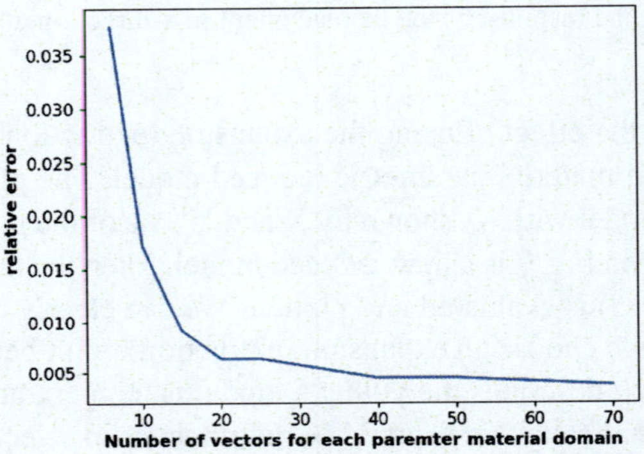

Fig. 8. The relation between the relative error and the size of the reduced space. On the horizontal axis is the number of vectors generated for each parameter matrix.

5 Conclusion and Outlook

In this work, we introduced a solution for reducing a parameterized finite element model, so that material parameters can be preserved in the symbolic form (provided, they can be factorized in front of the system matrices). In contrast to the previously studied parametric model order reduction, our finite element model produces a parametric system with parameter at the right and left hand sides. A new implementation had been carried out to accurately build the Krylov subspaces, which allow for moment matching between the transfer functions of the full- and reduced order model.

The results from this work build an intermediate stage towards reducing nonlinear reliability models. Material parameters are very important in the micro electronics studies and optimization processes. Parametric reduced order model based on Krylov subspace were generated for both single material domain parameter and multi-material domain parameters.

The optimal expansion point for such a system is a concern. This study shows that choosing the expansion point has a great influence on the accuracy of parametric reduced order model. In addition, the proper choice of the dimensions of the reduced space plays important role to build an efficient and accurate reduced model.

Thus, in this paper we have been able to achieve one of COMPAS project [17] goals, to preserve material properties in symbolic form within a reduced order model. Our next step is to reduce fully nonlinear reliability models.

Acknowledgments. This work was carried out as a part of the COMPAS project, which is supported by ITEA, the Eureka Cluster on software innovation, under project number 19037. COMPAS receives funding from 'Agentschap Innoveren en Ondernemen' (Belgium), 'Bundesministerium für Bildung und Forschung' (Germany), and 'Rijksdienst voor Ondernemend Nederland' (Netherlands).

References

1. Antoulas, A.C.: Approximation of Large-Scale Dynamical Systems. Society for Industrial and Applied Mathematics (2009)
2. Schilders, W., Van der Vorst, H., Rommes, J.: Model order reduction: theory, research aspects and applications. In: Mathematics in Industry Series, vol. 13 (2008). Springer, Germany-Model order reduction: theory, research aspects and applications. ISBN: 978-3-540-78840-9. https://doi.org/10.1007/978-3-540-78841-6
3. Benner, P.: Volume 1: System- and Data-Driven Methods and Algorithms. De Gruyter, Berlin (2021). ISBN:9783110498967. https://doi.org/10.1515/9783110498967
4. Benner, P.: Volume 2: Snapshot-Based Methods and Algorithms. De Gruyter, Berlin (2021). ISBN:9783110671490. https://doi.org/10.1515/9783110671490
5. Benner, P.: Volume 3: Applications. De Gruyter, Berlin (2021). ISBN:9783110499001. https://doi.org/10.1515/9783110499001
6. Gunupudi, P., Nakhla, M.: Multi-dimensional model reduction of VLSI interconnects. In: Proceedings of IEEE Custom Intergrated Circuits Conference, pp. 499–502 (2000)
7. Daniel, L., Siong, O. C., Chay, L.S., Lee, K.H., White, J.: A multiparameter moment-matching model-reduction approach for generating geometrically parameterized interconnect performance models. Comput.- Aided Des. Integr. Circuits Syst. **23**, 678–693 (2004)
8. Feng, L.H., Rudnyi, E.B., Korvink, J.G.: Preserving the film coefficient as a parameter in the compact thermal model for fast electrothermal simulation. IEEE Trans. Comput. Aided Design Integr. Circ. Syst. **24**, 1838–1847 (2005)
9. Rudnyi, E.B., Korvink, J.G.: Model order reduction for large scale finite element engineering models. In: European Conference on Computational Fluid Dynamics ECCOMAS CDF, TU Delft, Netherland (2006)
10. Lohmann, B., et al.: Model Order Reduction in Mechanical Engineering. Model order reduction. De Gruyter, Berlin (2020)
11. Bechtold, T., Hohlfeld, D., Rudnyi, E.B., Günther, M.: Efficient extraction of thin-film thermal parameters from numerical models via parametric model order reduction. J. Micromech. Microeng. **20**(4), 045030 (2010)
12. Gugercin, S., Antoulas, A.C., Beattie, C.: H_2 model reduction for large-scale linear dynamical systems. J. Matrix Anal. Appl. **30**, 609–638 (2008)
13. Baumann, M., Eberhard, P.: Interpolation-based parametric model order reduction for material removal in elastic multibody systems. Multibody Syst. Dyn. **39**, 21–36 (2017). https://doi.org/10.1007/s11044-016-9516-9
14. Yuan, C.D., Jadhav, O.S., Rudnyi, E.B., Hohlfeld, D., Bechtold, T.: parametric model order reduction and system-level simulation of a thermoelectric generator for electrically active implants. In: Accepted for Publication in Proceedings EuroSimE (2018)

15. Thoben, M., Xie, X., Silber, D., Wilde, J.: Reliability of Chip/DCB solder joints in AlSiC base plate power modules: influence of chip size. Microelectron. Reliabil. **41**(910), 121–1223 (2001)
16. Ansys®. Academic Research Mechanical, Release 21.2
17. ITEA 4. Compact modelling of high-tech systems for health management and optimization along the supply chain. https://itea4.org/project/compas.html
18. Yoo, E..: Parametric model order reduction for structural analysis and control. Ph.D. thesis, Technical University of Munich (2010). http://www.ct.tkk.fi/publications/dr-janne/main.html